Modelling Optimization and Control of Biomedical Systems

Modelling Optimization and Control of Biomedical Systems

Edited by

Efstratios N. Pistikopoulos
Texas A&M University, USA

Ioana Naşcu
Texas A&M University, USA

Eirini G. Velliou
Department of Chemical and Process Engineering
University of Surrey, UK

WILEY

This edition first published 2018

Registered Offices
John Wiley & Sons, Inc., 111 River Street, Hoboken, NJ 07030, USA
John Wiley & Sons Ltd, The Atrium, Southern Gate, Chichester, West Sussex, PO19 8SQ, UK

Editorial Office
The Atrium, Southern Gate, Chichester, West Sussex, PO19 8SQ, UK

For details of our global editorial offices, customer services, and more information about Wiley products visit us at www.wiley.com.

Wiley also publishes its books in a variety of electronic formats and by print-on-demand. Some content that appears in standard print versions of this book may not be available in other formats.

Library of Congress Cataloging-in-Publication data applied for

ISBN: 9781118965597

Cover design by Wiley
Cover image: Courtesy of Efstratios N. Pistikopoulos

Set in 10/12pt Warnock by SPi Global, Pondicherry, India

10 9 8 7 6 5 4 3 2 1

Contents

List of Contributors

Dr. Maria Fuentes-Gari
Process Systems Enterprise (PSE)
London
UK

Professor Michael C. Georgiadis
Laboratory of Process Systems Engineering
School of Chemical Engineering
Aristotle University of Thesaloniki
Greece

Dr. Alexandra Krieger
Jacobs Consultancy
Kreisfreie Stadt Aachen Area
Germany

Dr. Romain Lambert
Department of Chemical Engineering
Imperial College London
UK

Professor Athanasios Mantalaris
Department of Chemical Engineering
Imperial College London
UK

Dr. Ruth Misener
Department of Computing
Imperial College London
UK

Dr. Ioana Naşcu
Artie McFerrin Department of Chemical Engineering
Texas A&M University
College Station
USA

Dr. Richard Oberdieck
DONG energy A/S
Gentofte
Denmark

Dr. Nicki Panoskaltsis
Department of Medicine
Imperial College London
UK

Dr. Eleni Pefani
Clinical Pharmacology Modelling and Simulation
GSK
UK

Professor Efstratios N. Pistikopoulos
Texas A&M Energy Institute
Artie McFerrin Department of Chemical Engineering
Texas A&M University
USA

Dr. Pedro Rivotti
Department of Chemical Engineering
Imperial College London
UK

Susana Brito dos Santos
Department of Chemical
Engineering
Imperial College London
UK

Dr. Eirini G. Velliou
Department of Chemical and Process
Engineering
Faculty of Engineering and Physical
Sciences
University of Surrey
UK

Dr. Stamatina Zavitsanou
Paulson School of Engineering &
Applied Sciences
Harvard University
USA

Preface

A great challenge when dealing with severe diseases, such as cancer or diabetes, is the implementation of an appropriate treatment. Design of treatment protocols is not a trivial issue, especially since nowadays there is significant evidence that the type of treatment depends on specific characteristics of individual patients.

In silico design of high-fidelity mathematical models, which accurately describe a specific disease in terms of a well-defined biomedical network, will allow the optimisation of treatment through an accurate control of drug dosage and delivery. Within this context, the aim of the Modelling, Control and Optimisation of Biomedical Systems (MOBILE) project is to derive intelligent computer model-based systems for optimisation of biomedical drug delivery systems in the cases of diabetes, anaesthesia and blood cancer (i.e., leukaemia).

From a computational point of view, the newly developed algorithms will be able to be implemented on a single chip, which is ideal for biomedical applications that were previously off-limits for model-based control. Simpler hardware is adequate for the reduced on-line computational requirements, which will lead to lower costs and almost eliminate the software costs (e.g., licensed numerical solvers). Additionally, there is increased control power, since the new MPC approach can accommodate much larger – and more accurate – biomedical system models (the computational burden is shifted off-line).

From a practical point of view, the absence of complex software makes the implementation of the controller much easier, therefore allowing its usage as a diagnostic tool directly in the clinic by doctors, clinicians as well as patients without the requirement of specialised engineers, therefore progressively enhancing the confidence of medical teams and patients to use computer-aided practices. Additionally, the designed biomedical controllers increase treatment safety and efficiency, by carefully applying a "what-if" prior analysis that is tailored to the individual patient's needs and characteristics, therefore reducing treatment side effects and optimising the drug infusion rates. Flexibility of the device to adapt to changing patient characteristics and incorporation of the physician's performance criteria are additional great advantages.

There were several highly significant achievements of the project for all different diseases and biomedical cases under study (i.e., diabetes, leukaemia and anaesthesia). From a computational point of view, achievements include the construction of high-fidelity mathematical models as well as novel algorithm derivations. The methodology followed for the model design includes the following steps: (a) the derivation of a high-fidelity model, (b) the conduction of sensitivity analysis, (c) the application of parameter estimation techniques on the derived model in order to identify and estimate the sensitive model parameters and variables and (d) the conduction of extensive validation studies based on patient and clinical data. The validated model is then reduced to an approximate model suitable for optimisation and control via model reduction and/or system identification algorithms. The several theoretical (*in silico*) components are incorporated in a closed-loop (*in silico–in vitro*) framework that will be evaluated with *in vitro* trials (i.e., through experimental evaluation of the control-based optimised drug delivery). The outcome of the experiments will indicate the validity of the suggested closed-loop delivery of anaesthetics, chemotherapy dosages for leukaemia and insulin delivery doses in diabetes. It should be mentioned that this is the first closed-loop system including computational and experimental elements. The output of such a framework could be introduced, at a second step, in phase 1 clinical trials.

Chapter 1 is an overview of the framework for modelling, optimisation and control of biomedical systems. It describes the mathematical modelling of drug delivery systems that usually requires a pharmacokinetic part, a pharmacodynamic part and a link between the two. Model analysis, parameter estimation and approximation are used here in order to obtain an in-depth understanding of the model. Mathematical optimisation and control of the biomedical system could lead to a better prediction of the optimal drug and/or therapy treatment for a specific disease.

Chapter 2 presents in detail the theoretical background, computational tools and methods that are used in all the different biomedical systems analysed within the book. More specifically, Chapter 2 focuses on describing the computational tools, part of the developed multiparametric model predictive control framework presented in Chapter 1. It also presents the theory for multiparametric mixed-integer programming and explicit optimal control. This is part of the larger class of hybrid biomedical systems (i.e., biomedical systems featuring both discrete and continuous dynamics).

Chapters 3 and 4 aim at applying the presented framework to the process of anaesthesia: both volatile as well as intravenous. They present the procedure step by step from the model development to the design of a multiparametric model predictive controller for the control of depth of anaesthesia. Chapter 3 focuses on the process of volatile anaesthesia. A detailed physiologically based pharmacokinetic–pharmacodynamic patient model for volatile anaesthesia is presented where all relevant parameters and variables are analysed. A model predictive control (MPC) strategy is proposed to assure safe and robust control of anaesthesia by including an on-line parameter estimation step that accounts for patient variability. A Kalman filter is implemented to obtain an estimate of the states based on the measurement of the end-tidal concentration. An on-line estimator is added to the closed control loop for the estimation of the PD parameter C50 during the course of surgery. Closed-loop control simulations for the system for conventional MPC, explicit MPC and the on-line parameter estimation are presented for induction and disturbances during maintenance of anaesthesia.

In Chapter 4, we describe the process of intravenous anaesthesia. The mathematical model for intravenous anaesthesia is presented in detail, and sensitivity analysis is performed. The main objective is to develop explicit MPC strategies for the control of depth of anaesthesia in the induction and maintenance phases. State estimation techniques are designed and implemented simultaneously with mp-MPC strategies to estimate the state of each individual patient. Furthermore, a hybrid formulation of the patient model is performed, leading to a hybrid mp-MPC that is further implemented using several robust techniques.

Chapter 5 is focused on type 1 diabetes mellitus, more specifically on modelling, model analysis, optimisation and glucose regulation. The basic idea is to develop an automated insulin delivery system that would mimic the endocrine functionality of a healthy pancreas. The first level is the development of a high-fidelity mathematical model that represents in depth the complexity of the glucoregulatory system, presents adaptability to patient variability and demonstrates adequate capture of the dynamic response of the patient to various clinical conditions (normoglycaemia, hyperglycaemia and hypoglycaemia). This model is then used for detailed simulation and optimisation studies to gain a deep understanding of the system. The second level is the design of model-based predictive controllers by incorporating techniques appropriate for the specific demands of this problem.

The last three chapters are focused on the development of a systematic framework for the personalised study and optimisation of leukaemia (i.e., a severe cancer of the blood): from *in vivo* to *in vitro* and *in silico*. More specifically, Chapter 6 is a general description of the independent building blocks of the integrated framework, which are further analysed in the next chapters. Chapter 7 focuses on the detailed description of the *in vitro* building block of the framework. More specifically, it includes analysis of the disease, analysis of the experimental platform and environmental (stress) stimuli that are monitored within the platform, and a description of cellular biomarkers for monitoring the evolution of leukaemia *in vitro*. Chapter 8 focuses on the *in silico* building block of the framework. It describes the pharmacokinetic and pharmacodynamic models developed for the optimisation of chemotherapy treatment for leukaemia. Finally, the simulation results and analysis of a patient case study are presented.

The main outcome of this work is to develop models and model-based control and optimisation methods and tools for drug delivery systems, which would ensure: (a) reliable and fast calculation of the optimal drug dosage without the need for an on-line computer, while taking into account the specifics and constraints of the patient model (personalised health care); (b) flexibility to adapt to changing patient characteristics, and incorporation of the physician's performance criteria; and (c) safety of the patients, as optimisation of drug infusion rates would reduce the side effects of treatment. The major novelty introduced by mobile technology is that it is no longer necessary to trade off control performance against hardware and software costs in drug delivery systems. The parametric control technology will be able to offer state-of-the-art model-based optimal control performance in a wide range of drug delivery systems on the simplest of hardware. All of this will lead to some very important advantages, like: enhancing the confidence of medical teams to use computer-aided practices, increasing the confidence of patients to use such practices, enhancing safety by carefully applying a "what-if" prior analysis tailored made to patients' needs, a simple "look-up function," an optimal closed-loop response and cheap hardware implementation.

The book shows the newest developments in the field of multiparametric model predictive control and optimisation and their application for drug delivery systems.

This work was supported by the European Research Council (ERC), that is, by ERC-Mobile Project (no. 226462), ERC-BioBlood (no. 340719), the EU 7th Framework Programme (MULTIMOD Project FP7/2007-2013, no. 238013), the Engineering and Physical Sciences Research Council (EPSRC: EP/G059071/1 and EP/I014640), the Richard Thomas Leukaemia Research Fund and the Royal Academy of Engineering Research Fellowship (to Dr. Ruth Misener).

Part I

1

Framework and Tools: A Framework for Modelling, Optimization and Control of Biomedical Systems

Eirini G. Velliou[1], *Ioana Naşcu*[2], *Stamatina Zavitsanou*[3], *Eleni Pefani*[4], *Alexandra Krieger*[5], *Michael C. Georgiadis*[6], *and Efstratios N. Pistikopoulos*[7]

[1] *Department of Chemical and Process Engineering, Faculty of Engineering and Physical Sciences, University of Surrey, UK*
[2] *Artie McFerrin Department of Chemical Engineering, Texas A&M University, College Station, USA*
[3] *Paulson School of Engineering & Applied Sciences, Harvard University, USA*
[4] *Clinical Pharmacology Modelling and Simulation, GSK, UK*
[5] *Jacobs Consultancy, Kreisfreie Stadt Aachen Area, Germany*
[6] *Laboratory of Process Systems Engineering, School of Chemical Engineering, Aristotle University of Thesaloniki, Greece*
[7] *Texas A&M Energy Institute, Artie McFerrin Department of Chemical Engineering, Texas A&M University, USA*

1.1 Mathematical Modelling of Drug Delivery Systems

Drug delivery can be defined as the process of administering a pharmaceutical agent in the human body, including the consequent effects of this agent on the tissues and organs. Mathematical modelling of drug delivery can be divided into two different yet complementary approaches, the pharmacokinetic and pharmacodynamic approaches. Pharmacokinetics describes the effect of the drug in the body, by capturing absorption, distribution, diffusion and elimination of the drug. Pharmacodynamics describes the effects of a drug in the body, which are expressed mathematically by relations of drug dose–body responses. Usually, modelling of the drug delivery system requires a pharmacokinetic part, a pharmacodynamics part and a link between the two (Figure 1.1).

1.1.1 Pharmacokinetic Modelling

Two approaches for pharmacokinetic models dominate the literature, the compartmental models and the physiologically based pharmacokinetic models.

1.1.1.1 Compartmental Models

The basic idea of compartmental modelling is to group organs with similar properties, such as the well-perfused organs, in one compartment and describe the uptake based on these tissues' properties (e.g. drug solubility and perfusion). The basic assumptions of compartmental modelling are: (a) homogeneity: uniform distribution and instant mixing within the compartment; (b) conservation of mass;

Modelling Optimization and Control of Biomedical Systems, First Edition.
Edited by Efstratios N. Pistikopoulos, Ioana Naşcu, and Eirini G. Velliou.

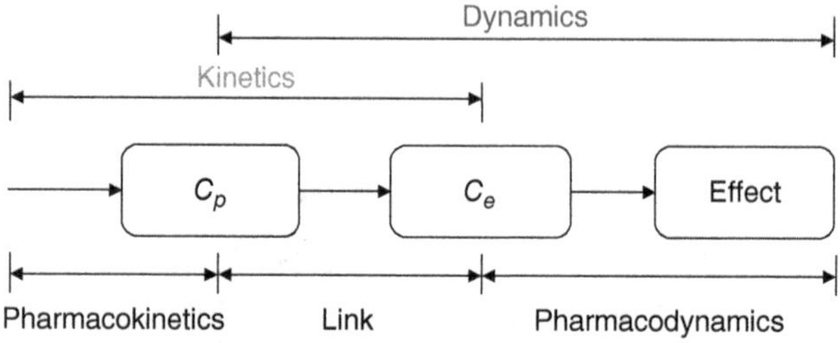

Figure 1.1 Mathematical representation of a drug delivery system. *Source*: Ette and Willliams (2007). Reproduced with permission of John Wiley and Sons.

(c) the intrinsic properties are constant (e.g. temperature and volume); (d) there are no time delays between compartments; and (e) all exiting fluxes are linearly proportional to the drug concentration in the compartment.

The simplest approach is to consider the whole body as one single compartment in which the drug is administered and also eliminated. Usually, this mathematical approach is used for the description of drugs that are intravenously injected and well diffused, the elimination of which follows first-order kinetics. Practically, within the human body, usually more than one compartment is considered due to the slow diffusion of the drug to the peripheral tissues (Figure 1.2).

There are several challenges related to compartmental model development, such as the correlation of the model parameters (e.g. transfer coefficients) to physiological parameters, as well as difficulties related to the determination of the appropriate number of compartments that should be used in order to represent the pharmacokinetics of a population. Furthermore, the ability of these models to give a valid estimation of the drug profile of a newly studied patient is rather questionable. The major source of model uncertainty is due to the fact that the values of the variables are based on the interpretation of the mean concentration profile of a group of patients. This *mean concentration profile* in most of the cases is not representative of the behaviour of patients in the group studied, let alone the whole patient population. These drawbacks are satisfied to a certain extent by the physiologically based pharmacokinetic models.

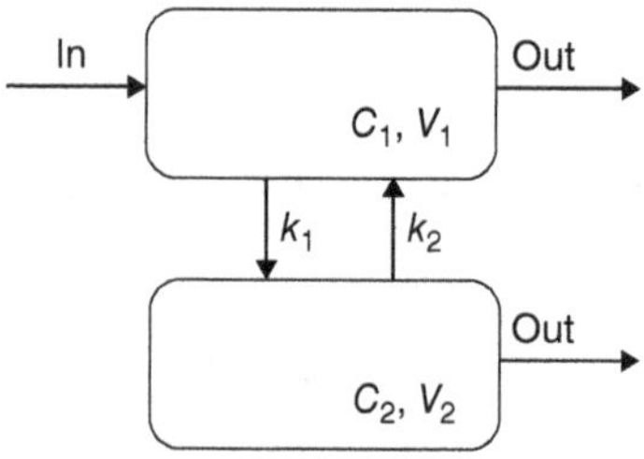

Figure 1.2 Schematic of a two-compartment pharmacokinetic model. *Source*: Saltzman (2001). Reproduced with permission of Oxford University Press.

1.1.1.2 Physiologically Based Pharmacokinetic Models

Physiological models are high compartmental models that use existing knowledge of the physiological mechanisms which regulate the drug action. These models capture the administration, diffusion and elimination of a drug in body organs that react with the drug. The drug mass balance for each organ can be described by Equation 1.1 (Saltzman, 2001):

$$V_i \frac{dC_i}{dt} = Flow_{in} - Flow_{out} - q_{el,i} V_i \tag{1.1}$$

where i is each specific organ/compartment, V_i is the organ volume, C_i is the drug concentration in the organ/compartment i, $q_{el,i}$ is the rate of drug metabolism in the organ/compartment i, and $Flow_{in}$ and $Flow_{out}$ are the inflow and outflow of the drug in the organ/compartment i.

A schematic overview of a physiological pharmacokinetic model, where each body organ is considered an independent compartment, is shown in Figure 1.3.

This modelling approach requires an in-depth understanding of the physiology, but it describes more accurately than empirical compartmental models the drug delivery system. The advantages of physiologically based models over empirical compartmental models lie in the ability to be extrapolated between different species and different drug dosages (Cashman *et al.*, 1996; Saltzman, 2001). The main drawback of physiologically based models is that, sometimes, certain parameters cannot be measured, and their values are difficult to be accurately predicted.

The description of one compartment itself in either of the previously mentioned approaches can be described by complex interactions and flows between, for example, blood cells, plasma, intestinal fluid, a rapid interactive pool and a slow interactive pool.

Both compartmental and physiological models range from simple to more detailed models that are based on fewer assumptions. Simplifications in the previous scheme can be made, depending on the exact system which is studied. In Figure 1.3, organs which do not contain important amounts of the drug agent can be neglected (Saltzman, 2001). However, the level of detail added to the model depends on the data availability and the purpose of the model.

1.1.2 Pharmacodynamic Modelling

Pharmacodynamic models describe the effect of a drug in the body (i.e. the impact of a drug that enters the cell on the cellular function). Due to the high complexity of the drug mechanism of action that enables precise measurements of the drug effect, detailed pharmacodynamic models are not in use and empirical expressions which correlate the drug concentration with the drug effect are

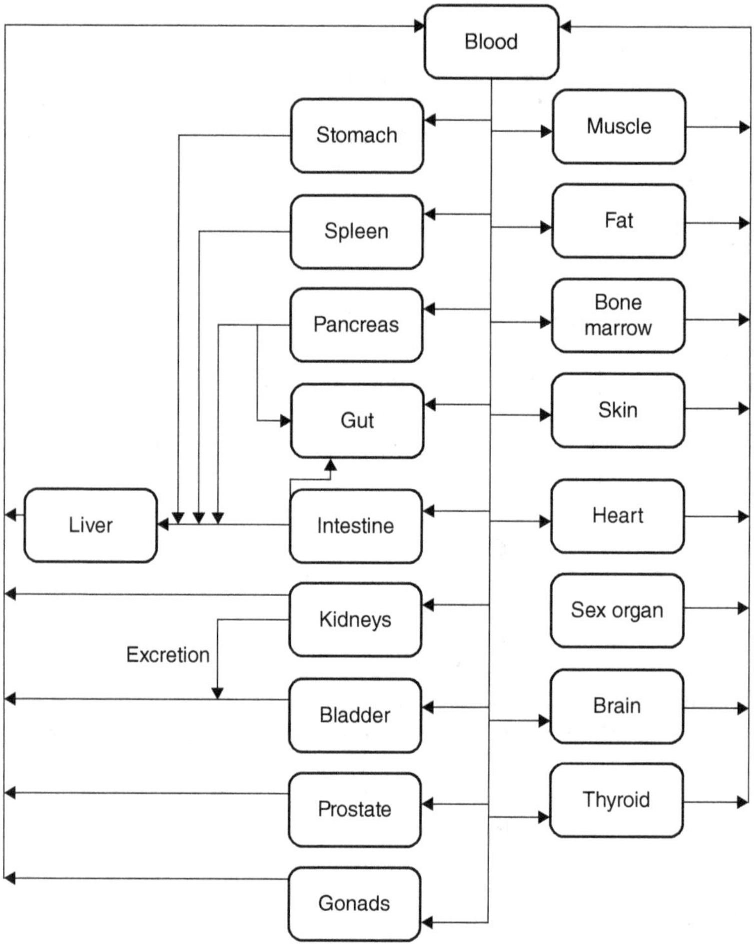

Figure 1.3 Schematic of a physiological pharmacokinetic model. *Source*: Saltzman (2001). Reproduced with permission of Oxford University Press.

more preferable (Holford & Sheiner, 1982). Practically, the pharmacodynamic model is determined by testing potential models and estimating the parameters when a reference pharmacokinetic model is used, and the accuracy of the pharmacodynamic model is highly dependent on precision of the pharmacokinetic model. The usage of a pharmacokinetic model is essential for the valuable expression of a pharmacodynamic model, as the latter assumes that the concentration of the drug is in equilibrium with the effect site, which might be the case only in the steady state.

Table 1.1 The most common types of empirical pharmacodynamic models.

Model	Model equations	Description
Fixed-effect model	–	Effect: present (1) or absent (0), or degree of effect
Linear model	$E = S \cdot C + E_o$	E = drug effect, C = drug concentration, S = slope parameter, E_o = initial drug effect
Log-linear model	$E = S \cdot \log C + I$	E = drug effect, C = drug concentration, S = slope parameter, I = constant
$E_{\max}$ model	$E = E_o - \frac{E_{\max} \cdot C}{EC_{50} + C}$	E = drug effect, C = drug concentration, E_{max} = maximum drug effect, E_o = initial drug effect from previous application, EC_{50} = concentration producing half of the maximum drug effect
Sigmoid $E_{\max}$ model	$E = \frac{E_{\max} \cdot C^n}{EC_{50}^n + C^n}$	E = drug effect, C = drug concentration, E_{max} = maximum drug effect, E_o = initial drug effect from previous application, EC_{50} = concentration producing half of the maximum drug effect, n = constant affecting the shape of the drug effect–concentration curve

Source: Holford and Sheiner (1982). Reproduced with permission of Elsevier.

In general, pharmacodynamics is the study of dose–response relationships. For the development of pharmacodynamic models, target cells are exposed *in vitro* in different drug concentrations, and drug effect curves are obtained. These data are then used to fit empirical pharmacodynamic models (Table 1.1). An example of a common dose–response curve is presented in Figure 1.4. The drug effect curves are of crucial importance, especially for the early clinical trial phases, for the determination of maximal dose effect as well as for estimation of the effective drug dosing window.

1.2 Model analysis, Parameter Estimation and Approximation

Model analysis includes analysis of parameters and variables of the developed pharmacokinetic model, in order to define uncertainty of parameters. This uncertainty usually originates from inter-patient or experimental variability. In a consecutive step, the model is analysed towards its most influential parameters and variables. The methods that are usually used in order to obtain in-depth understanding of the model are global sensitivity analysis, variability analysis, parameter estimation and parameters correlation.

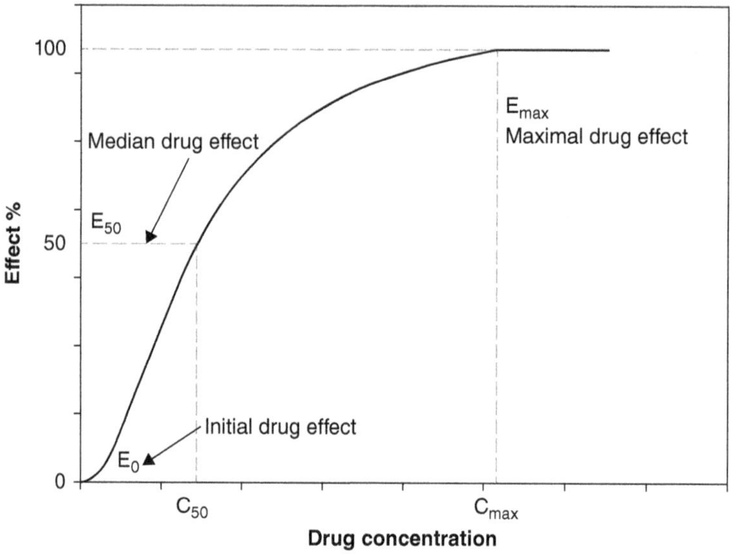

Figure 1.4 Illustration of a pharmacodynamic dose–response curve.

1.2.1 Global Sensitivity Analysis

Global sensitivity analysis allows the understanding and identification of crucial model parameters that affect the model output. In the case of mathematical models that describe biomedical systems, global sensitivity analysis enables the identification of the relative influence of parameters of the pharmacokinetic and/or pharmacodynamic part of the model, on the model output. Performance analysis is conducted in the graphical user interface/high-dimensional model representation (GUI-HDMR) software, which uses random sampling HDMR (RS-HDMR) to construct an expression for the output as a function of the parameters with orthogonal polynomials. This expression accounts for up to second-order interactions and corresponds to the ANOVA decomposition truncated to the second order. From the coefficients of the representation, the sensitivity index is derived. The sensitivity indices are calculated based on partial variances, which themselves are calculated from the approximation of the model by orthonormal polynomials (Li *et al.*, 2002; Ziehn and Tomlin, 2009).

1.2.2 Variability Analysis

Variability analysis focuses on the identification of the influence of the individual parameters and variables on the model outputs. Global sensitivity analysis gives a measure of the relative influence of each parameter on the output. However, that approach does not incorporate whether a higher or lower value of the parameter or variable of interest is increasing or decreasing the model output. Variability analysis enables the detection of the influence of each parameter and variable on the output, therefore facilitating the understanding of the actual physical influence of the pharmacokinetic and pharmacodynamic variables and parameters. In particular, when performing variability analysis, an investigation of whether an increase in the pharmacokinetic and/or pharmacodynamic variable or parameter increases or decreases the model output, y, takes place (Equation 1.2):

$$P_{\%,i} = \frac{y_{max,i} - y_{min,i}}{y_{nom}}. \tag{1.2}$$

where $P_{\%,i}$ is the percentage of change due to an increase in variable or parameter i, $y_{max,i}$ is the upper bound model output, $y_{min,i}$ is the lower bound model output and y_{nom} is the calculated nominal model output.

1.2.3 Parameter Estimation and Correlation

Parameter estimation is the process of fitting the model parameters to clinical data. If the parameters are estimated with high precision, then the model's response is closer to reality. The parameter estimation problem is evaluated by the correlation matrix C of the estimated parameters. An entry in the off-diagonal elements of the correlation matrix C close to one ($|Cij| \approx 1$) indicates a high correlation of the corresponding parameters i and j, whereas an entry of zero ($C_{ij} \approx 0$) indicates no correlation. The entries of the correlation matrix are calculated based on the variance–covariance matrix V, the variance of a parameter is given on the diagonal (V_{ii}) and the covariance of two parameters i and j is given on the off-diagonal elements (V_{ij}).

$$C_{i,j} = \frac{V_{i,j}}{\sqrt{V_{ii}V_{jj}}}, i \neq j \tag{1.3}$$

$$C_{ii} = 1 \tag{1.4}$$

1.3 Optimization and Control

Mathematical optimization and control of biomedical systems could lead to a better prediction of the optimal drug and/or therapy treatment for a specific disease. Advanced mathematical and computational techniques such as multiparametric predictive control, sensitivity analysis and model reduction are extensively discussed in Chapter 2. Moreover, those techniques are applied in a variety of diseases (i.e. anaesthesia, diabetes and leukaemia) that are further discussed in the following chapters.

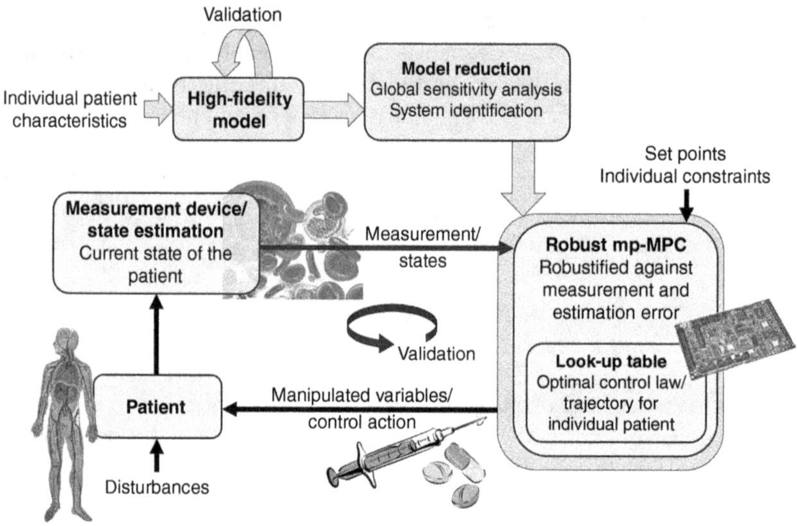

Figure 1.5 Framework towards optimal drug delivery systems.

Anaesthesia (see Chapters 3 and 4) is a process which provides hypnosis, analgesia and muscle relaxation while maintaining the vital functions of a living organism. For efficient prediction and control of this bio-process, a model predictive controller (see Chapter 2) is required.

In *type 1 diabetes* (see Chapter 5), the goal is to maintain the blood's glucose concentration within normal levels. From a mathematical point of view, this can be formulated as a model predictive control problem.

In acute myeloid leukaemia (see Chapters 6, 7, and 8), the ultimate goal is to determine the optimal chemotherapy dose that would lead to minimization of the cancerous population while maintaining the normal/healthy population above a minimum acceptable level. From a computational point of view, this is a scheduling problem.

For all of these diseases, the overall framework for the design of optimal drug delivery systems is presented in Figure 1.5.

In order to move towards the design of optimal delivery systems, development of a high-fidelity model able to describe the biomedical problem for an individual patient has to take place. Identification of parameters that crucially affect the model output enables the model reduction and, at a second step, the predictive control of the drug dose, therefore leading to process optimization.

Especially in the case of acute myeloid leukaemia, we have developed an appropriate *in vitro* system which allows *ex vivo* experimentation of leukaemic patient cells for the more efficient understanding and further identification of parameters that crucially affect the model output (i.e. the drug dose determination). Moreover, experimental data serve as an input for our mathematical model, allowing validation and improvement (Chapters 6, 7, and 8). Therefore, this *in vivo–in vitro–in silico* closed loop enables the accurate study and further determination of the optimal drug dose for an individual/specific patient (Velliou *et al.*, 2014).

References

Cashman, J.R., Perotti, B.Y., Berkman, C.E., & Lin, J. (1996). Pharmacokinetics and molecular detoxication. *Environmental Health Perspectives*, 104(Suppl. 1), 23–40.

Ette, E.I., & Williams, P.J. (2007). *Pharmacometrics: the science of quantitative pharmacology*. Hoboken, NJ: John Wiley & Sons.

Holford, N.H.G., & Sheiner, L B. (1982). Kinetics of pharmacologic response. *Pharmacology & Therapeutics*, 16, 143–166.

Li, G., Wang, S.W., Rabitz, H., Wang, S., & Jaffe, F. (2002). Global un-certainty assessments by high dimensional model representations (HDMR). *Chemical Engineering Science*, 57, 4445–4460.

Saltzman, W.M. (2001). *Drug delivery: engineering principles for drug therapy*. Oxford: Oxford University Press.

Velliou, E., Fuentes-Garí, M., Misener, R., Pefani, E., Rende, M., Panoskaltsis, N., Pistikopoulos, E.N., & Mantalaris, A. (2014). A framework for the design, modeling and optimization of biomedical systems. In M. Eden, J.D. Siirola, & G.P. Towler (Eds.), *Proceedings of the 8th International Conference on Foundations of Computer-Aided Process Design – FOCAPD*. Amsterdam: Elsevier.

Zichn, T., & Tomlin, A.S. (2009). Environmental modelling & software GUI-HDMR – a software tool for global sensitivity analysis of complex models. *Environmental Modeling & Software*, 24(7), 775–785.

2

Draft Computational Tools and Methods

Ioana Naşcu[1], Richard Oberdieck[2], Romain Lambert[3], Pedro Rivotti[3], and Efstratios N. Pistikopoulos[4]

[1] *Artie McFerrin Department of Chemical Engineering, Texas A&M University, College Station, USA*
[2] *DONG energy A/S, Gentofte, Denmark*
[3] *Department of Chemical Engineering, Imperial College London, UK*
[4] *Texas A&M Energy Institute, Artie McFerrin Department of Chemical Engineering, Texas A&M University, USA*

2.1 Introduction

This chapter focuses on describing the computational tools that are part of the developed multiparametric model predictive control (MPC) framework presented in Chapter 1. The framework enables the solution of demanding optimization and control problems through a step-by-step procedure presented in this chapter. The key advantage of this is that it follows a multiparametric approach for the controller design that transfers the computational burden offline (Pistikopoulos 2000). Furthermore, the proposed procedure is not process dependent and can be adapted to any process at hand. All the steps included in the framework are realized through the developing software platform PAROC (PARametric Optimization and Control). PAROC is a user-friendly software platform that utilizes the communication between gPROMS ModelBuilder and MATLAB. Through this software interoperability, the multiple steps are realized in a way that convenient for the user and, most importantly, tractable.

A comprehensive schematic representation of the framework is shown in Figure 2.1, and a thorough explanation of the computational tool required for the steps is provided within this chapter.

The high-fidelity model developed in the *modelling and design optimization* step usually results in *d*ifferential-*a*lgebraic *e*quation (DAE) systems of high complexity. The DAE systems are approximated by discrete time models in state-space representation. In order to do that, complex model–order reduction techniques as well as identification methods and toolboxes are employed. The key objectives are to simplify the representation of the system without compromising the accuracy of the high-fidelity model. Although there is a

Modelling Optimization and Control of Biomedical Systems, First Edition.
Edited by Efstratios N. Pistikopoulos, Ioana Naşcu, and Eirini G. Velliou.

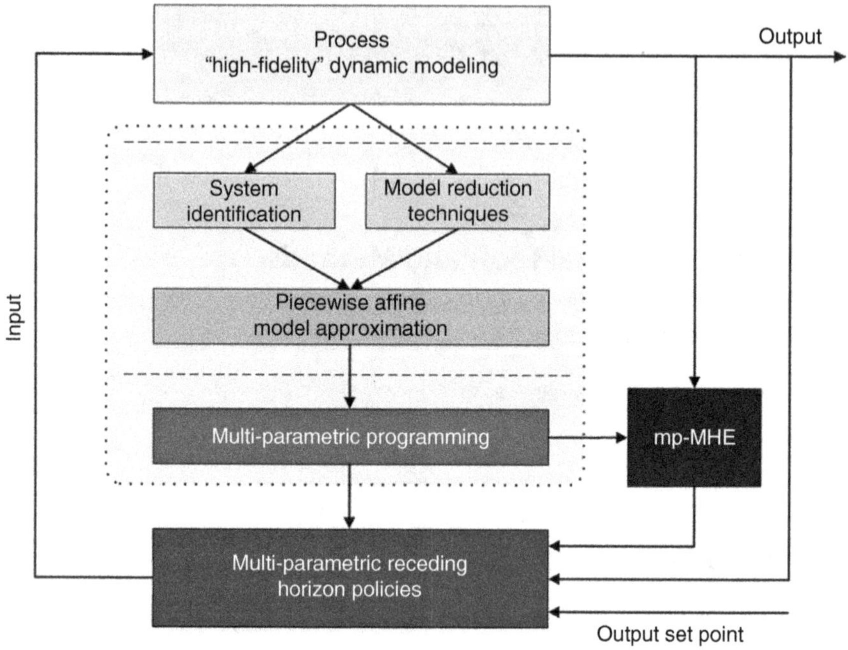

Figure 2.1 A framework for explicit/multiparametric model predictive control and moving horizon estimation. *Source*: Naşcu *et al.* (2016). Reproduced with permission of Elsevier.

variety of model reduction and approximation techniques (Lambert *et al.* [2013] and references therein), the System Identification Toolbox of MATLAB is also commonly used. In this chapter, we will focus on model reduction techniques as a method of model approximation.

2.2 Sensitivity Analysis and Model Reduction

2.2.1 Sensitivity Analysis

The use of sensitivity analysis in the context of biomedical engineering is of critical importance. Sensitivity analysis has been increasingly used for the assessment of the robustness of complex biological and biomedical models and in uncertainty quantification (Kontoravdi *et al.* 2005, 2010; Yue *et al.* 2008; Kiparissides *et al.* 2009; Kucherenko *et al.* 2009). This is particularly relevant in the field of pharmacometrics when trying to estimate the relative influence of pharmacokinetics, pharmacodynamics and other uncertain parameters. Sensitivity analysis is also used in model simplification as an approach to decrease the parametric dimensionality of biological systems. On some occasions, it might be possible to remove some parts of a model that do not significantly affect its response. This is usually done by fixing non-essential parameters to their mean value, so that more attention can be dedicated to critically important factors to perform tasks like parameter estimation or optimal design of experiment. In recent years, global sensitivity analysis has gained considerable attention due to its advantages over local sensitivity analysis approaches (Homma and Saltelli 1996; Saltelli 2004). Global sensitivity is model-independent by design and can detect parametric interactions, unlike one factor at a time (OAT) local methods (Saltelli *et al.* 2010). An eminent class of global sensitivity analysis techniques is that of variance-based method, which includes the well-known Sobol method of sensitivity indices (SIs) (Sobol 1993, 2001). One of the disadvantages of such methods that are based on Monte Carlo sampling is the necessity to repeatedly run potentially expensive simulations. This is exacerbated in the case of high-dimensional input spaces for which exploration may become computationally intractable. One way to reduce the computational expense of performing sensitivity analyses has been the use of surrogate models or meta-models. This approach consists of using relatively simpler models that emulate the dynamic behaviour of the original computationally intensive models. Various surrogate modelling approaches have been suggested, such as Gaussian process modelling, polynomial chaos expansion (PCE) (Sudret 2008), radial basis function (Buhmann 2003) and high-dimensional model representation (actually, a particular instance of PCE) (Li *et al.* 2002). The two main difficulties of these approaches are: the ability to handle higher dimension spaces and the sampling requirements to achieve convergence. For example, regression-based PCE approaches are better suited for systems with no more than 10 input variables (Blatman and Sudret 2010). Methods based on numerical integration like high-dimensional model representation (HDMR) are able to perform in high-dimensional spaces but may require a significant amount of sampling realization in order to achieve convergence. An efficient solution is the combined use of low computational screening methods to discard non-essential variables prior to the use of a variance-based method on the remaining parameters. One of the most commonly used screening methods is the Morris method (Morris 1991). A very powerful set of data-driven approaches is the class of inductive modelling methods, in particular the group method of data handling (GMDH) (Ivakhnenko and Muller 1995). The GMDH is based on the cybernetic principle of self-organization and has the ability to perform with limited data samples and in very high-dimensional spaces, by selecting important parameters in an adaptive fashion. Another advantage of the approach is its immunity to noise. This is a very relevant aspect, as in many cases the sensitivity analysis practitioner does not necessarily have access to a model but only noisy field data. The objective of this study is twofold: firstly, we demonstrate the screening capabilities of GMDH in combination with the HDMR approach; and, secondly, the noise immunity capabilities are evaluated on numerical examples. The mathematical fundamentals to HDMR and GMDH are introduced, and a simple methodology combining both techniques is presented. This methodology is then applied to first principle biomedical models.

2.2.1.1 Sobol's Sensitivity Analysis

Sobol's sensitivity analysis method is a variance-based approach based on the ANOVA decomposition (Sobol 2001; Sobol & Kucherenko 2005). If f is an integrable function defined on the unit hypercube I^n and $x \epsilon I^n$, $x = (x_1, \ldots, x_n)$ the input variables, the output $f(x)$ of the function may be expressed as:

$$f(x) = f_0 + \sum_{s=1}^{n} \sum_{i_1 < \ldots < i_s}^{n} f_{i_1 \ldots i_s}(x_{i_1}, \ldots, x_{i_s}) \tag{2.1}$$

where f_0 is the mean response of f; and the terms $f_i(x_i)$ and $f_{ij}(x_i, x_j)$ represent the first-order and second-order terms, and so on. The formula above is termed ANOVA decomposition. The component functions may then be expressed as integrals of f:

$$\begin{aligned} &\int f(x)dx = f_0 \\ &\int f(x) \prod_{k \neq i} dx_k = f_0 + f_i(x_i) \\ &\int f(x) \prod_{k \neq i,j} dx_k = f_0 + f_i(x_i) + f_{ij}(x_i, x_j) \end{aligned} \tag{2.2}$$

One of the best known global sensitivity analysis methods was introduced by Sobol (2001). If it is assumed that f is square integrable over I^n, we have:

$$\int f^2(x)dx - f_0 = \sum_{s=1}^{n} \sum_{i_1 < \ldots < i_s}^{n} \int f_{i_1 \ldots i_s} dx_1 \ldots dx_s \tag{2.3}$$

$$D = \int f^2(x)dx - f_0 \text{ and } D_{(i_1 \ldots i_s)} = \int f_{(i_1 \ldots i_s)} dx_1 \ldots dx_s \tag{2.4}$$

The terms represent the variance and partial variance, respectively. Sobol's SIs can be given by:

$$S_{i_1 \ldots i_s} = \frac{D_{i_1 \ldots i_s}}{D} \tag{2.5}$$

where:

$$\sum_{s=1}^{n} \sum_{i_1 < \ldots < i_s}^{n} S_{i_1 \ldots i_s} = 1 \tag{2.6}$$

If a set of variables $y = (x_1, \ldots, x_s)$ is considered and z is a set of the complementary variables, we note $x = (y, z)$. Using the previous definition of the variance, the total variance of the subset y can be computed as:

$$D_y^{tot} = D - D_z \tag{2.7}$$

and:

$$S_y^{tot} = \frac{D_y^{tot}}{D} \tag{2.8}$$

The following inequality holds:

$$0 \leq S_y \leq S_y^{tot} \leq 1 \tag{2.9}$$

If $S_y = S_y^{tot} = 0$, then *f* does not depend on *y*.

If $S_y = S_y^{tot} = 1$, then *f* only depends on *y*.

The indices enable us to rank variables and discard unessential variables. Sensitivity analysis indices are usually computed through Monte Carlo numerical integration (Sobol, 2001).

$$D_y = \int f(x) f(y,z)\, dxdz - f_0^2 \tag{2.10}$$

Using low-discrepancy sequences has been shown to increase the efficiency of the technique, especially Sobol's sequence for uniform sampling.

2.2.1.2 High-Dimensional Model Representation

In order to efficiently build the map of the input–output behaviour of a model function involving high-dimensional inputs (typically, $n \sim 10^2 - 10^3$), the HDMR approach was introduced as a set of quantitative tools. In most engineering problems, the expansion of functions can be truncated to the second-order component function by Li *et al.* (2002, 2006):

$$f(x) \approx h(x) = f_0 + \sum_{i=1}^{n} f_i(x_i) + \sum_{1 \leq i < j \leq n} f_{ij}(x_i, x_j) \tag{2.11}$$

A particular way of deriving an HDMR representation through Monte Carlo sampling is the random sampling HDMR (RS-HDMR) technique. Since the computation of multidimensional integrals may become prohibitive (Sobol, 1993), an alternative technique based on the use of interpolation of over families over low-order component functions has been introduced by Rabitz and co-workers (Li *et al.*, 2002). If a set of piecewise continuous component functions $\{\varphi\}$ is considered, we can derive:

$$\begin{aligned} f_i(x_i) &= \sum_{r=1}^{k} \alpha_r^i \varphi_r(x_i) \\ f_{ij}(x_i, x_j) &= \sum_{p=1}^{l} \sum_{q=1}^{m} \beta_{pq}^{ij} \varphi_p(x_i) \varphi_q(x_j) \end{aligned} \tag{2.12}$$

Once a family of component functions has been selected, the coefficients

$$\begin{aligned}\forall r \in [1,k], \alpha_r^i &= \int_0^1 f_i(x_i)\varphi_r(x_i)dx_i \\ \forall p \in [1,l], \forall q \in [1,m], \beta_{pq}^{ij} &= \int_0^1\int_0^1 f_{ij}(x_i,x_j)\varphi_p(x_i)\varphi_q(x_j)dx_i dx_j\end{aligned} \tag{2.13}$$

In practice, these calculations are done through Monte Carlo integration. There is a direct relationship between the HDMR expansion coefficients and Sobol's sensitivity analysis technique:

$$D_i = \sum_{r=1}^{k_i}\left(\alpha_r^i\right)^2 \tag{2.14}$$

and

$$D_{ij} = \sum_{p=1}^{l_i}\sum_{q=1}^{l_j}\left(\beta_{pq}^{ij}\right)^2 \tag{2.15}$$

The sensitivity indices are obtained by dividing with the total variance, even though the total effect coefficients and the total variance involving interaction orders greater than three will still require the use of Sobol's original approach.

Although HDMR has been very successful in a number of sensitivity analysis studies, it can be problematic in the case of a large number of parameters. The calculation of its component often requires large sampling sets, even though the method is able to present high-dimensional input–output relationships. In the case of computationally intensive simulation models, this may become very impractical.

2.2.1.3 Group Method of Data Handling

GMDH is based on the principle of self-organization and is sometimes referred to as *polynomial neural networks.* This technique is based on representing complex functions through networks of elementary expressions, like other advanced surrogate-modelling approaches such as neural networks or the HDMR approach. Lorentz (1966) and Kolmogorov (1957) have shown that any continuous function $f(x_1, \ldots, x_d)$ of dimension d on $[0,1]^d$ can be exactly represented as a composition of sums and continuous one-dimensional functions. The GMDH approach is very efficient in data-driven modelling of complex systems, with several advantages over conventional neural networks. We can refer to Ivakhnenko and Muller (1995) and Lemke (1997) for more ample theoretical description of the method. An advantage over the classical neural networks is that GMDH is inductive, adaptively creating models from data under the form of networks of optimized active neurons in an evolutionary manner. The aim is to estimate an optimal structure of a network that self-organizes itself during training, making this a combined structure and parameter estimation procedure that starts from a basic structure of the mean value of the time series output data.

A first layer is built by considering all possible variable pairs and inductively self-constructing and validating neurons made of simple expressions, usually within linear or second-order polynomials. This will result in a set of transfer functions for the first network layer. A number of fittest and best generalizing models consisting of neurons are then selected via an external criterion. After each single induction step, model validation is performed as an integrated critical part of model self-organization. In the classical approach, in order to create a new layer, the selected neurons are subsequently used as inputs, while other neurons are discarded. More complex organizations can be generated by using the selection criterion and using the cybernetics inheritance principle. The final optimal complex structure consists of a single network. There is no need to predefine the number of neurons or layers to be used since they are adaptively determined through the learning process.

The model self-organization stops itself when an optimal complex model has been found (i.e., further increasing model complexity would result in over-fitting the design data by starting to adapt to noise). This is an important advantage over the RS-HDMR approach or regression-based PCE, which require the computation of a full set of predefined parameters. HDMR requires the computation of a large number of α_r^i coefficients through numerical integration β_{pq}^{ij}, for many combinations of parameters (x_i, x_j) and polynomial orders, and unessential parameters can only be weeded out *a posteriori* upon calculation of these coefficients.

2.2.1.4 GMDH-HDMR

As shown in this chapter, GMDH holds a number of advantages that are essential to global sensitivity analysis. The method is able to handle high dimensionalities, this being important in the context of biomedical engineering. Moreover, GMDH is, by design, a very efficient screening procedure in itself by adaptively weeding out unessential parameters in a computationally tractable manner. Also, it has good performance for small data samples. The presented method is based on the direct construction of the HDMR expansion by using GMDH inductive modelling. If a set of parameters $(x_i)_{i\in[\![1,n]\!]}$ is considered, additional 'synthetic' variables are built. These correspond to Legendre orthogonal polynomials of up to a predefined order n and evaluated on the original variables: $X_{r,i} = \varphi_r(x_i), r = 1..n$.

The GMDH algorithm is performed only on these variables, imposing a multilinear relationship between the variables. For the calculation of Sobol's SIs, the coefficient of the GMDH expression is used.

The main advantage of this method is its inductive ability to eliminate unessential parameters during the modelling process, leading to the elimination of the calculation of coefficients for parameters that do not contribute to the variance of the output. The method, indeed, incorporates the screening step and calculation of SIs in a single procedure.

2.2.2 Model Reduction

Model order reduction (MOR) describes a methodology intended to reduce the dimensionality of a dynamical system while preserving its input–output behaviour (Figure 2.2). The main purpose of MOR originally stemmed from a need to derive approximations of large-scale dynamical systems for simulation purposes. One major area of application has concerned the reduction of finite element models originating from the discretization of large-scale systems of ordinary differential equations (ODEs), differential algebraic equations (DAEs), partial differential equations (PDEs) and partial differential algebraic equations (PDAEs). In effect, sophisticated discretization techniques yield computationally prohibitive high-dimensional systems. These discretized systems tend to be extremely complex and sometimes intractable for the purposes of prediction and simulation, and even more so in the case of the resolution of inverse problems characterizing optimization, parameter estimation and MPC. In the context of multiparametric/explicit MPC, this complexity takes a very specific meaning. Indeed, complexity directly materializes in a steep increase in the

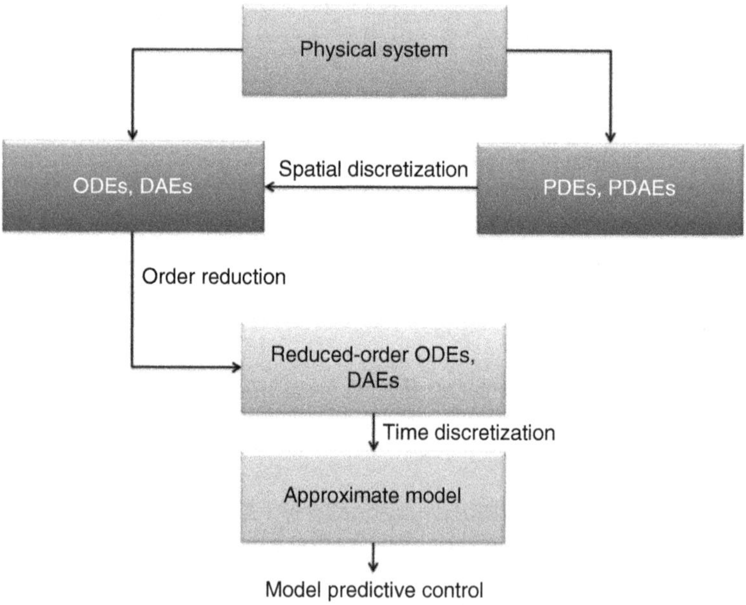

Figure 2.2 Schematic representation of the MOR approximation procedure.

number of critical regions, which results from the compounded effect of a high number of state variables (parameters) and constraints (dependent on the length of the prediction horizon).

2.2.2.1 Linear Model Order Reduction

An important class of model reduction techniques concerns linear systems. A major area of application of this class of problem has been the reduction of large-scale microelectromechanical systems (Antoulas, 2005). Most MOR techniques are projection based (i.e. they consist of projecting the dynamics of the original system on a lower dimensional subspace). One major class of methods is singular value decomposition (SVD) methods, which are based on the more general concept of principal component analysis (PCA). PCA is a procedure concerned with inferring the covariance structure of a system by converting a set of observations of possibly correlated variables into a set of values of linearly uncorrelated variables called the *principal components*. The transformation results in a hierarchized set of principal components ordered by decreasing variance. In particular, it allows the identification of the principal directions (e.g. state variables) in which the data vary. The two main classes of MOR techniques are SVD methods and moment-matching approaches. In balanced truncation, a transformation is operated that projects the system dynamics in a space where the most observable systems correspond to the most controllable ones. Following the procedure described in Antoulas (2005), we formulate a dynamical system in an equivalent balanced form:

$$\begin{aligned} x_{t+1} &= Ax_t + Bu_t \\ y_t &= Cx_t \end{aligned} \tag{2.16}$$

The linear gramians' controllability and observability gramians, W_C and W_O, are defined as the unique positive definite solution to the Lyapunov equations:

$$\begin{aligned} AW_C + W_C A^T &= -BB^T \\ A^T W_O + W_O A &= -C^T C \end{aligned} \tag{2.17}$$

Finding a balanced form for these gramians consists of finding a diagonal matrix Σ such that:

$$\bar{W}_O = \bar{W}_C = \Sigma = diag\left[(\sigma_i)_{i\in[1,n]}\right] \tag{2.18}$$

where:

$$\begin{aligned} \bar{W}_C &= TW_C T_T \\ \bar{W}_O &= \left(T^{-1}\right)^T W_O T^{-1} \end{aligned} \tag{2.19}$$

where T is a transformation matrix, and the σ_i are the Hankel singular values. The transformation matrix is then used to reformulate the dynamical system in an equivalent balanced form:

$$\begin{aligned}\bar{x}_{t+1} &= TAT^{-1}\bar{x}_t + TBu_t \\ y_t &= CT^{-1}x_t\end{aligned} \tag{2.20}$$

It is possible to truncate the system by retaining the states accounting for most of its dynamical behaviour by partitioning the balanced system: noting $\bar{A} = TAT^{-1}$ and $\bar{B} = TB$, a reduced-order LTI is obtained:

$$\begin{aligned}\bar{x}_{1t+1}A_{11}\bar{x}_{1t} &+ B_1u_t \\ y_t &= C_1\bar{x}_{1t}\end{aligned} \tag{2.21}$$

where:

$$\bar{A} = \begin{bmatrix} A_{11} & A_{12} \\ A_{21} & A_{22} \end{bmatrix}, \bar{B}\begin{bmatrix} B_1 \\ B_2 \end{bmatrix}, \bar{C} = \begin{bmatrix} C_1 & C_2 \end{bmatrix}, \bar{x} = \begin{bmatrix} \bar{x}_1 \\ \bar{x}_2 \end{bmatrix} \tag{2.22}$$

Those synthetic (i.e. physically meaningless) states form an ordered set of decreasing controllability and observability. Another very important class of linear MOR techniques is that of moment-matching approaches. This class of method consists of the interpolation of the transfer function of a system, usually via the Pade approximation (Gallivan *et al.* 1994). It also belongs to the wider class of projection techniques known as *Krylov subspace methods* (Krylov 1931). Two widely used moment-matching methods are those of Arnoldi (1951) and Lanczos (1950). Current research concerns the combination of the two paradigms (Antoulas and Sorensen 2001). These techniques are commonly referred to as *SVD-Krylov methods.* For a thorough overview of linear MOR techniques, the reader will refer to Antoulas (2005). In some cases, a linear system is not sufficient to accurately capture the dynamics of a dynamical system. As linearization potentially leads to a significant loss of information, nonlinear model reduction approaches are introduced.

2.2.2.2 Nonlinear Model Reduction

The second approach employed is nonlinear balanced truncation, which is a snapshot-based technique and an empirical extension of the linear balanced truncation technique. Consider a nonlinear system of ODEs of the following form:

$$\begin{aligned}\dot{x}(t) &= f(x(t), u(t)) \\ y(t) &= h(x(t), u(t))\end{aligned} \tag{2.23}$$

As in linear balanced truncation, the method consists of finding a transformation matrix T in order to project the state vector on a lower order subspace $\bar{x} = Tx$. In order to compute these matrices, empirical gramians or covariance matrices are derived from simulation data from the system.

Defining the following sets:

$$\begin{aligned}
&T^n = \left\{T_1,...,T_r; T_i \in \Re^{n\times n}, T_i^T T_i = I, i = 1,...,r\right\} \\
&M = \left\{c_1,...,c_s; c_i \in \Re, c_i \geq 0 = I, i = 1,...,s\right\} \\
&E^n = \left\{e_1,...,e_n\right\} \text{standard unit vectors in } \Re^n
\end{aligned} \tag{2.24}$$

where r is the number of matrices for perturbation directions, s the number of different perturbation sizes for each direction, and n the number of inputs of the system. Using the sets above, it is possible to derive empirical controllability and observability gramians as follows:

$$W_C = \sum_{l=1}^{r}\sum_{m=1}^{s}\sum_{i=1}^{p}\frac{1}{rsc_m{}^2}\int_0^\infty \Phi^{ilm}(t)dt \tag{2.25}$$

$\Phi^{ilm}(t) \in \Re^{n\times n}$ is given by $\Phi^{ilm}(t) = \left(x^{ilm}(t) - x_0{}^{ilm}\right)\left(x^{ilm}(t) - x_0{}^{ilm}\right)^T$, where $x^{ilm}(t)$ is the state of the nonlinear system corresponding to the impulse input, $u(t) = c_m T_l e_i \delta(t) + u_0$; and $x_0{}^{ilm}$ corresponds to the steady state of the system. Similarly, an empirical observability gramian is defined by:

$$W_C = \sum_{l=1}^{r}\sum_{m=1}^{s}\frac{1}{rsc_m{}^2}\int_0^\infty T_l \Psi^{lm}(t) T_l^T dt \tag{2.26}$$

$\Psi^{lm}(t) \in \Re^{n\times n}$ is defined as $\Psi^{lm}{}_{ij}(t) = \left(y^{ilm}(t) - y_0^{ilm}\right)\left(y^{jlm}(t) - y_0^{jlm}\right)$, where $y^{ilm}(t)$ is the output of the system corresponding to the initial condition $x_0 = c_m T_l e_i + x_0$. The y_0^{ilm} corresponds to the output measurement when the system is at steady state. A balanced system is then obtained from the previously defined empirical gramians as:

$$\begin{aligned}
&\dot{\bar{x}}(t) = Tf(T^{-1}\bar{x}(t), u(t)) \\
&y(t) = g(T^{-1}\bar{x}, u)
\end{aligned} \tag{2.27}$$

Using a Garlekin projection $P = [I, 0]$ matrix with the same rank as the reduced system, the unimportant states may be set at a nominal steady-state value and the nonlinear reduced-order model:

$$\begin{aligned}
&\dot{\bar{x}}_1(t) = PTf(T^{-1}P^T\bar{x}(t), u(t)) \\
&\dot{\bar{x}}_2(t) = \bar{x}_{2ss}(0) \\
&y = g(T^{-1}\bar{x}, u)
\end{aligned} \tag{2.28}$$

Note that in the case of the presence of parametric uncertainty, the system may be reduced by treating the parameters as exogenous inputs in a similar way as the method described above:

$$\begin{aligned}
&\dot{x}(t) = f(x(t), u(t), \theta(t)) \\
&y(t) = h(x(t), u(t), \theta(t))
\end{aligned} \tag{2.29}$$

Table 2.1 Classification of the main order reduction techniques

Linear systems (Antoulas 2000)		Nonlinear systems	
Moment-matching methods	**SVD methods**	**SVD methods**	**Linearization based**
Arnoldi (Arnoldi 1951) Lanczos (Lanczos 1950)	Balanced truncation (Moore 1982) Singular perturbation method (Kokotovix 1976) Hankel approximation (Adamjan, Arov *et al.* 1971; Antoulas and Sorensen 2001)	POD (Wong 1971; Astrid 2004) Empirical balanced truncation (Lall, Marsden *et al.* 1999; Hahn and Edgar 2002)	TPWL (Rewieński and White 2001)

Source: P. V. Kokotovic, R. E. O'Malley, P. Sannuti, Singular Perturbations and Order Reduction in Control Theory - an Overview, Automatica, 12: 123–132, 1976.

Simply by posing $\tilde{u} = \begin{pmatrix} u \\ \theta \end{pmatrix}$.

A classification of linear and nonlinear model reduction techniques can be found in Table 2.1, and Table 2.2 presents a summary of the literature on MOR for multiparametric model predictive control (mp-MPC) applications.

2.3 Multiparametric Programming and Model Predictive Control

Multiparametric programming is a technique for solving any optimization problem, where the objective is to minimize or maximize a performance criterion subject to a given set of constraints and where some of the parameters vary between specified lower and upper bounds. The main characteristic of multiparametric programming is its ability to obtain: (1) the objective and optimization variable as a function of the varying parameters, and (2) the regions in the space of the parameters where these functions are valid.

The advantage of using multiparametric programming to address these problems is that for problems pertaining to plant operations, such as for process planning, scheduling and control, one can obtain a complete map of all the optimal solutions. Hence, as the operating conditions vary, one does not have to reoptimize for the new set of conditions (Pistikopoulos *et al.* 2007).

A general multiparametric programming problem may be formulated as follows:

$$\begin{aligned} & \min f(x,\theta) \\ & s.t.\ g_i(x,\theta) \leq 0, \quad \forall i = 1,..,p \\ & \quad h_j(x,\theta) = 0, \quad \forall j = 1,...,q, \\ & \quad x \in X \subseteq \Re^n, \\ & \quad \theta \in \Theta \subseteq \Re^m, \end{aligned} \tag{2.30}$$

Table 2.2 Summary of the literature on model order reduction for mp-MPC applications.

Authors	Methodologies	Key features
Narciso and Pistikopoulos (2008)	Balanced truncation, mp-MPC	Combines linear balanced truncation and explicit MPC, incorporating the error bound into the control formulation
Singh and Hahn (2005)	Empirical balanced truncation, Luenberg-type observers	State estimation on nonlinear reduced-order models obtains through empirical balanced truncation.
Hovland *et al.* (2008)	POD, mp-MPC, Kalman filters	Implementation of a 'goal-oriented' model constrained optimization framework to determine the optimal POD reduction projection basis Simultaneous use of Kalman state estimation on the reduced-order systems
Bonis *et al.* (2012)	Successive linearization, Krylov methods	'Equation-free' successive linearization of nonlinear systems of ODEs to which an Arnoldi order reduction scheme is applied
Agarwal and Biegler (2013)	POD	Implementation of a trust-region framework to guarantee optimality conditions with respect to the original system in optimization problems defined on reduced-order POD models
Hedengren and Edgar (2005)	Empirical balanced truncation, ISAT	Order reduction through empirical balanced truncation coupled to complexity reduction and linearization via ISAT
Xie *et al.* (2012)	ANNs, POD	A hybrid, data-driven approach, constructing POD approximate models with $a_i(t)$ time-varying coefficient(s) determined via ANN black-box models and the basis function in POD from data plant 'snapshots'
Lambert *et al.* (2013)	Empirical balanced truncation	Empirical balanced balance truncation combined with linearization and balanced truncation for application of mp-MHE
Rivotti *et al.* (2012)	Empirical balanced truncation	Empirical balanced truncation combined with nonlinear mp-NMPC
Lambert *et al.* (2013)	Variance-based model reduction	Use numerical integration for a variance-based approximation technique using global sensitivity analysis principles.
Xie *et al.* (2011)	POD, TPWL, mp-MPC	POD model order reduction of the dimensionality with respect to the spatial coordinate and use of TPWL to linearize the time-dependent coefficients in the POD expansion

where *f*, *g* and *h* are twice continuously differentiable in x and θ.

If (2.30) has a quadratic objective function and linear constraints, and the parameters appear on the right-hand side of the constraints, the equation will have the following form:

$$\begin{aligned} & z(\theta) = \min_x fc^T x + \frac{1}{2} x^T Qx \\ \text{s.t.} \; & Ax \leq b + F\theta \\ & x \in X \subseteq \Re^n, \\ & \theta \in \Theta \subseteq \Re^m, \end{aligned} \tag{2.31}$$

where c is a constant vector of dimension n, Q is an $(n \times n)$ symmetric positive definite constant matrix, A is a $(p \times n)$ constant matrix, F is a $(p \times m)$ constant matrix, b is a constant vector of dimension p, and X and Θ are compact polyhedral convex sets of dimensions n and m, respectively. Note that a term of the form $\theta^T Px$ in the objection function can also be addressed in the following formulation, as it can be transformed into the form given in (2.31) by substituting $x = s - Q^{-1}P^T\theta$, where s is a vector of arbitrary variables of dimensions n, and P is a constant matrix of dimension $(m \times n)$.

If we apply the basic sensitivity theorem (Floudas 1995) to (2.31) at $[x(\theta_Q), \theta_Q]$, we will obtain the following result:

$$\begin{pmatrix} \dfrac{dx(\theta_Q)}{d\theta} \\ \dfrac{d\lambda(\theta_Q)}{d\theta} \end{pmatrix} = -(M_Q)^{-1} N_Q, \tag{2.32}$$

where:

$$\begin{aligned} & M_Q = \begin{bmatrix} Q & A_1^T & \cdot & \cdot & \cdot & A_p^T \\ -\lambda_1 & A_1 - V_1 & & & & \\ \cdot & & & \cdot & & \\ \cdot & & & & \cdot & \\ \cdot & & & & & \cdot \\ -\lambda_p A_p & & & & & -V_p \end{bmatrix}, \\ & N_Q = [Y, \lambda_1 F_1, \ldots, \lambda_p F_p]^T, \\ & V_i = A_i x(\theta_Q) - b_i - F_i \theta_Q, \end{aligned} \tag{2.33}$$

and Y is a null matrix of dimension $(n \times m)$. Thus, in the linear-quadratic optimization problem, the Jacobian reduces to a mere algebraic manipulation of the matrices declared in (2.31).

$$\begin{bmatrix} x_Q(\theta) \\ \lambda_Q(\theta) \end{bmatrix} = -(M_Q)^{-1} N_Q(\theta - \theta_Q) + \begin{bmatrix} x_Q(\theta_Q) \\ \lambda_Q(\theta_Q) \end{bmatrix} \tag{2.34}$$

The space of θ where this solution remains optimal is defined as the critical region, CR^Q, and can be obtained by using feasibility and optimality conditions. The notation *CR* will be used to denote the set of points in the space of θ that lie in *CR*, as well as to denote the set of inequalities which define *CR*. Feasibility is ensured by substituting $x_Q(\theta)$ into the inactive inequalities given in (2.31), whereas the optimality condition is given by $\tilde{\lambda}_Q(\theta) \geq 0$, where $\tilde{\lambda}_Q(\theta)$ corresponds to the vector of active inequalities, resulting in a set of parametric constraints. This is represented by:

$$CR^R = \left\{ \breve{A} x_Q(\theta) \leq \breve{b} + \breve{F}\theta, \tilde{\lambda}_Q(\theta) \geq 0, CR^{IG} \right\}, \tag{2.35}$$

where $\breve{A}$, $\breve{b}$ and $\breve{F}$ correspond to the inactive inequalities; and CR^{IG} represents a set of linear inequalities defining an initial given region. A compact representation of CR^Q is obtained from the parametric inequalities by removing the redundant inequalities:

$$CR^Q = \Delta\left\{ CR^R \right\}, \tag{2.36}$$

where Δ is an operator which removes redundant constraints (Pistikopoulos *et al.* 2007). Once CR^Q, which is a polyhedral region, has been defined for a solution, $[x(\theta_Q), \theta_Q]$, the next step is to define the rest of the region, CR^{rest}, as proposed in Pistikopoulos *et al.* (2007):

$$CR^{rest} = CR^{IG} - CR^Q \tag{2.37}$$

We then obtain another set of solutions in each of these regions and their corresponding *CR*s. The algorithm terminates when there are no more regions to be explored, namely, when the solution of the differential equation (2.32) has been fully approximated by first-order expansions.

The main steps of the algorithm are presented in Table 2.3. While defining the rest of the regions, some of the regions are split, and hence the same optimal solutions may be obtained in more than one region. Therefore, the regions with the same optimal solutions are united, and a compact representation of the final solution is obtained.

The design of mp-MPC is based on the validated procedure described in Pistikopoulos *et al.* (2002). The standard state-space representation is used to formulate a model predictive controller of the generic form subjected to the state-space model and constraints imposed by the boundary values of the input, output, state and measured disturbance vectors. The resulting multiparametric quadratic programming problem is solved with standard

Table 2.3 mp-QP algorithm.

Step 1	In a given region, solve (2.31) by treating $\boldsymbol{\theta}$ as a free variable to obtain a feasible point $[\boldsymbol{\theta}_Q]$.
Step 2	Fix $\theta = \theta_Q$ and solve (2.31) to obtain $[x(\theta_Q), \lambda(\theta_Q)]$.
Step 3	Compute $[-(M_Q)^{-1}N_Q]$ from (2.32).
Step 4	Obtain $[x_Q(\theta), \lambda_Q(\theta)]$ from (2.34).
Step 5	Form a set of inequalities, CR^R, as described in (2.35).
Step 6	Remove redundant inequalities from this set of inequalities, and define the corresponding CR^Q as given in (2.36).
Step 7	Define the rest of the region, CR^{rest}, as given in (2.37).
Step 8	If there are no more regions to explore, go to the next step; otherwise, go to Step 1.
Step 9	Collect all the solutions, and unify the regions having the same solution to obtain a compact representation.

multiparametric techniques using the POP toolbox in MATLAB, and the map of optimal control actions is acquired.

2.3.1 Dynamic Programming and Robust Control

The combination of multiparametric programming and dynamic programming (Bellman 1957) has been reported as a method suitable for reducing the complexity of the optimization problem involved in multistage decision processes, such as explicit model predictive control (Borrelli *et al.* 2005; Faisca *et al.* 2008; Kouramas *et al.* 2011). By using this method, the original problem is disassembled into a set of smaller sub-problems with lower dimensionality, which are sequentially solved in a recursive manner. Recently, this approach has been extended to the problem of constrained dynamic programming of mixed-integer linear problems (Rivotti and Pistikopoulos 2013).

A brief review on parametric programming algorithms that handle problems with solely continuous optimization variables is presented in Table 2.4. Table 2.5 shows the most important algorithms for parametric optimization problems with mixed discrete and continous optimization decisions.

In Table 2.6, we present the most important works on sensitivity analysis and parametric programming of infinite-dimensional dynamic optimization problems.

Despite the significant amount of publications and algorithms proposed for the multiparametric problems presented in Table 2.6, most classes of problems remain active subjects of research. Even for well-established classes of problems, such as multiparametric linear programming problems, there is continued interest in further improving the efficiency of the algorithm, reducing the complexity of exploring large-dimensional parameter spaces, and extending the approach to wider ranges of uncertainty descriptions in the cost function or constraints of the problem.

Table 2.4 Literature review on continuous parametric programming techniques for static problems.

Authors	Theory development
Multiparametric linear programming	
Gal and Nedoma (1972)	Linear objective – constraints
Gal (1995)	Extension of simplex algorithm – basis exchange
Dua *et al.* (2002) Bemporad *et al.* (2000)	Solve LP and use sensitivity analysis to derive expressions. Use inactive inequalities and Lagrange multipliers to define regions. Avoid degenerate solutions.
Filippi and Romanin-Jacur (2002)	Enhancement of Gal's mp-LP approach. Overcome dual degeneracy via lexicographic pivoting.
mp-Nonlinear Programming	
Dua *et al.* (1999, 2004)	Nonconvex optimization Generate convex over-estimator and under-estimator. Linearization of convex functions mp-NLP on the functions Spatial B&B to obtain solution
Jonker *et al.* (2001)	p-Linear/quadratic optimization Distinguish between 5 types of generalized critical points g.c. according to whether: (a) linear-independent constraint qualification, (b) strict complementary conditions or (c) non-degeneracy of Hessian and Lagrangian hold. Switch between regions according to what condition is fulfilled, to derive optimal mapping. Complete enumeration of constraint space.
Dua *et al.* (2002)	Quadratic objective linear constraints (mp-QP) Uses sensitivity analysis from Fiacco and Kyparisis (1986) to get the expressions. Convex assumptions; no linearization
Tøndel *et al.* (2003)	mp-QP Exploit the relation between polyhedral critical regions and sets of active constraints to improve the off-line computation time of the explicit solution of Dua *et al.* (2002).
Berkelaar *et al.* (1999)	Quadratic/linear complementarity program – QP/LCP Use maximally complementarity conditions of interior point methods to characterize complete parametric region. Key features: pivot step as in LP for switching basis Use optimal basis identification/optimal partition identification.

(*Continued*)

Table 2.4 (Continued)

Authors	Theory development
Zafiriou (1990)	Quadratic programs Obtain solutions for different sets of active constraints.
Fiacco and Ishizuka (1990) Fiacco and Kyparisis (1988)	Convex problem Sensitivity analysis around optimal Linearization to obtain expressions
Benson (1982)	Non-convex problem Convex over- and under-estimators

Table 2.5 Literature review on mixed-integer parametric programming techniques for static problems.

Authors	Theory development
mp- Mixed-integer linear programming	
Acevedo and Pistikopoulos (1996, 1999)	Modified branch-and-bound Comparison procedure between parametric solutions
Ohtake and Nishida (1985)	B&B algorithm for p-MILP LB: relaxed integers; UB: fixed integers No formal comparison procedure
Dua and Pistikopoulos (2000)	Decomposition between (m)p-LP and MILP
Crema (2002)	mp-ILP: iterate between a MILP with free θ and ILP with fixed θ. Stop when no improved realization is found.
mp- Mixed-integer nonlinear programming	
Dua *et al.* (1999)	Convex systems Iterate between an mp-NLP and a MINLP
Acevedo and Pistikopoulos (1996) Pertsinidis *et al.* (1998) Papalexandri and Dimkou (1998) Dua *et al.* (1999)	Convex systems Iterate between an (m)p-NLP and na (m) p-MILP

The benefits of using explicit model predictive control have motivated research aimed at extending the theory to several different classes of control problems. Table 2.7 presents references for problems that are common in the control literature which have been addressed using explicit model predictive control.

Table 2.6 Literature review on parametric programming/sensitivity analysis in dynamic optimization.

Authors	Theory development
Ito and Kunisch (1992)	Optimal control. Inequality constraints. Linear dynamics. Derive functions of control, state and adjoint variables with respect to problem parameters.
Diehl *et al.* (2002)	On-line dynamic optimization. Use derivatives of objectives and constraints with respect to parameters to derive a perturbation manifold. Use it as an approximation to the dependence of the optimal conditions. Not complete profile of optimal conditions in terms of parameters
Dontchev *et al.* (2000)	Optimal control. Solution stability under perturbations. Expressions for neighbourhood around the optimal solution. Do not consider range of variations.
Büskens and Maurer (2000)	Optimal control. Detailed sensitivity analysis of optimal solution. Use first- and second-order Taylor expansion. Find sensitivity of adjoints and states. Employ SQP methods. No exploration of full parameter space
Malanowski and Maurer (2001) Augustin and Maurer (2001)	Sensitivity analysis of nonlinear optimal control problems. Compute derivatives of optimal conditions as a function of parameters. Problems with high-order path constraints
Solís-Daun *et al.* (1999)	Stability of a dynamic system subject to uncertainty. Optimal feedback control as a function of parameters. No constraints included. Cannot be used for a range of variations
Alt (1991)	Optimal control. Objective: stability of optimal solution. Constraints included. Subject to perturbations. Restricted to neighbourhood around optimal

Robust control can be defined as the solution of an optimal trajectory of the system under the presence of uncertainty and/or disturbances, which guarantees constraint satisfaction for all admissible values of uncertainty, and optimally steers the system to the target reference points (Bemporad and Morari 1999; Rawlings and Mayne 2009). The uncertainty can originate from model-mismatch, non-captured hidden process dynamics and/or input or output disturbances (Muske and Badgwell 2002).

Table 2.7 Relevant methods for designing robust model-based controllers.

Authors	On-line/off-line	Remarks
Campo and Morari (1987)	On-line	Min worst-case ∞ norm
Lee and Yu (1997)	On-line	Min worst-case quadratic cost, use of dynamic programming for closed loop
Schwarm and Nikolaou (1999) Badgwell (1997)	On-line	Min nominal objective s.t. robustness quadratic/linear constraints. Large number of combinations
Kassmann *et al.* (2000)	On-line	Apply robustness constraints to steady-state target calculation.
Scokaert and Mayne (1998)	On-line	Min worst-case quadratic, invariant set for stability
Kothare *et al.* (1996) Casavola and Mosca (2007)	On-line	Min upper bound on worst-case quadratic cost Convert min-max optimization to LMI-based optimization. Use also terminal cost and finite horizon tuning for stability.
Lee and Cooley (2000)	On-line	Min worst-case quadratic cost s.t. quadratic constraints for stability
Zafiriou (1990)	Semi off-line	Min quadratic cost. Add stability constraints, no feasibility guaranteed. Large number of constraints combinations
Mayne and Schroeder (1997)	Off-line	Min settling time; use invariant set.
Kouvaritakis *et al.* (2000)	Semi off-line	Off-line: obtain the gain of the feedback controller via min quadratic robust perf. s.t. linear matrix inequalities. On-line: quadratic min to compute bias. Suboptimal strategy
Bemporad *et al.* (2003)	Off-line	Min worst-case – ∞ norm. Use of dynamic programming. Solve consecutively mp-LPs.
Ramírez and Camacho (2001, 2002)	Semi off-line	Min worst-case quadratic cost. CARIMA models, PWA solution via geometric arguments. No constraint considered min worst-case quadratic cost. Rigorous on-line LP to compare minima and locate solution

2.4 Estimation Techniques

Estimation techniques are vital for obtaining information about a system's state and condition but also for realizing control of a system based on the state information (Rao 2000; Rawlings and Mayne 2009). The purpose of estimation is hence often to reconstruct this state information from a possibly noisy set of measurements or when state estimation is not directly available. A long-existing model-based technique for state estimation is the Kalman filter (Welch and Bishop 2001), which is an unconstrained method. The Kalman filter is an algorithm that uses a series of measurements observed over time, containing noise and other inaccuracies, and produces estimates of unknown variables that tend to be more precise than those based on a single measurement alone. The use of constrained estimation techniques such as the moving horizon estimator (MHE) can lead to significant improvements of the estimation result (Rao 2000; Darby and Nikolaou 2007) by adding system knowledge. MHE is an estimation method that obtains the estimates by solving a constrained optimization problem given a number, or *horizon*, of past measurements. It obtains not only the state information but also the noise sequence over the horizon.

The advantages of MHE over the Kalman filter are that, firstly, the MHE can handle constraints that could improve the estimation result significantly. Secondly, it estimates not only the state values but also the values of the disturbance over the horizon.

The obtained estimates can then be used for the control of the system using MPC, which is a model-based control method that solves a constrained optimization problem to obtain the optimal control action to fulfil the purpose of the system, for example a certain quality of the final product for as little production cost as possible. The main advantages of MPC are the handling of multiple-input/multiple-output systems and the handling of system constraints that for example represent limitations to guarantee the safety of the system.

Estimation techniques are often embedded in a control structure that relies on estimated values such as MPC (Rawlings and Mayne 2009). MPC solves a constrained optimization to obtain the best control action and requires the current state information, which often cannot be measured directly but is only available as the result of an estimator. If noise is present in the available measurements, there will inevitably be an error in the estimator's results. The MPC should hence not simply assume that the estimated values are correct but needs to consider the estimation error in order to obtain the best performance (Rawlings and Mayne 2009). While the determination of this error set is rather straightforward for unconstrained estimators such as the Luenberger observer (Raković and Kouramas 2006; Mayne *et al.* 2009) or the unconstrained MHE (Sui *et al.* 2008), no method exists for constrained MHE. Since the constrained MHE can give significantly better estimates than unconstrained methods, the controller based on those better estimates will also perform better (Rawlings and Mayne 2009).

2.4.1 Kalman Filter

The Kalman filter is a standard method for unconstrained state estimations, and it follows a two-step procedure to calculate the maximum *a posteriori* Bayesian estimate (Rao 2000; Welch and Bishop 2001; Rawlings and Mayne 2009). The first step is the time update, which uses the system model to predict the current state of the system based on the last estimate. The second step is the measurement update. The prediction from the previous step is updated by using the sensor information. Therefore, we can say that the Kalman filter is a predictor-corrector type estimator that is optimal in the sense that it minimizes the estimated error covariance.

2.4.1.1 Time Update (Prediction Step)

Prediction of the state:

$$\hat{x}_{k/k-1} = A\hat{x}_{k-1/k-1} + Bu_{k-1} \tag{2.38}$$

Projection of the error covariance:

$$P_{k/k-1} = AP_{k-1}A^T + Q_{kal} \tag{2.39}$$

2.4.1.2 Measurement Update (Correction Step)

Computation of the Kalman gain:

$$K_k = P_{k/k-1}C^T(CP_{k/k-1}C^T + R_{kal})^{-1} \tag{2.40}$$

Update of the estimate with measurement:

$$\hat{x}_{k/k} = \hat{x}_{k/k-1} + K_k(y_k - Cx_{k/k-1}) \tag{2.41}$$

Update of the error covariance:

$$P_k = (I - K_kC)P_{k/k-1} \tag{2.42}$$

In Equations (2.38) through (2.42), Q_{kal} and R_{kal} represent the measure of confidence in the model and the measurement, respectively.

The solution of the discrete algebraic Riccati equation (Söderström 2002):

$$P = A^T PA - A^T PB(B^T PB + R_{kal})^{-1} B^T PA + Q_{kal} \tag{2.43}$$

can be used for the calculation of the steady-state gain which makes $\mathrm{P}_{k+1} = \mathrm{P}_k = \mathrm{P}$ constant.

2.4.2 Moving Horizon Estimation

The idea of moving horizon estimation is to estimate the state using a moving and fixed-size window of data. Once a new measurement becomes available, the oldest measurement is discarded, and the new measurement is added. The concept is to penalize deviations between measurement data and predicted outputs. In addition – for theoretical reasons – a regularization term on the initial state estimate is added to the objective function. There are two main characteristics that distinguish MHE from other estimation strategies, such as the Kalman filter: (a) prior information in the form of constraints on the states, disturbances and parameters can be included; and (b) since MHE is optimization based, it is able to handle explicitly nonlinear system dynamics through the use of approximative nonlinear optimization algorithms. In Haseltine and Rawlings (2005), MHE was shown to possess superior estimation properties compared to the extended Kalman filter.

The Kalman filter considers only one set of measurements at a time. Rawlings and Mayne (2009) have shown that the Kalman filter is the algebraic solution to the following unconstrained least-squares optimization problem:

$$\Phi(x_0,\ \{w\}) = \left\|\hat{x}_0 - \underline{x}_0\right\|^2_{P_0^{-1}} + \sum_{k=0}^{T-1}\left\|\hat{w}_k\right\|^2_{Q_k^{-1}} + \sum_{k=0}^{T}\left\|\hat{v}_k\right\|^2_{R_k^{-1}} \tag{2.44}$$

where:

$$\begin{aligned} x_{k-1} &= Ax_k + Bu_k + Gw_k \\ y_k &= Cx_k + v_k \end{aligned} \tag{2.45}$$

and $Qk > 0$, $Rk > 0$ and $P0 > 0$ are positive definite matrices. This optimization problem now opens the possibility to add system knowledge in the form of constraints. The constraints might for example capture the fact that a leak is always an outflowing stream, or account for non-zero non-Gaussian noise (Robertson and Lee 2002).

The optimization problem (2.44) is then not equivalent to the Kalman filter anymore. If all the available past measurements are used for the estimation, as in (2.44), the estimation problem grows unbounded with time. This is referred to as the *full information estimator* (Rao 2000). The derivation is based on the maximization of the *a posteriori* Bayesian estimate. In order to keep the estimation problem computationally tractable, it is necessary to limit the processed data, for example by discarding the oldest measurement once a new one becomes available. This essentially slides a window over the data, leading to the MHE. The data that are not considered anymore can be accounted for by the so-called *arrival cost* so that the information is not lost. The MHE then considers only a limited amount of data, so that the constrained optimization problem becomes:

$$\begin{aligned} &\min_{\hat{x}_{T-N/T},\hat{W}_T} \left\|\hat{x}_{T-N/T} - \underline{x}_{T-N/T}\right\|_{P^{-1}_{T-N/T-1}} + \left\|Y_{T-N}^{T-1} - O\hat{x}_{T-N/T} - \overline{c}bU_{T-N}^{T-2}\right\|^2_{P-1} \\ &+ \sum_{k=T-N}^{T-1}\left\|\hat{w}_k\right\|_{Q_k^{-1}} + \sum_{k=T-N}^{T-1}\left\|\hat{v}_k\right\|_{R_k^{-1}} \\ &\text{st:} \\ &\qquad \hat{x}_{k+1} = A\hat{x}_k + Bu_k + G\hat{w}_k \\ &\qquad y_k = C\hat{x}_k + \hat{v}_k \\ &\qquad \hat{x}_k \in X, \hat{w}_k \in \Theta, \hat{v}_k \in V \end{aligned} \tag{2.46}$$

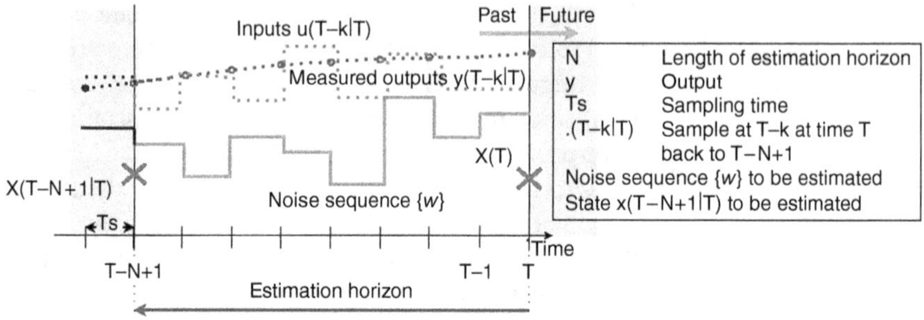

Figure 2.3 Concept of MHE.

where T is the current time; $Q_k \succ 0, R_k \succ 0, P_{T-N/T-1} \succ 0$ are the covariances of $w_k v_k x_{T-N}$ assumed to be symmetric; N is the horizon length of the MHE; $Y^{T-1}_{T-N} = \left[y^T_{T-N}, \ldots, y^T_T \right]^T$ is a vector containing the past $N+1$ measurements; $U^{T-1}_{T-N} = \left[u^T_{T-N}, \ldots, u^T_{T-1} \right]^T$ is a vector containing the past N inputs; x, v, w denote the variables of the system; $\hat{x}, \hat{v}, \hat{w}$ denote the estimated variables of the system; and $\hat{x}_{T/T-N}$ and $\hat{W}_T = W^{T-1}_{T-N} = \{\hat{w}\}^{T-1}_{T/T-N}$ $\hat{x}_{T-N/T}, \hat{W}_T$ denote the decision variables of the optimization problem, respectively the estimated state variable and the noise sequence.

$$\left\| \hat{x}_{T-N/T} - \underline{x_{T-N/T}} \right\|^2_{P_{T-N/T-1}} + \left\| Y^{T-1}_{T-N} - O\hat{x}_{T-N/T} - \overline{c}bU^{T-2}_{T-N} \right\|^2_{P-1} \tag{2.47}$$

is described as the smoothed arrival cost. For steady-state MHE, $Q_k = Q, R_k = R$ and $P_{T-N|T-1} = P$ are time invariant.

The current state of the system can be calculated from the initial state $x_{T|T-N}$ by forward programming using the discrete time linear system if the deterministic input $U^{T-1}{}_{T-N}$ and the noise sequence W^{T-1}_{T-N} are known. It is thus sufficient to estimate the initial state $\hat{}x^*{}_{T|T-N}$ and the noise $\hat{}W^*{}_T$. The concept of MHE is illustrated in Figure 2.3, where $(\cdot)_{T-k|T}$ denotes the sample at time $T-k$ obtained at time T.

The MHE is applied with the following steps:

1) The optimization problem (2.40) is solved to obtain $\hat{x}^*_{T|T-N}$ and $\hat{W}^*_T$.
2) The current state estimate $\hat{x}^*_{T|T}$ is obtained by substituting $\hat{x}^*_{T|T-N}$ and $\hat{W}^*_T$ into the system dynamics $\hat{x}_{k+1} = A\hat{x}_k + Bu_k + G\hat{w}_k$ and projecting the state values forward from time $T-N$ to the current time T by $\hat{x}^*_{T/T} = A^N \hat{x}^*_{T-N/T} + \sum_{j=T-N}^{T-1} A^{T-1-j} G\hat{w}_k$.
3) When the next measurement becomes available (at the next sampling instance), steps 1 and 2 are repeated.

Remark 2.1 *In the case of* $T \le N$, *the full information estimator is solved using the arrival cost* $\left\| \hat{x}_{T-N/T} - \underline{x}_{T-N/T} \right\|^2_{P_0^{-1}}$. *The horizon 'fills up,' and no data are discarded (Rao* 2000).

Remark 2.2 *Rao (2000) points out that wrongly posed constraints might lead to an infeasible optimization problem and that constraints on* $\hat{v}_k$ *could be problematic due to the possibility of outliers in the measurement. Any constraints posed in* (2.40) *should hence be chosen such that the real system does not violate them (Rao 2000).*

This observation leads to the following assumption for the work in this thesis:

Assumption 2.1 *The real system does not violate the constraints posed to the MHE –* $x \in \hat{X}, v \in \hat{V}, w \in \hat{W}$ *– which contain the origin in their interior.*

The formulation of the MHE with the smoothed arrival cost is still an open issue which will be addressed here. In order to formulate and solve the constrained MHE with multiparametric programming, the optimization problem needs to be reformulated into the standard multiparametric quadratic form.

Darby and Nikolaou (2007) have reformulated the MHE with the filtered arrival cost.

The multiparametric formulation of the constrained MHE is obtained by substituting the state-space formulation of the estimated $\hat{x}_{k+1} = A\hat{x}_k + Bu_k + G\hat{w}_k$, $\hat{v}_k = y_k - C\hat{x}_k$ into:

$$[\hat{x}_{T-N+1}, \{\hat{w}\}_{T-N+1}^{T-1}] - \min_{x_{T-N+1}, \{w\}_{T-N+1}^{T-N}} \sum_{k=T-N+1}^{T-N} (w_k - \overline{w})^T \cdot Q^{-1} \cdot (w_k - \overline{w})$$
$$+ \sum_{k=T-N+1}^{T} (v_k^T \cdot R^{-1} \cdot v_k) + (x_{T-N+1} - \overline{x}_{T-N+1/T-N})^T \cdot P_{SS}^{-1} \cdot (x_{T-N+1} - \overline{x}_{T-N+1/T-N}) \tag{2.48}$$

subject to:

$x_{k+1} = A \cdot x_k + B \cdot u_k + G \cdot w_k$	discrete state-space formulation
$y_k = C \cdot x_k + v_k$	$\rightarrow v_k = y_k - C \cdot x_k$
$Hx \cdot x \le h$	path constraints on state variables
$Kw \cdot w \le k$	path constraints on the noise w
$Lv \cdot v \le l$	path constraints on the noise v[1]
$\overline{x}_{T-N+1/T-N} = A \cdot \hat{x}_{T-N/T-N} + B \cdot u_k + G \cdot \overline{w}$	update of the cost to arrive

where N is the length of the horizon, T is the current point in time, Q and R are positive definite diagonal weighting matrices on the noises, and P_{SS} is the steady-state solution for the Kalman filter. $\{\hat{w}\}_{T-N+1}^{T-N}$ and $\{v\}_{T-N+1}^{T}$ are sequences of independent, normally distributed random numbers with mean values $\bar{w}$ for $\{w\}$ and zero-mean for $\{v\}$. $\bar{x}_{T-N+1/T-N}$ is the arrival cost, which captures the previous measurements that are not considered anymore. $\hat{x}_{T-N+1}$ is the solution of the MHE at the previous time step.

$$\min_{\hat{x}_{T-N+1},\{\hat{w}\}_{T-N+1}^{T-1}} \frac{1}{2}\left[\hat{x}_{T-N+1},\{\hat{w}\}_{T-N+1}^{T-1}{}^{T}\right]H\begin{bmatrix}\hat{x}_{T-N+1}\\ \{\hat{w}\}_{T-N+1}^{T-1}\end{bmatrix}+\theta.f.\begin{bmatrix}\hat{x}_{T-N+1}\\ \{\hat{w}\}_{T-N+1}^{T-1}\end{bmatrix} \tag{2.49}$$

where:

$$\theta=\left[\underline{x}_{T-N/T}^{T},y_{T-N}^{T},\ldots,y_{T}^{T},u_{T-N}^{T},\ldots,u_{T-1}^{T}\right]^{T},$$

$$H\begin{bmatrix}2P^{-1}-2O^{T}\cdot W^{-1}\cdot O+2ca^{T}\cdot diag\,(R^{-1})\cdot ca & 2ca\cdot diag\,(R^{-1})\cdot cg\\ 2cg^{T}\cdot diag\,(R^{-1})\cdot ca & 2diag\,(Q^{-1})+2cg^{T}\cdot diag(R^{-1})\cdot cg\end{bmatrix},$$

$$f=\begin{bmatrix}[-2P^{-1},0]\\ 2diag\,(R^{-1}\cdot C)\cdot[a,g]\\ 2b^{T}\cdot diag\,(C^{T}\cdot R^{-1}\cdot C)\cdot[a,g]\end{bmatrix}, \tag{13}$$

$$a=\begin{bmatrix}I\\ A\\ A^{2}\\ \vdots\\ A^{N-1}\end{bmatrix}, g\begin{bmatrix}0 & 0 & \ldots & 0\\ G & 0 & \ldots & 0\\ A\cdot G & G & \ldots & 0\\ A^{2}\cdot G & G & \ldots & 0\\ \vdots & & \ddots & \\ A^{N-2}\cdot G & A^{N-2}\cdot G & \ldots & G\end{bmatrix},$$

$$b=\begin{bmatrix}0 & 0 & \ldots & 0\\ B & 0 & \ldots & 0\\ A\cdot B & B & \ldots & 0\\ A^{2}\cdot B & A\cdot G & \ldots & 0\\ \vdots & & \ddots & \\ A^{N-2}\cdot B & A^{N-3}\cdot B & \ldots & B\end{bmatrix}, \begin{matrix}\mathrm{ca}=\mathrm{diag(C)}\cdot\mathrm{a},\\ \mathrm{ca}=\mathrm{diag(C)}\cdot\mathrm{g},\\ \mathrm{ca}=\mathrm{diag(C)}\cdot\mathrm{b},\end{matrix} \tag{2.50}$$

where *diag*(·) denotes a matrix of appropriate size, with (·) on its main diagonal and zero everywhere else. Further details can be found in Darby and Nikolaou (2007) and Voelker *et al.* (2013).

The optimal solution to the MHE optimization (2.49) $\left[\hat{x}_{T-N+1}^{*}(\theta),\{\hat{w}\}_{T-N+1}^{T-1}{}^{*}(\theta)\right]^{T}$ (where * denotes the the optimizer of problem [2.49]) is a piecewise affine function of the measurements, inputs and arrival cost.

$$\begin{bmatrix} \hat{x}^*_{T-N+1}(\theta) \\ \{\hat{w}\}_{T-N+1}^{T-1\,*}(\theta) \end{bmatrix} = F(\theta) = \begin{cases} K_1\theta + c_1 & if\ \theta \epsilon CR^1 \\ \vdots & \vdots \\ K_l\theta + c_l & if\ \theta \epsilon CR^l \end{cases} \tag{2.51}$$

where l is the number of critical regions (CRs).

Remark 2.3 *Often, only the state estimate* $\hat{x}_{T/T}$ *is of interest, and hence it might be sufficient if only* $\check{x}_{T/T}$ *has to fulfil the constraints rather than all* $\hat{x}_{k/T}$ *for all* k = T − N,…, T. *It might therefore suffice to pose constraints only on* $\hat{x}_{T/T}$ *rather than the whole horizon by replacing the matrix* col $(\hat{d}_x)$ *with* dx *and the matrix* diag $(\hat{D}_x)$ *with* $\begin{bmatrix} 0 & \cdots & \hat{D}_x \end{bmatrix}$. *This reduces the number of parameters in the optimization problem.*

2.5 Explicit Hybrid Control

Mixed-logical dynamical (MLD) systems are a well-studied class of systems (Bemporad and Morari 1999; Heemels *et al.* 2001). Their basic principle is that, additionally to the commonly encountered continuous parts, discrete elements are present in the problem formulation, as either inputs, states, variables or outputs.

In particular, in the case where the discrete elements are switches, this immediately leads to bilinear terms. However, it was shown how these bilinear terms can be avoided, and via auxiliary variables a linear formulation can be obtained according to:

$$x_{k+1} = Ax_k + B_1u_k + B_2\delta_k + B_3z_k \tag{2.52a}$$

$$y_k = Cx_k + D_1u_k + D_2\delta_k + D_3z_k \tag{2.52b}$$

$$E_1u_k + E_2\delta_k \le E_3z_k + E_4x_k + E_5, \tag{2.52c}$$

where (2.52a)–(2.52b) represent the state-space model, while (2.52c) describes any type of constraints of the system (Bemporad and Morari 1999). The states at time k are thereby denoted by x_k, the outputs by y_k, the inputs by u_k, the discrete (auxiliary) variables by δ_k and the continuous auxiliary variables by z_k. Note that all matrices have appropriate dimensions.

Consequently, in order to obtain an optimal, model-based control strategy, the following optimization problem needs to be solved:

$$\begin{aligned} J^*(x_0) = &\min_U J(U,X,Y,\Delta,Z) \\ &\text{s.t. Eq.}(3.4)\ \forall k = 0,\dots,N-1 \\ &x_N \in \Omega \end{aligned} \tag{2.53}$$

where N is the number of steps considered (also referred to as the *horizon*), $U \triangleq [u_0, u_1, \ldots, u_{N-1}]$, $X \triangleq [x_0, x_1, \ldots, x_N]$, $Y \triangleq [y_1, y_2, \ldots, y_{N-1}]$, $\Delta \triangleq [\delta_0, \delta_1, \ldots, \delta_{N-1}]$, $Z \triangleq [z_0, z_1, \ldots, z_{N-1}]$ and Ω is a terminal weight to ensure stability.

Remark 2.4 *It is convention that no output y_k is recorded at $k = 0$. Thus, the mathematical representation of problem (2.48) is slightly inaccurate. However, this is accepted due to the increased meaning of the formulation due to its conciseness.*

Problems of type (2.53) are referred to as *mixed-integer programming* (MIP) problems. In the case of classical MPC, problem (2.53) is solved as soon as the nominal value for x_0 becomes available. In the case of explicit MPC, problem (2.53) is solved offline over a certain bounded range of x_0, and thus requires the solution of a multiparametric mixed-integer programming (mp-MIP) problem.

2.5.1 Multiparametric Mixed-Integer Programming

The solution of mp-MIP problems is very challenging. The additional complications introduced by the presence of integer variables are (a) combinatorial complexity and (b) nonconvexity. In this section, the problem and solution characteristics of mp-MIP problems are described, before approaches are reviewed which have been proposed to solve certain mp-MIP problems. Based upon these, a general framework for the solution of a certain class of mp-MIP problems is presented, as well as a novel strategy on how to handle the presence of nonconvexity in the description of the critical regions.

Remark 2.5 *The most common mp-MIP problems, and thus the ones considered here, are multiparametric mixed-integer linear and quadratic programming (mp-MILP and mp-MIQP, respectively) problems; mp-MILP problems are a subclass of the mp-MIQP problems with a linear objective function.*

2.5.1.1 Problem and Solution Characterization

Here, the following mp-MIQP problem is considered:

$$
\begin{aligned}
z(\theta) = \min_{x,y}\ & (Q\omega + H\theta + c)^T \omega \\
\text{s.t.}\quad & Ax + Ey \le b + F\theta \\
& x \in R^n, y \in \{0,1\}^p, \omega \in [x^T, y^T]^T \\
& \theta \in \Theta := \{\theta \in R^q \mid \theta_l^{min} \le \theta_l \le \theta_l^{max}, l = 1, \ldots, q\}
\end{aligned}
\tag{2.54}
$$

where Θ is a polyhedral subset of the parameter space, and the matrices have appropriate dimensions. Note that this formulation only includes binary variables. Furthermore, only $Q > 0$ is considered.

When the integer combination $y = \bar{y}$ is fixed in problem (2.54), the following multiparametric quadratic programming (mp-QP) problem results:

$$
\begin{aligned}
z(\theta) = {} & \min_x (Q_x x + H_x\theta + \tilde{c}_x)^T x + f(\theta) \\
\text{s.t.} \quad & Ax \le (b - E\bar{y}) + F\theta \\
& x \in R^n, \\
& \theta \in \Theta := \left\{\theta \in R^q \,|\, \theta_l^{min} \le \theta_l \le \theta_l^{max}, l = 1,\ldots,q\right\},
\end{aligned}
\tag{2.55}
$$

where the index x denotes the part of the matrix or variable associated with the continuous variable x, and $\tilde{c}_x = c_x + Q_{xy}\bar{y}$, where Q_{xy} is the part of Q associated with the combination of continuous and binary variables. Note that $f(\theta)$ does not influence the solution $x(\theta)$, as it is a scaling factor of the objective function value.

The solution of problems of type (2.55) has been studied extensively, and its solution characteristics are reported in Definitions 2.1 and 2.2 and Theorem 2.1.

Definition 2.1 Piecewise affinity and critical regions (Bemporad *et al.* 2002) *A function $x(\theta): \Theta \mapsto R^n$, where $\Theta \subseteq R^q$ is a polyhedral set, is piecewise affine if it is possible to partition Θ into convex polyhedral regions, called critical regions, CR_i, and $x(\theta) = K_i\theta + r_i, \forall \theta \in CR_i$.*

Remark 2.6 *The definition of a piecewise quadratic function is analogous.*

Definition 2.2 Parametric profile *The solution of an mp-P problem is referred to as a* parametric profile, *and it consists of the closure of the critical regions and the solutions associated with them.*

Theorem 2.1 Properties of mp-QP solution (Bemporad *et al.* 2002; Dua *et al.* 2002) *Consider the optimal solution mp-QP problem, and let Q_x be positive definite, Θ convex. Then, the set of feasible parameters $\Theta_f \subseteq \Theta$ is convex, the solution $x(\theta)$ is piecewise affine and the optimizer solution $z(\theta): \Theta_f \mapsto R$ is continuous, convex and piecewise quadratic.*

In the case of mp-MIQP problems of type (2.54), the solution properties are given by Theorem 2.2 and Lemma 2.1.

Theorem 2.2 Properties of mp-MIQP solution (Borrelli *et al.* 2005) *Consider the optimal solution of problem (2.54), and let Q be positive definite. Then, the solution $x(\theta)$ is piecewise affine, and the set CR_i has the following form:*

$$CR_i = \{\theta : \theta^T G_{i,j}\theta + h_{i,j}^T\theta \le w_{i,j}, j = 1,\ldots,t_i\} \tag{2.56}$$

where t_i is the number of constraints that describe CR_i.

Lemma 2.1 Quadratic boundaries (Borrelli *et al.* 2005) *Consider the solution of problem (2.54) as a combination of solutions of the mp-QP problems associated with all possible combinations of binary variables. Then, quadratic boundaries arise from the comparison of quadratic objective functions associated with feasible combinations of binary variables.*

2.5.1.2 Literature Review

The different approaches for the solution of mp-MIP problems can be distinguished based on their strategy to handle the integer variables: branch-and-bound methods, decomposition-type approaches and exhaustive enumeration.

Branch-and-Bound Methods The first solution approach for mp-MILP problems was presented by Acevedo and Pistikopoulos in 1997 (Acevedo and Pistikopoulos 1997): in order to handle the integer variables y, they are relaxed in the original problem by treating them as continuous variables bounded between 0 and 1 (i.e. $y \in [0,1]^p$). Based on this root node, a binary search tree is generated and the binary variables were then successively fixed at each node of the search tree (Figure 2.4).

The resulting multiparametric linear programming (mp-LP) problems were solved at each node. The solution to the mp-LP problem was then compared to the upper bound according to:

$$\Delta z(\theta) = z^{*}(\theta) - z_{UB}(\theta) \leq 0 \tag{2.57}$$

where $z^{*}(\theta)$ is the objective function value of the solution of the current node, and $z_{UB}(\theta)$ is the objective function value of the upper bound. Note that all objective function values are linear, and thus $\Delta z(\theta)$ is always linear.

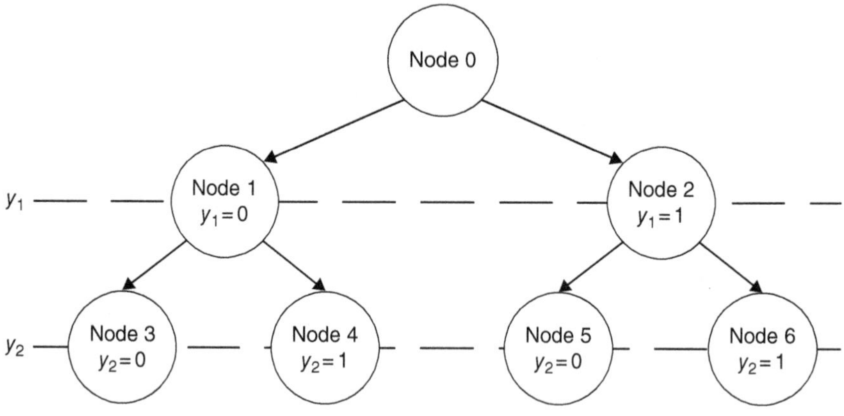

Figure 2.4 A schematic representation of a binary search tree used in branch-and-bound methods for the solution of certain mp-MIP problems.

Remark 2.7 *According to Theorem 2.2, the objective function value of an mp-MILP without bilinear terms is piecewise affine. Thus, (2.54) has to be considered for every intersection CR_{int} of the polyhedral regions associated with $z^*(\theta)$ and $z_{UB}(\theta)$, respectively.*

Consequently, Equation (3.9) is used to partition CR_{int} according to:

$$CR_1 = CR_{int} \cap \{\Delta z(\theta) \leq 0\} \tag{2.58a}$$

$$CR_2 = CR_{int} \cap \{\Delta z(\theta) \geq 0\} \tag{2.58b}$$

where in CR_1, $z^*(\theta)$ is optimal; while in CR_2, $z_{UB}(\theta)$ remains optimal. Note that due to the linearity of $\Delta z(\theta)$, CR_1 and CR_2 are polyhedral.

As $z^*(\theta)$ is a solution at a node of the binary search tree, it is possible that the associated solution $y^*(\theta)$ is either integer (i.e. all the relaxed integer variables, if any, have a value of either 0 or 1) or not. If $z^*(\theta)$ corresponds to an integer solution, then the upper bound in CR_{int} is updated according to the conclusions drawn from (2.58). If $z^*(\theta)$ does not correspond to an integer solution, then $z^*(\theta)$ is a lower bound of an integer solution – if feasible – as it relaxes the integer variables, and thus allows them to adhere to values which, albeit not integer, yield a lower objective function value $z^*(\theta)$ at the current node.

As it does not correspond to an integer solution, this current node has to be branched further, and (2.58) can be used to fathom parts of the parameter space considered, namely CR_2 according to (2.58b), as in that region the current best upper bound has a lower value than the lower bound provided by the solution $z^*(\theta)$. A schematic representation of the comparison procedure is shown in Figure 2.55.

Remark 2.8 *It is possible that CR_1 or CR_2 are empty sets (see Figure 2.5).*

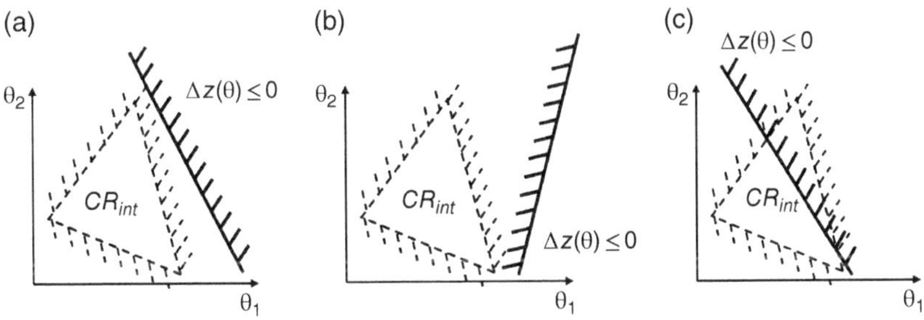

Figure 2.5 A schematic representation of the comparison procedure employed in Acevedo and Pistikopoulos (1997). According to Equation (3.10), in case (a), $CR_{int} = CR_1$; in case (b), $CR_{int} = CR_2$; and in case (c), CR_{int} is split into CR_1 and CR_2. *Source*: Acevedo and Pistikopoulos (1997). Reproduced with permission of American Chemical Society.

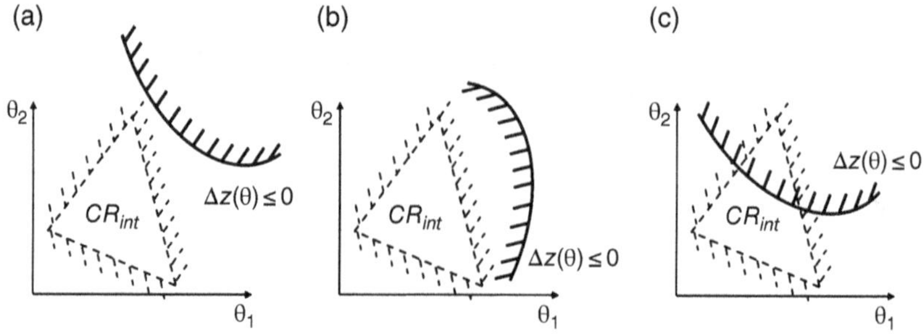

Figure 2.6 A schematic representation of different scenarios for a comparison procedure of objective functions featuring bilinear and/or quadratic terms, an issue considered in Axehill *et al.* (2011, 2014) and Oberdieck *et al.* (2014).

Once the entire solution at the current node has been considered, the union of all the parameter spaces that are passed on to the child nodes is considered. If the node is a terminal node or the union is empty, then no branching is performed. Otherwise, two new nodes according to Figure 2.4 are created.

This approach only considered the purely linear case, and thus the objective function is always linear (Lemma 2.5.1 and Figure 2.5). However, in the case of quadratic and/or bilinear terms in the objective function, this might lead to nonconvexity when applying (2.58), as shown in Figure 2.6.

Thus, (2.55) cannot be applied in a straightforward way, as in Acevedo and Pistikopoulos (1997), and novel comparison procedures are needed. So far, two approaches have been presented in the open literature, and both rely on a concept called an *envelope of solutions.*

Definition 2.3 Envelope of solutions *An* envelope of solutions *describes the case where more than one solution is associated with one critical region. The optimal solution is thereby always present among those solutions associated with the critical region.*

The first approach, presented by Oberdieck *et al.* in 2014 (Oberdieck *et al.* 2014), creates McCormick relaxations (McCormick 1976) of the nonconvex parts of $\Delta z(\theta)$, that is:

$$g(\theta) \leq \Delta z(\theta) \leq h(\theta) \tag{2.59}$$

where $g(\theta)$ and $h(\theta)$ are affine functions. These relaxations are used to partition CR_{int} according to:

$$CR_1 = CR_{int} \cap \{h(\theta) \leq 0\} \tag{2.60a}$$

$$CR_2 = CR_{int} \cap \{g(\theta) \geq 0\} \tag{2.60b}$$

$$CR_3 = CR_{int} \cap \{g(\theta) \leq 0, h(\theta) \geq 0\} \tag{2.60c}$$

where in CR_1 $z^*(\theta)$ is optimal, while in CR_2 $z_{UB}(\theta)$ remains optimal, and in CR_3 an envelope of solutions containing $z^*(\theta)$ and $z_{UB}(\theta)$ is created (see Figure 2.7a).

Remark 2.9 *In this case, it was assumed that only one $z_{UB}(\theta)$ is associated with the critical region. However, due to the nonconvexity, this is not necessarily the case. However, all the stated results also hold for the case where the upper bound consists of an envelope of solutions. Thus, the simplest representation without an envelope of solutions in the upper bound is used.*

In the second approach, presented independently by Axehill *et al.* (2011, 2014), no direct comparison using $\Delta z(\theta)$ is made. Firstly, the comparison is classified by solving the following optimization problem:

$$\delta_{max} = \max_{\theta \in CR_{\text{int}}} \Delta z(\theta) \tag{2.61a}$$

$$\delta_{min} = \min_{\theta \in CR_{\text{int}}} \Delta z(\theta) \tag{2.61b}$$

The solution of problem (2.61) allows for the following classification based on:

$$\begin{aligned} &\text{Case(a)}: \delta_{max} \leq 0 \\ &\text{Case(b)}: \delta_{min} \geq 0 \\ &\text{Case(c)}: \delta_{min} \leq 0 \wedge \delta_{max} \geq 0 \end{aligned} \tag{2.62}$$

If cases (a) or (b) are realized, then $z^*(\theta)$ and $z_{UB}(\theta)$, respectively, are assigned to CR_{int}. However, if case (c) is realized, then an envelope of solutions, containing $z^*(\theta)$ and $z_{UB}(\theta)$, is created in CR_{int}.

Remark 2.10 *Solving the optimization problems in (2.61) might be challenging as $\Delta z(\theta)$ might be nonconvex. In Axehill et al. (2011, 2014), these problems are solved via a spatial branch-and-bound algorithm implemented in YALMIP (Löfberg 2004). Also note that in Axehill et al. (2011, 2014), the concept of suboptimality was introduced, which however is not discussed further here.*

A schematic representation of both comparison approaches is shown in Figure 2.7. In particular, it shows that while the approach presented in Oberdieck *et al.* (2014) reduces the volume of the critical region associated with an envelope of solutions (grey area in Figure 2.7), the number of critical regions created is smaller in the approach presented in Axehill *et al.* (2011, 2014).

Decomposition-Type Methods A different solution approach for problems of type (2.54) was presented in 2002 by Dua *et al.* (Dua *et al.* 2002). Instead of gradually fixing the integer variables as proposed with the branch-and-bound

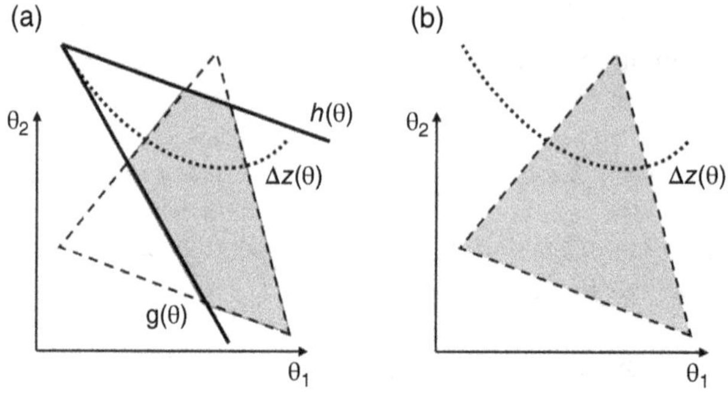

Figure 2.7 A schematic representation of two comparison procedures presented for the solution of mp-MIQP problems: in case (a), McCormick relaxations (McCormick 1976) are used to divide $\boldsymbol{CR_{int}}$ into three regions, one of which contains an envelope of solutions (grey area) (Oberdieck *et al.* 2014), while in case (b) the entire $\boldsymbol{CR_{int}}$ is regarded as an envelope of solution (grey area) (Axehill *et al.* 2011, 2014).

approach, the key idea was to decompose the original problem (2.54) into two subproblems: in the first subproblem, an integer candidate solution is found via global optimization, while in the second subproblem this integer candidate solution is substituted into the original problem (2.54), thus resulting in an mp-QP of form (2.55).

Here, each step of the presented algorithm is considered in detail:

Step 0 – Initialization: The mp-MIQP problem (2.54) is considered. A candidate solution for the binary variables is found by solving the following MIQP problem:

$$
\begin{aligned}
z_{global} = & \min_{x,y,\theta}(Q\omega + H\theta + c)^T \omega \\
s.t. \quad & Ax + Ey \le b + F\theta \\
& x \in \mathbb{R}^n, y \in \{0,1\}^p, \omega = [x^T\, y^T]^T \\
& \theta \in \Theta := \{\theta \in \mathbb{R}^q \mid \theta_l^{min} \le \theta_l \le \theta_l^{max}, l = 1,\ldots,q\}
\end{aligned}
\tag{2.63}
$$

where the parameter θ is treated as an optimization variable, and the problem is solved using available MIQP solvers. If the problem is infeasible, problem (2.54) is also infeasible. Otherwise, an integer solution y^* is obtained and subsequently fixed in problem (2.54), thus resulting in an mp-QP of the form (2.55). This problem can be solved using one of the approaches presented in the literature, which results in an initial partitioning of the parameter space according to Theorem 2.1 and provides a parametric upper bound to the solution.

Step 1 – Candidate solution for binary variables: In the first step of the algorithm, in each critical region CR_i of the current upper bound, the parameter θ is treated as an optimization variable and the following optimization problem is solved:

$$\begin{aligned} z_{global} &= \min_{x,y,\theta} (Q\omega + H\theta + c)^T \omega \\ s.t. \quad & Ax + Ey \leq b + F\theta \\ & (Q\omega + H\theta + c)^T \omega - z_{UB,i}(\theta) \leq 0 \\ & \sum_{k \in J_i} y_k - \sum_{k \in T_i} y_k \leq |J_i| - 1 \\ & x \in \mathbb{R}^n, y \in \{0,1\}^p, \omega = [x^T y^T]^T \\ & \theta \in CR_i \end{aligned} \tag{2.64}$$

where $i = 1,\ldots,v$, where v is the number of critical regions that constitute the upper bound; $z_{UB,i}(\theta)$ is the objective function value of the upper bound associated with the critical region CR_i; $|\cdot|$ is the cardinality; and J_i and T_i are the sets containing the indices of the integer variables of the integer combination $y_{UB,i}$ associated with the upper bound $z_{UB,i}(\theta)$ that attain the values 0 and 1, respectively, that is:

$$J_i = \{k | y_{UB,i}^k = 1\} \tag{2.65a}$$

$$T_i = \{k \mid y_{UB,i}^k = 0\} \tag{2.65b}$$

The two additional constraints in problem (2.64) compared to problem (2.63) are called *parametric* and *integer cut*, respectively. They ensure that the new candidate integer solution has a lower objective function value than the upper bound in at least one point, and that previously considered integer solutions are not considered again.

Step 2 – Solution of mp-QP problem: Similarly to the initialization step, the candidate integer solution y^* is substituted into problem (2.54), thus resulting in an mp-QP of the form (2.55). This mp-QP is then solved using an approach from the open literature (e.g. Bemporad *et al.* 2002; Dua *et al.* 2002; Tøndel *et al.* 2003; Spjøtvold *et al.* 2006; Gupta *et al.* 2011; Feller *et al.* 2013), which results in an initial partitioning of the parameter space according to Theorem 2.1.

Step 3 – Comparison with upper bound: In order to avoid computational burden, no comparison procedure is performed in Dua *et al.* (2002). Instead, an envelope of solution is immediately created. Note that, due to the parametric cut in problem (2.64), there will be at least one point in the originally considered critical region CR_i where the objective function value of the new solution is lower than the one of the upper bound.

Step 4 –Termination: If problem (2.64) is infeasible for critical region CR_i, then CR_i is not considered further. If problem (2.64) is infeasible for all critical regions, then the algorithm terminates.

Exhaustive Enumeration A third method to solve problems of type (2.54) was presented by Borrelli (2003, ch. 8). Thereby, all possible combinations of binary variables, in total 2^p according to Equation (2.54), are evaluated. Similar to the decomposition-type approach, each combination is fixed in problem (2.54), resulting in the corresponding mp-QP which is solved. The solution to this mp-QP is then combined with the upper bound, as no comparison procedure is carried out. The algorithm terminates when every integer combination has been evaluated.

Remark 2.11 *Albeit quite simplistic, this method has proved to be numerically very stable, as it only relies on a suitable solver for the mp-QP problem, and does not require the solution of bigger mp-QP problems (branch-and-bound methods) or the use of global optimization (decomposition-type methods).*

Comments Based on the literature review, several comments can be made:

- Binary and continuous variables are treated separately in all solution approaches presented so far.
- The comparison between two parametric profiles is not straightforward, and might lead to envelopes of solutions (i.e. that more than one solution is associated with a critical region).
- All approaches require the solution of mp-QP problems, in order to solve for the continuous variables in the problem formulation.
- The approaches of handling the binary variables differ in their termination criterion: while branch-and-bound and decomposition-type methods are exploratory (i.e. during the solution procedure, new problems are formulated), the complexity of the exhaustive enumeration methods is fixed *a priori*.

2.5.1.3 A General Framework for the Solution of mp-MIQP Problems

Based on the similarities highlighted in the comments, it is possible to formulate a unified framework for the solution of mp-MIQP problems of type (2.54). This framework incorporates all approaches presented so far and shows their underlying similarities. This not only leads to a good theoretical understanding, but also is the basis for a unified software implementation of all the approaches.

This framework is based on five key components:

Initialization: The algorithm is initialized.

Integer handling: A candidate integer solution is found which is fixed in the original problem, thus transforming it into an mp-QP problem. The three options to find a suitable candidate are (a) global optimization (Dua *et al.* 2002), (b) branch-and-bound (Acevedo and Pistikopoulos 1997; Axehill *et al.* 2011, 2014; Oberdieck *et al.* 2014) and (c) exhaustive enumeration (Borrelli 2003).

mp-QP solution: The mp-QP problem is solved using available solvers (e.g. Bemporad *et al.* 2002; Dua *et al.* 2002; Tøndel *et al.* 2003; Spjøtvold *et al.* 2006).

Comparison procedure: The objective function values of the mp-QP problem and the upper bound in the critical region considered are compared against each other to form a new, tighter upper bound. The four comparison procedures are (a) no comparison of the objective function (Dua *et al.* 2002; Borrelli 2003), (b) comparison of the objective function over the entire critical region considered (Axehill *et al.* 2011, 2014), (c) linearization of the nonlinearities in the objective function using McCormick relaxations (Oberdieck *et al.* 2014) and (d) calculation of the exact solution via piecewise outer approximation of quadratically constrained critical regions, a procedure which will be explained in detail in Section 3.2.5. Note that the approaches (a–c) might result in envelopes of solutions.

Termination: The algorithm terminates if a termination criterion is reached.

A schematic representation of the framework is shown in Figure 2.8.

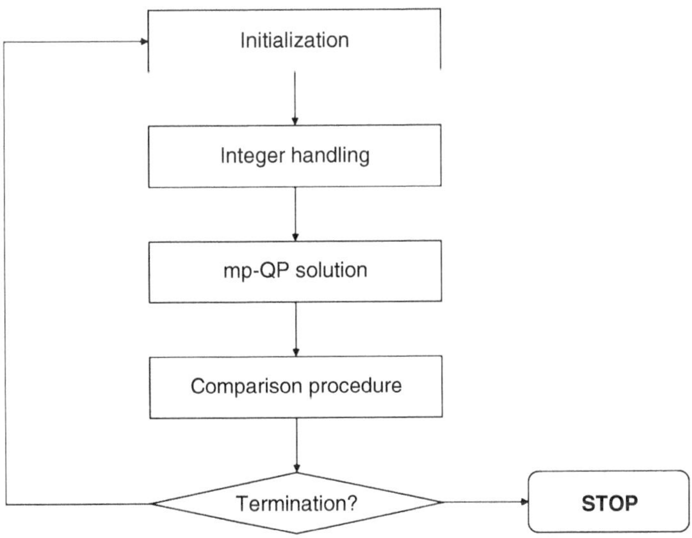

Figure 2.8 The general framework for the solution of mp-MIQP problems.

The framework presents several options for the ways that integer variables are handled and the comparison procedures are done. Since the integer handling has a greater influence on the structure of the algorithm, as it changes how the problem is approached, this will be used to distinguish the way the unit operations are approached by the integer-handling method chosen.

In this section, each aspect of the framework is discussed, and suggestions for their implementation are presented.

2.5.1.4 Detailed Analysis of the General Framework

Initialization The aim of the initialization procedure is to convert the input data efficiently into a suitable form such that the main algorithm can solve the problem. The initialization procedures for mp-MIQP problems presented so far involve initializing the binary search tree (in the case of a branch-and-bound procedure) (Acevedo and Pistikopoulos 1997; Axehill *et al.* 2011, 2014; Oberdieck *et al.* 2014), creating a list of all possible integer variable combinations (in the case of an exhaustive enumeration procedure) (Borrelli 2003) and setting the upper bound to ∞ (Acevedo and Pistikopoulos 1997; Dua *et al.* 2002; Oberdieck *et al.* 2014).

Remark 2.12 *Notably, this part could also be used for other tasks, such as upper bound creation of initialization of parallelization strategies. These topics are subject to ongoing research and provide possibilities of lowering the computational burden.*

Integer Handling The aim of this unit operation is to (a) find a candidate integer variable combination and (b) fix it in the original mp-MIQP, resulting in an mp-QP. As mentioned here, three different approaches have been presented:

Global optimization (Dua *et al.* 2002): A candidate binary variable solution is found by solving the global optimization problem (2.58).

Branch-and-bound (Acevedo and Pistikopoulos 1997; Axehill *et al.* 2011, 2014; Oberdieck *et al.* 2014): The branch-and-bound procedure relies on relaxing the binary variables $y \in \{0,1\}^p$ to continuous variables $\bar{y} \in [0,1]^p$, which results in a mp-QP problem. Based on this relaxation, a binary search tree (see Figure 2.4), and at each node one binary variable, is fixed to either 0 or 1. Additionally, each node inherits a parameter space from its parent node, where the mp-QP is solved. This parameter space is based on the original parameter space Θ, after the following fathoming criteria have been applied:

- The problem is infeasible.
- An integer solution is found.

– The optimal objective function value of the parent node is greater than the current best upper bound in the entire parameter space considered.

Exhaustive enumeration (Borrelli 2003): All possible combinations of binary variables, a total of 2^p, are considered exhaustively. Note that this approach is identical to considering the final depth of a binary search tree.

Solution of the mp-QP Problem The solution of the mp-QP problem has been discussed extensively in the open literature. The two main approaches are:

- *Geometrical approach* (Bemporad *et al.* 2002; Dua *et al.* 2002; Tøndel *et al.* 2003; Spjøtvold *et al.* 2006; Patrinos and Sarimveis 2010, 2011): A geometrical interpretation of the polyhedral regions as half spaces is used to successively explore the initial parameter space. Different techniques discussing different exploration and redundancy checks have thereby been discussed.
- *Combinatorial approach* (Gupta *et al.* 2011; Feller and Johansen 2013; Feller *et al.* 2013): The solution of the mp-QP problem is viewed via the Karush–Kuhn–Tucker conditions, where a certain combination of the inequality constraints is active and inactive, thus leading to a system of linear equations. The consideration of all combinations of all constraints thereby exhaustively solves the problem. The different approaches presented aim at limiting this combinatorial complexity.

Comparison Procedure After the solution to the mp-QP problem has been obtained, it has to be compared against the current upper bound. The need for such a comparison procedure arises from the fact that it is not possible to explore the parameter space like performed for the solution of the mp-QP problem due to the presence of the integer variables.

Mathematically, this comparison procedure can be described as:

$$PP_{sol} = \min_{\theta \in \Theta} PP_1 \cup PP_2 \tag{2.66}$$

where PP_i refers to the i-th parametric profile (see Definition 2.2); and $PP_1 \cup PP_2$ represents the union of the two parametric profiles. The main challenges to solve (2.65) are thereby:

Combinatorial complexity: In order to obtain PP_{Sol}, each critical region of PP_1 has to be compared to each critical region of PP_2, thus leading to combinatorial complexity.

Nonconvexity: The possibly quadratic nature of the objective functions might lead to nonconvexities. As the handling of nonconvex critical regions is a challenging problem, other ways have to be found to deal with this issue.

Remark 2.13 *All comparison procedures presented to date focus on how to deal with the nonconvexity, while little attention has been given to the combinatorial complexity of the problem. At the end of this section, we will present a first analysis of the problem. Additionally, in the next section, a new comparison procedure resulting in the exact partitioning of the parameter space is presented.*

In the open literature, three different comparison procedures have been presented:

- **No objective function comparison (Dua *et al.* 2002):** This approach, first presented in Dua and Pistikopoulos (1999), does not compare the objective functions of the two parametric profiles, but directly creates an envelope of solutions (see Definition 2.3).
- **Objective function comparison over entire CR (Axehill *et al.* 2011, 2014):** This approach was first presented in the realms of multiparametric dynamic programming (mp-DP) problems (Borrelli *et al.* 2005), and later on applied to mp-MIQP problems. It consists of classifying the situation in the currently considered intersect of critical regions according to Figure 2.6. If case (c) is realized, then an envelope of solutions (Definition 2.3) is created over the entire critical region.
- **Direct objective function comparison via McCormick relaxations (Oberdieck *et al.* 2014):** In this procedure, the difference $\Delta z(\theta)$ between the objective functions in the considered intersect of critical regions is directly considered. If quadratic terms are present in $\Delta z(\theta)$, then affine under- and over-estimators using McCormick relaxations (McCormick 1976) are calculated. These estimators generate polyhedral regions, in which either the objective function of one of the parametric profiles is optimal or an envelope of solutions is created.

In order to lessen the combinatorial complexity of the system, it is necessary to incorporate additional information into the solution of the problem. In particular, it is necessary to classify each critical region CR_k of parametric profile i according to the following criteria:

Case (a): CR_k does not overlap with any critical region from the other parametric profile, PP_j.
Case (b): CR_k partially overlaps with one or multiple critical regions from the other parametric profile, PP_j.
Case (c): CR_k completely overlaps with one or multiple critical regions from the other parametric profile, PP_j.

This situation is schematically depicted in Figure 2.9.

Any part of CR_k that does not overlap with PP_j does not have to be considered in the comparison procedure, as it cannot be compared to anything. Thus,

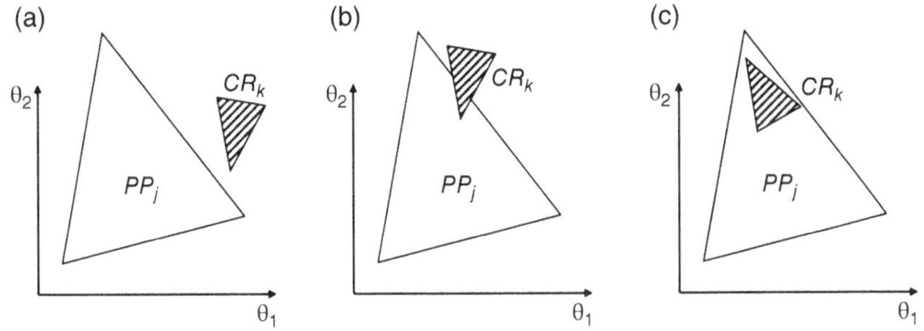

Figure 2.9 The three classifications of overlap between $\boldsymbol{CR_k}$ and $\boldsymbol{PP_j}$.

the number, or volume, of critical regions considered can be reduced, thus lessening the impact of the combinatorial complexity. However, to perform the classification according to Figure 2.9, the union of all critical regions in PP_j has to be calculated, a procedure which also is subject to combinatorial complexity. However, this complexity is lower than the previous one, as it only requires the comparison of all critical regions associated with one parametric profile, not with two.

Termination In order to conclude the framework description, a suitable way of defining the termination criterion has to be outlined. The structure of this termination criterion depends on the integer-handling approach chosen:

- **Global optimization**: If the problem is infeasible in every critical region
- **Branch-and-bound**: If the list of nodes N is empty
- **Exhaustive enumeration**: If all combinations of binary variables have been explored.

As mentioned, the exhaustive enumeration approach is thereby the only approach which allows for an *a priori* determination of the number of problems that will be solved in the course of the algorithm.

Comments Based on the proposed framework, the following comments are made:

- Any of the comparison procedures presented can be combined with any of the integer-handling strategies to create an algorithm which is most suitable to one's needs. Among these combinations are also the mp-MIQP algorithms that have been presented so far. Thus, this framework sheds a light on the development of these algorithms and gives great theoretical insight.

- It is possible to include relative and absolute sub-optimality into this framework (the first description of which was presented in Axehill *et al.* 2011, 2014). However, it does not significantly add to the scientific content presented and is thus omitted.
- If any advances are made in any of the components of the framework (e.g. initialization, integer handling or comparison procedure), then these advances can be readily incorporated into the existing framework. Thus, it is a flexible account not of a specific algorithm, but rather of a general solution approach for this class of problems.

Based on the last comment, in Section 2.5.1.5, a novel comparison procedure is presented which allows for the exact solution of mp-MIQP problems, thus resulting in nonconvex critical regions.

2.5.1.5 Description of an Exact Comparison Procedure

To exactly solve a mp-MIQP problem, possibly resulting in nonconvex critical regions, two main questions have to be answered:

- How can nonconvex critical regions be handled in the integer handling and the mp-QP solution part of the framework (Figure 2.8)?
- Is it possible to solve mp-MIQP problems exactly without employing global optimization?

In order to answer these questions, the difference between two objective functions according to (3.9) is considered.

$$\Delta z(\theta) = z^*(\theta) - z_{UB}(\theta) \tag{2.67}$$

Assuming the same classification as in Figure 2.6, the exact comparison comes into effect in case (c), that is, when there is a transition of optimality between $z^*(\theta)$ and $z_{UB}(\theta)$ within the critical region CR_{int}.

Therefore, if this case is realized, the following critical regions are created:

$$CR_1 = CR_{int} \cap \{\Delta z(\theta) \le 0\} \tag{2.68a}$$

$$CR_2 = CR_{int} \cap \{\Delta z(\theta) \ge 0\} \tag{2.68b}$$

Note that this formulation is identical to the one introduced by Acevedo and Pistikopoulos (1997), however resulting in quadratically constrained regions CR_1 and CR_2. In addition to (3.20), affine over- and under-estimators are created according to Equation (3.11), that is:

$$g(\theta) \le \Delta z(\theta) \le h(\theta). \tag{2.69}$$

In the next iteration of the algorithm, a new combination of integer variables has to be fixed:

- **Global optimization**: The quadratically constrained critical region can readily be incorporated into the problem, as the parameters are treated as optimization variables and the quadratic constraints of the critical region are conceptually of the same complexity as the parametric cuts.
- **Branch-and-bound**: The selection of a new integer variable to fix is regardless of the form of one of the critical regions in the parameter space considered; thus, it does not influence the integer-handling procedure.
- **Exhaustive enumeration**: As this problem is predefined from the start due to the exhaustive enumeration of all integer combinations, a quadratically constrained upper bound region does not influence this integer-handling method.

However, the next component of the framework, the solution of the mp-QP solution, cannot be readily executed. For example, if global optimization is used to find a suitable combination of binary variables, then it would be necessary to solve a mp-QP of form (2.55), where the initial parameter space Θ is quadratically constrained.

In order to deal with this situation, the affine estimators of (2.69) are used to construct a polyhedron Ξ such that:

$$CR \subseteq \Xi \tag{2.70}$$

where CR denotes the quadratically constrained critical region.

Thus, any mp-QP problem with an initial quadratically constrained parameter space CR is solved in Ξ, which by definition also solves the mp-QP in the space covered by CR (and beyond).

As a next step, the solution to this mp-QP with augmented parameter space is compared to the current upper bound. If the upper bound only consists of polyhedral critical regions, then no additional action is necessary. However, if the upper bound consists of a quadratically constrained critical region, then two issues arise:

- How can an empty intersect (case (a) of Figure 2.9) be detected?
- How can the classification according to Figure 2.6 be performed?

The detection of an empty intersect between a polyhedral and a quadratically constrained critical region corresponds to the solution of the following feasibility problem:

$$\begin{aligned} &\min_{\theta} 0 \\ &s.t. \theta \in CR \cap \Phi \end{aligned} \tag{2.71}$$

where CR and Φ denote a quadratically constrained and a polyhedral critical region, respectively. One way to approach problem (2.71) is the use of a spatial branch-and-bound algorithm, where the quadratically constrained critical region is approximated by a polyhedral critical region and a corresponding linear program (LP) is solved until a predefined tolerance is reached.

Remark 2.14 *Spatial branch-and-bound techniques are, among others, commonly encountered in global optimization. However, problem (2.71) is not a (nonconvex) optimization problem, as it only requires to find one feasible point $\theta \in CR \cap \Phi$, regardless of the objective function value associated with it. Thus, only if exactly one point θ exists (within the predefined tolerance), the complexity of problem (2.71) is the same as the corresponding optimization problem, since, if only one point is feasible, this point is also optimal.*

In order to perform a classification according to Figure 2.6, a similar argument to one just presented can be made: let CR_{int} be a quadratically constrained critical region, and let $\Delta z(\theta)$ be quadratic. Then the following regions are defined:

$$CR_1 = CR_{int} \cap \{\Delta z(\theta) \le 0\} \tag{2.72a}$$

$$CR_2 = CR_{int} \cap \{\Delta z(\theta) \ge 0\} \tag{2.72b}$$

At this point, it is possible to perform a feasibility test according to (2.71) on CR_1 and CR_2, and thus perform the classification:

$$\begin{aligned} CR_2 = \varnothing &\rightarrow \text{Case (a)} \\ CR_1 = \varnothing &\rightarrow \text{Case (b)} \\ CR_1 \ne \varnothing \wedge CR_2 \ne \varnothing &\rightarrow \text{Case (c)} \end{aligned} \tag{2.73}$$

where $\varnothing$ denotes an empty set.

Remark 2.15 *As this comparison procedure is performed for the solution of the mp-QP problem obtained in $\Xi \supseteq CR$, an additional step has to be undertaken, in which the original quadratic constraints are added to the newly formed critical regions (e.g. CR_1 and CR_2) and the problem has to be solved again.*

A schematic representation of the exact comparison procedure is shown in Figure 2.10.

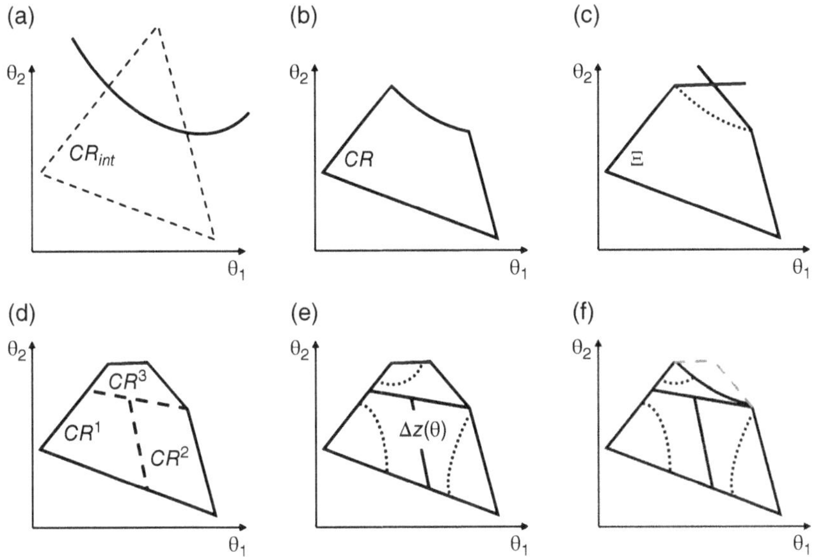

Figure 2.10 A schematic representation for the exact comparison procedure for the solution of mp-MIQP problems. In part (a), the quadratic boundary resulting from the exact use of $\boldsymbol{\Delta z(\theta)}$ is combined with $\boldsymbol{CR_{int}}$ to form a quadratically constrained critical region $\boldsymbol{CR}$ (part (b)). For this region, a polyhedral outer approximation Ξ is calculated such that $\boldsymbol{CR} \subseteq \Xi$ (part [c]). In part (d), the corresponding mp-QP problem is solved in Ξ, resulting in a partition of the parameter space. Each of these critical regions, $\boldsymbol{CR^{1-3}}$, is compared against the current upper bound, thus resulting in a new set of $\boldsymbol{\Delta z(\theta)}$ (part [e]). Lastly, in part (f), the original quadratic constraints from $\boldsymbol{CR}$ are reintroduced, thus closing the loop.

References

Acevedo, J. and E. N. Pistikopoulos (1996). "A parametric MINLP algorithm for process synthesis problems under uncertainty." *Industrial and Engineering Chemistry Research* **35**(1): 147–158.

Acevedo, J. and E. N. Pistikopoulos (1997). "A multiparametric programming approach for linear process engineering problems under uncertainty." *Industrial & Engineering Chemistry Research* **36**(3): 717–728.

Acevedo, N. and E. N. Pistikopoulos (1999). "An algorithm for multiparametric mixed-integer linear programming problems." *Oper. Res. Lett.* **24**(3): 139–148.

Adamjan, V. M., D. Z. Arov and M. G. Krein (1971). "Analytic properties of Schmidt pairs for a Hankel operator and the generalized Schur-Takagi problem." *Math. USSR Sbornik* **15**: 31–73.

Agarwal, A. and L. T. Biegler (2013). "A trust-region framework for constrained optimization using reduced order modeling." *Optimization and Engineering* **14**(1): 3–35.

Alt, W. (1991). "Local stability of solutions to differentiable optimization problems in banach spaces." *Journal of Optimization Theory and Applications* **70**(3): 443–466.

Antoulas, A. C. (2005). "An overview of approximation methods for large-scale dynamical systems." *Annual Reviews in Control* **29**(2): 181–190.

Antoulas, A. C. and D. C. Sorensen (2001). "Lyapunov, Lanczos, and inertia." *Linear Algebra and Its Applications* **326**(1–3): 137–150.

Arnoldi, W. E. (1951). "The principle of minimized iterations in the solution of the matrix eigenvalue problem." *Quarterly of Applied Mathematics* **9**: 17–29.

Augustin, D. and H. Maurer (2001). "Computational sensitivity analysis for state constrained optimal control problems." *Annals of Operations Research* **101**(1–4): 75–99.

Axehill, D., T. Besselmann, D. M. Raimondo and M. Morari (2011). Suboptimal explicit hybrid mpc via branch and bound. *IFAC World Congress*. Milano.

Axehill, D., T. Besselmann, D. M. Raimondo and M. Morari (2014). "A parametric branch and bound approach to suboptimal explicit hybrid MPC." *Automatica* **50**(1): 240–246.

Badgwell, T. A. (1997). "Robust model predictive control of stable linear systems." *International Journal of Control* **68**(4): 797–818.

Bellman, R. (1957). *Dynamic Programming*. Princeton, NJ, USA, Princeton University Press.

Bemporad, A., F. Borrelli and M. Morari (2000). The explicit solution of constrained LP-based receding horizon control. In *Proceedings of the IEEE Conference on Decision and Control*.

Bemporad, A., F. Borrelli and M. Morari (2003). "Min-max control of constrained uncertain discrete-time linear systems." *IEEE Transactions on Automatic Control* **48**(9): 1600–1606.

Bemporad, A. and M. Morari (1999). "Control of systems integrating logic, dynamics, and constraints." *Automatica* **35**(3): 407–427.

Bemporad, A., M. Morari, V. Dua and E. N. Pistikopoulos (2002). "The explicit linear quadratic regulator for constrained systems." *Automatica* **38**(1): 3–20.

Benson, H. P. (1982). "Algorithms for parametric nonconvex programming." *Journal of Optimization Theory and Applications* **38**(3): 319–340.

Berkelaar, A. B., B. Jansen, K. Roos and T. Terlaky (1999). "Basis- and partition identification for quadratic programming and linear complementarity problems." *Mathematical Programming, Series B* **86**(2): 261–282.

Blatman, G. and B. Sudret (2010). "An adaptive algorithm to build up sparse polynomial chaos expansions for stochastic finite element analysis." *Probabilistic Engineering Mechanics* **25**(2): 183–197.

Bonis, I., W. Xie and C. Theodoropoulos (2012). "A linear model predictive control algorithm for nonlinear large-scale distributed parameter systems." *AIChE Journal* **58**(3): 801–811.

Borrelli, F. (2003). *Constrained optimal control of linear and hybrid systems.* Springer, Berlin.

Borrelli, F., M. Baotic, A. Bemporad and M. Morari (2005). "Dynamic programming for constrained optimal control of discrete-time linear hybrid systems." *Automatica* **41**(10): 1709–1721.

Buhmann, M. D. (2003). *Radial Basis Functions: Theory and Implementations.* Cambridge University Press, Cambridge.

Büskens, C. and H. Maurer (2000). "SQP-methods for solving optimal control problems with control and state constraints: adjoint variables, sensitivity analysis and real-time control." *Journal of Computational and Applied Mathematics* **120**(1–2): 85–108.

Campo, P. J. and M. Morari (1987). "Robust model predictive control." *American Control Conference. ACC. Minneapolis*: pp. 1021–1026.

Casavola, A. and E. Mosca (2007). "A correction to "Min-max predictive control strategies for input-saturated polytopic uncertain systems" [Automatica 36 (2000) 125–133] (DOI:10.1016/S0005-1098(99)00112-0)." *Automatica* **43**(2): 384–385.

Crema, A. (2002). "The multiparametric 0-1-integer linear programming problem: a unified approach." *European Journal of Operational Research* **139**(3): 511–520.

Darby, M. L. and M. Nikolaou (2007). "A parametric programming approach to moving-horizon state estimation." *Automatica* **43**(5): 885–891.

Diehl, M., H. G. Bock, J. P. Schlöder, R. Findeisen, Z. Nagy and F. Allgöwer (2002). "Real-time optimization and nonlinear model predictive control of processes governed by differential-algebraic equations." *Journal of Process Control* **12**(4): 577–585.

Dontchev, A. L., W. W. Hager, K. Malanowski and V. M. Veliov (2000). "On quantitative stability in fpo optimization and optimal control." *Set-Valued Analysis* **8**(1–2): 31–50.

Dua, V., N. A. Bozinis and E. N. Pistikopoulos (2002). "A multiparametric programming approach for mixed-integer quadratic engineering problems." *Computers and Chemical Engineering* **26**(4–5): 715–733.

Dua, V., K. P. Papalexandri and E. N. Pistikopoulos (1999). "A parametric mixed-integer global optimization framework for the solution of process engineering problems under uncertainty." *Computers and Chemical Engineering* **23**(Suppl. 1): S19–S22.

Dua, V., K. P. Papalexandri and E. N. Pistikopoulos (2004). "Global optimization issues in multiparametric continuous and mixed-integer optimization problems." *Journal of Global Optimization* **30**(1): 59–89.

Dua, V. and E. N. Pistikopoulos (1999). "Algorithms for the solution of multiparametric mixed-integer nonlinear optimization problems." *Industrial & Engineering Chemistry Research* **38**(10): 3976–3987.

Dua, V. and E. N. Pistikopoulos (2000). "An algorithm for the solution of multiparametric mixed integer linear programming problems." *Annals of Operations Research* **99**(1–4): 123–139.

Faisca, N. P., K. I. Kouramas, P. M. Saraiva, B. C. C. Rustem and E. N. Pistikopoulos (2008). "A multi-parametric programming approach for constrained dynamic programming problems." *Optimization Letters* **2**(2): 267–280.

Feller, C. and T. A. Johansen (2013). Explicit MPC of higher-order linear processes via combinatorial multi-parametric quadratic programming. *Control Conference (ECC), 2013 European*: 536–541.

Feller, C., T. A. Johansen and S. Olaru (2013). "An improved algorithm for combinatorial multi-parametric quadratic programming." *Automatica* **49**(5): 1370–1376.

Fiacco, A. V. and Y. Ishizuka (1990). "Sensitivity and stability analysis for nonlinear programming." *Annals of Operations Research* **27**(1): 215–235.

Fiacco, A. V. and J. Kyparisis (1988). "Computable bounds on parametric solutions of convex problems." *Mathematical Programming* **40**(1–3): 213–221.

Filippi, C. and G. Romanin-Jacur (2002). "Multiparametric demand transportation problem." *European Journal of Operational Research* **139**(2): 206–219.

Floudas, C. A. (1995). "Nonlinear and mixed integer optimisation." *Oxford University Press, New York.*

Gal, T. (1995). "Postoptimal Analyses, Parametric Programming and Related Topics." *second edition ed. Walter de Cruyter. Berlin, New York.*

Gal, T. and J. Nedoma (1972). "Multiparametric Linear Programming." *Management Science* **18**(7): 406–422.

Gallivan, K., E. Grimme and P. Van Dooren (1994). *Pade approximation of large-scale dynamic systems with Lanczos methods.* Proceedings of the IEEE Conference on Decision and Control.

Gupta, A., S. Bhartiya and P. S. V. Nataraj (2011). "A novel approach to multiparametric quadratic programming." *Automatica* **47**(9): 2112–2117.

Hahn, J. and T. F. Edgar (2002). "An improved method for nonlinear model reduction using balancing of empirical gramians." *Computers and Chemical Engineering* **26**(10): 1379–1397.

Hedengren, J. D. and T. F. Edgar (2005). "In situ adaptive tabulation for real-time control." *Industrial and Engineering Chemistry Research* **44**(8): 2716–2724.

Heemels, W. P. M. H., B. d. Schutter and A. Bemporad (2001). "Equivalence of hybrid dynamical models." *Automatica* **37**(7): 1085–1091.

Homma, T. and A. Saltelli (1996). "Importance measures in global sensitivity analysis of nonlinear models." *Reliability Engineering and System Safety* **52**(1): 1–17.

Hovland, S., J. T. Gravdahl and K. E. Willcox (2008). “Explicit model predictive control for large-scale systems via model reduction.” *Journal of Guidance, Control, and Dynamics* **31**(4): 918–926.

Ito, K. and K. Kunisch (1992). “Sensitivity analysis of solutions to optimization problems in Hilbert spaces with applications to optimal control and estimation.” *Journal of Differential Equations* **99**(1): 1–40.

Ivakhnenko, A. G. and J. A. Muller (1995). “Self-organization of nets of active neurons.” *SAMS* **20**: 93–106.

Jonker, P., G. Still and F. Twilt (2001). “One-Parametric Linear-Quadratic Optimization Problems.” *Annals of Operations Research* **101**(1–4): 221–253.

Kassmann, D. E., T. A. Badgwell and R. B. Hawkins (2000). “Robust steady-state target calculation for model predictive control.” *AIChE Journal* **46**(5): 1007–1024.

Kiparissides, A., S. S. Kucherenko, A. Mantalaris and E. N. Pistikopoulos (2009). “Global sensitivity analysis challenges in biological systems modeling.” *Industrial and Engineering Chemistry Research* **48**(15): 7168–7180.

Kolmogorov, A. N. (1957). “On the representation of continuous functions of many variables by superpositions of continuous functions of one variable and addition.” *Doklay Akademii Nauk USSR* **14**(5): 953–956.

Kontoravdi, C., S. P. Asprey, E. N. Pistikopoulos and A. Mantalaris (2005). “Application of global sensitivity analysis to determine goals for design of experiments: An example study on antibody-producing cell cultures.” *Biotechnology Progress* **21**(4): 1128–1135.

Kontoravdi, C., E. N. Pistikopoulos and A. Mantalaris (2010). “Systematic development of predictive mathematical models for animal cell cultures.” *Computers and Chemical Engineering* **34**(8): 1192–1198.

Kothare, M. V., V. Balakrishnan and M. Morari (1996). “Robust constrained model predictive control using linear matrix inequalities.” *Automatica* **32**(10): 1361–1379.

Kouramas, K. I., N. P. Faisca, C. Panos and E. N. Pistikopoulos (2011). “Explicit/multi-parametric model predictive control (MPC) of linear discrete-time systems by dynamic and multi-parametric programming.” *Automatica* **47**(8): 1638–1645.

Kouvaritakis, B., J. A. Rossiter and J. Schuurmans (2000). “Efficient robust predictive control.” *IEEE Transactions on Automatic Control* **45**(8): 1545–1549.

Krylov, A. N. (1931). “On the numerical solution of equation by which are determined in technical problems the frequencies of small vibrations of material systems.” *Izvestija AN SSSR (News of Academy of Sciences of the USSR), Otdel. mat. i estest. nauk* 7(4): 491–539.

Kucherenko, S., M. Rodriguez-Fernandez, C. Pantelides and N. Shah (2009). “Monte Carlo evaluation of derivative-based global sensitivity measures.” *Reliability Engineering and System Safety* **94**(7): 1135–1148.

Lall, S., J. E. Marsden and S. Glavaski (1999). "Empirical model reduction for controlled nonlinear systems." *Proceedings of the IFAC World Congress*: 473–478.

Lambert, R. S. C., P. Rivotti and E. N. Pistikopoulos (2013). "A Monte-Carlo based model approximation technique for linear model predictive control of nonlinear systems." *Computers and Chemical Engineering* **54**: 60–67.

Lanczos, C. (1950). "An iteration method for the solution of the eigenvalue problem of linear differential and integral operators." *Res. Nat. Bureau Stan* **45**: 255–282.

Lee, J. H. and B. L. Cooley (2000). "Min-max predictive control techniques for a linear state-space system with a bounded set of input matrices." *Automatica* **36**(3): 463–473.

Lee, J. H. and Z. Yu (1997). "Worst-case formulations of model predictive control for systems with bounded parameters." *Automatica* **33**(5): 763–781.

Lemke, F. (1997). "Knowledge extraction from data using self-organizing modeling technologies." *eSEAM'97 conference, MacSciTech organization*

Li, G., J. Hu, S. W. Wang, P. G. Georgopoulos, J. Schoendorf and H. Rabitz (2006). "Random sampling-high dimensional model representation (RS-HDMR) and orthogonality of its different order component functions." *Journal of Physical Chemistry A* **110**(7): 2474–2485.

Li, G., S. W. Wang and H. Rabitz (2002). "Practical approaches to construct RS-HDMR component functions." *Journal of Physical Chemistry A* **106**(37): 8721–8733.

Löfberg, J. (2004). YALMIP: a toolbox for modeling and optimization in MATLAB. *CCA/ISIC/CACSD*.

Lorentz, G. (1966). *Approximation of Functions*. Syracuse University Press, Syracuse.

Malanowski, K. and H. Maurer (2001). "Sensitivity analysis for optimal control problems subject to higher order state constraints." *Annals of Operations Research* **101**(1–4): 43–73.

Mayne, D. Q., S. V. Raković, R. Findeisen and F. Allgöwer (2009). "Robust output feedback model predictive control of constrained linear systems: time varying case." *Automatica* **45**(9): 2082–2087.

Mayne, D. Q. and W. R. Schroeder (1997). "Robust time-optimal control of constrained linear system." *Automatica* **33**(12): 2103–2118.

McCormick, G. P. (1976). "Computability of global solutions to factorable nonconvex programs: part i — convex underestimating problems." *Mathematical Programming* **10**(1): 147–175.

Morris, M. D. (1991). "Factorial sampling plans for preliminary computational experiments." *Factorial Sampling Plans for Preliminary Computational Experiments* **33**.

Narciso, D. and E. Pistikopoulos (2008). "A combined balanced truncation and multi-parametric programming approach for linear model predictive control." *Computer Aided Chemical Engineering* **25**: 405–410.

Naşcu, I., R. Oberdieck and E. N. Pistikopoulos (2016). "A framework for simultaneous state estimation and robust hybrid model predictive control in intravenous anaesthesia." *Computer Aided Chemical Engineering* **38**: 1057–1062.

Oberdieck, R., M. Wittmann-Hohlbein and E. N. Pistikopoulos (2014). "A branch and bound method for the solution of multiparametric mixed integer linear programming problems." *Journal of Global Optimization* **59**(2–3): 527–543.

Ohtake, Y. and N. Nishida (1985). "A branch-and-bound algorithm for 0-1 parametric mixed integer programming." *Operations Research Letters* **4**(1): 41–45.

Papalexandri, K. P. and T. I. Dimkou (1998). "A parametric mixed-integer optimization algorithm for multiobjective engineering problems involving discrete decisions." *Industrial and Engineering Chemistry Research* **37**(5): 1866–1882.

Patrinos, P. and H. Sarimveis (2010). "A new algorithm for solving convex parametric quadratic programs based on graphical derivatives of solution mappings." *Automatica* **46**(9): 1405–1418.

Patrinos, P. and H. Sarimveis (2011). "Convex parametric piecewise quadratic optimization: Theory and algorithms." *Automatica* **47**(8): 1770–1777.

Pertsinidis, A., I. E. Grossmann and G. J. McRae (1998). "Parametric optimization of MILP programs and a framework for the parametric optimization of MINLPs." *Computers and Chemical Engineering* **22**(SUPPL.1): S205–S212.

Pistikopoulos, E. N. (2000). *On-line Optimization via Off-line Optimization: A Guided Tour to Parametric Programming*. AspenWorld.

Pistikopoulos, E. N., V. Dua, N. A. Bozinis, A. Bemporad and M. Morari (2002). "On-line optimization via off-line parametric optimization tools." *Computers and Chemical Engineering* **26**(2): 175–185.

Pistikopoulos, E., M. Georgiadis and V. Dua (2007). *Multi-parametric Programming: Theory, Algorithms and Applications*. Wiley-VCH, Weinheim.

Raković, S. V. and K. I. Kouramas (2006). "The minimal robust positively invariant set for linear discrete time systems: approximation methods and control applications." In *Proceedings of the IEEE Conference on Decision and Control*.

Ramírez, D. R. and E. F. Camacho (2001). "On the piecewise linear nature of min-max model predictive control with bounded uncertainties." In *Proceedings of the IEEE Conference on Decision and Control*.

Ramírez, D. R. and E. F. Camacho (2002). "Characterization of min-max MPC with bounded uncertainties and a quadratic criterion." In *Proceedings of the American Control Conference*.

Rao, C. V. (2000). *Moving Horizon Strategies for the Constrained Monitoring and Control of Nonlinear Discrete-Time Systems*. PhD thesis, University of Wisconsin–Madison.

Rawlings, J. B. and D. Q. Mayne (2009). *Model Predictive Control: Theory and Design*. Nob Hill Publishing, Madison, WI.

Rewieński, M. and J. White (2001). A trajectory piecewise-linear approach to model order reduction and fast simulation of nonlinear circuits and micromachined devices. In *IEEE/ACM International Conference on Computer-Aided Design, Digest of Technical Papers.*

Rivotti, P., R. S. C. Lambert and E. N. Pistikopoulos (2012). "Combined model approximation techniques and multiparametric programming for explicit nonlinear model predictive control." *Computers and Chemical Engineering* **42**: 277–287.

Rivotti, P. and E. N. Pistikopoulos (2013). "Constrained dynamic programming of mixed-integer linear problems by multi-parametric programming." *Computer and Chemical Engineering*, submitted.

Robertson, D. G. and J. H. Lee (2002). "On the use of constraints in least squares estimation and control." *Automatica* **38**(7): 1113–1123.

Saltelli, A. (2004). "Global sensitivity analysis: an introduction." In *Proceeding of 4th International Conference on Sensitivity Analysis of Model Output (SAMO '04).*

Saltelli, A., P. Annoni, I. Azzini, F. Campolongo, M. Ratto and S. Tarantola (2010). "Variance based sensitivity analysis of model output: design and estimator for the total sensitivity index." *Computer Physics Communications* **181**(2): 259–270.

Schwarm, A. T. and M. Nikolaou (1999). "Chance-constrained model predictive control." *AIChE Journal* **45**(8): 1743–1752.

Scokaert, P. O. M. and D. Q. Mayne (1998). "Min-max feedback model predictive control for constrained linear systems." *IEEE Transactions on Automatic Control* **43**(8): 1136–1142.

Singh, A. K. and J. Hahn (2005). "State estimation for high-dimensional chemical processes." *Computers and Chemical Engineering* **29**(11–12 Spec. Iss.): 2326–2334.

Sobol, I. M. (1993). "Global sensitivity indices for nonlinear mathematical models and their Monte Carlo estimates." *Mathematical Modeling Computation* 407–414.

Sobol, I. M. (2001). "Global sensitivity indices for nonlinear mathematical models and their Monte Carlo estimates." *Mathematics and Computers in Simulation* **55**(1–3): 271–280.

Sobol, I. M. and S. S. Kucherenko (2005). "On global sensitivity analysis of quasi-Monte Carlo algorithms." *Monte Carlo Methods and Applications* **11**(1): 83–92.

Söderström, T. (2002). *Discrete-Time Stochastic Systems.* Springer, Berline.

Solís-Daun, J., J. Álvarez-Ramírez and R. Suárez (1999). "Semiglobal stabilization of linear systems using constrained controls: a parametric optimization approach." *International Journal of Robust and Nonlinear Control* **9**(8): 461–484.

Spjøtvold, J., E. C. Kerrigan, C. N. Jones, P. Tøndel and T. A. Johansen (2006). "On the facet-to-facet property of solutions to convex parametric quadratic programs." *Automatica* **42**(12): 2209–2214.

Sudret, B. (2008). "Global sensitivity analysis using polynomial chaos expansions." *Reliability Engineering and System Safety* **93**(7): 964–979.

Sui, D., L. Feng and M. Hovd (2008). Robust output feedback model predictive control for linear systems via moving horizon estimation. In *Proceedings of the American Control Conference.*

Tøndel, P., T. A. Johansen and A. Bemporad (2003). "An algorithm for multi-parametric quadratic programming and explicit MPC solutions." *Automatica* **39**(3): 489–497.

Voelker, A., K. Kouramas and E. N. Pistikopoulos (2013). "Simultaneous design of explicit/multi-parametric constrained moving horizon estimation and robust model predictive control." *Computers and Chemical Engineering* **54**: 24–33.

Welch, G. and G. Bishop (2001). "An introduction to the Kalman filter: SIGGRAPH 2001 Course 8." In *Proc. Comput. Graph., Annu. Conf. Comput Graph., Interact. Techn.*, 12–17.

Wong, E. (1971). *Stochastic Process in Information and Dynamical Systems.* McGraw-Hill, New York.

Xie, W., I. Bonis and C. Theodoropoulos (2011). "Off-line model reduction for on-line linear MPC of nonlinear large-scale distributed systems." *Computers and Chemical Engineering* **35**(5): 750–757.

Xie, W., I. Bonis and C. Theodoropoulos (2012). Linear MPC based on data-driven artificial neural networks for large-scale nonlinear distributed parameter systems. *Computer Aided Chemical Engineering* **30**: 1212–1216.

Yue, H., M. Brown, F. He, J. Jia and D. B. Kell (2008). "Sensitivity analysis and robust experimental design of a signal transduction pathway system." *International Journal of Chemical Kinetics* **40**(11): 730–741.

Zafiriou, E. (1990). "Robust model predictive control of processes with hard constraints." *Computers and Chemical Engineering* **14**(4–5): 359–371.

3

Volatile Anaesthesia

Alexandra Krieger[1], Ioana Naşcu[2], Nicki Panoskaltsis[3], Athanasios Mantalaris[4], Michael C. Georgiadis[5], and Efstratios N. Pistikopoulos[6]

[1] *Jacobs Consultancy, Kreisfreie Stadt Aachen Area, Germany*
[2] *Artie McFerrin Department of Chemical Engineering, Texas A&M University, College Station, USA*
[3] *Department of Medicine, Imperial College London, UK*
[4] *Department of Chemical Engineering, Imperial College London, UK*
[5] *Laboratory of Process Systems Engineering, School of Chemical Engineering, Aristotle University of Thesaloniki, Greece*
[6] *Texas A&M Energy Institute, Artie McFerrin Department of Chemical Engineering, Texas A&M University, USA*

3.1 Introduction

Mathematical modelling and model-based control of anaesthesia attracted the attention of researchers during the past decades and are believed to benefit the safety of the patient undergoing surgery and provide anaesthetists with valuable insights. Currently, anaesthetists rely on common practice and their personal experience to determine drug infusion rates and drug inhalation concentrations for the individual patient. The expectations are that detailed modelling and optimized regulation of anaesthesia could pave the way for personalized healthcare, where the individual patient characteristics are taken into account for optimal and flexible drug administration. This will improve the safety of the patient by minimizing side effects, the risk of awareness and overdosing during anaesthesia (Dua *et al.* 2010). When in need of an adequate mathematical model describing the process of drug distribution (pharmacokinetics [PK]) and drug effects (pharmacodynamics [PD]), employing model predictive control is the method of choice. With respect to model predictive control, one of the key challenges is the high inter-patient and intra-patient variability, which introduces a high degree of uncertainty into the system.

The anaesthetist faces the task of providing sufficient anaesthesia to the individual patient during the ongoing surgery, while maintaining their vital functions. The available drugs lead to the desired effects of hypnosis, amnesia, analgesia and muscle relaxation, while the side effects on the cardiovascular system, the respiratory system and the central nervous system, if not supervised accordingly, can have such a high impact that they are life-threatening.

Modelling Optimization and Control of Biomedical Systems, First Edition.
Edited by Efstratios N. Pistikopoulos, Ioana Naşcu, and Eirini G. Velliou.

Volatile anaesthetics are preferred by most anaesthetists to provide hypnosis, because the concentration of the anaesthetic in the expired air is directly related to the partial pressure in the brain and thus gives a direct feedback of the hypnotic state of the patient. The uptake and the elimination of volatile anaesthetics occur in the lungs. Intravenous drugs (e.g. intravenous anaesthetics or analgesics) are administered through a venous catheter and enter the cardiovascular system directly. From there, the drug is distributed to its side of action in the brain and/or the other tissues and organs.

Sufficient anaesthesia is provided by a combined administration of anaesthetics, analgesics, muscle relaxants and cardioactive drugs. The anaesthetic practice is based on the synergetic drug effects of anaesthetics and analgesics. One can assume that any administered drugs are interacting on a PK or PD basis with each other; therefore, a fundamental knowledge of drug interactions and possible side effects is mandatory for the anaesthetist.

The first essential step is to derive an adequate PK model describing the drug distribution. Physiologically based models for drug distribution go back to the 1920s and the pioneering work of Teorell for drug uptake, distribution and elimination (Teorell 1937a, 1937b). Specific to volatile anaesthesia, an in-depth analysis of the uptake of specific tissues and the influence of the partition coefficient and physiological variables such as the cardiac output is given by Kety (1951). In the first physiologically based models, the tissues with similar properties, such as the well-perfused organs, were grouped together, as presented by Mapleson (1963) and extended by Fiserova-Bergerova (1992) and Lerou *et al.* (1991a, 1991b), who derived a 14-compartment model for teaching and research purposes. This so-called physiologically based pharmacokinetic (PBPK) modelling is the focus of researchers for drug delivery, uptake and distribution models (Hall *et al.* 2012). Pioneering work towards individual patient variables of volatile agents, applying a model mapping the circulation for the uptake of ether in a dog, goes back to Haggard (1924).

Individualized modelling of the drug effect is the more challenging task, because of higher variability (Mertens and Vuyk 1998). To determine the drug effect (i.e. the hypnotic depth during anaesthesia), anaesthetists use the minimum alveolar concentration (MAC) as a guideline. MAC is defined as the concentration required to prevent movement in response to surgical incision in 50% of patients. Usually, 1.3 MAC of volatile anaesthetic is administered during anaesthesia to assure sufficient anaesthesia in 90% of patients (Miller 2015). Hence, by definition, this guideline of the MAC is based on probability. This highlights the challenge of identifying the individual patient's sensitivity and hypnotic depth to avoid awareness or overdosing. Individual factors influencing the individual patient's sensitivity are, for example, age. Studies by Brunner *et al.* (1994) investigated the correlation of MAC with patient characteristics or analgesics administered simultaneously during anaesthesia (e.g. elderly patients are more sensitive to anaesthetics); hence, MAC decreases with age (Mapleson 1996). Furthermore, the patient's sensitivity towards the anaesthetic agent changes depending on the surgical stimulation or the combination of drugs administered during anaesthesia, such as muscle relaxants and analgesics (Glass *et al.* 1997; Rosow 1997). Recent advances investigate the pharmacogenomic variability as an indicator of individual patients' sensitivity to anaesthetic agents (Searle and Hopkins 2009).

In this work, PBPK modelling is applied to describe the PK and address patient variability by including patient-specific characteristics in the mathematical description (Hall *et al.* 2012). The variability of the PK uncertainty is included analogously to Fiserova-Bergerova (1992), where all volumes for blood tissue and gas compartments are assigned specific to the individual patient's weight, height and age. By including these factors, the aim is to reduce significantly the variability in the PK, which is estimated to be around 60–80% (Mertens and Vuyk 1998). However, the challenge is the PD variability, which is assumed to be considerably higher, around 300–400% (Mertens and Vuyk 1998). Including the full PD variability in the mathematical model representation might be very complex, if not impossible. The consideration of all patient-specific factors might still not capture the full uncertainty, and a control strategy based on such a model might not be safe for the patient if used as the underlying model of a model predictive controller.

Motivated by this challenge of the implying uncertainty, a detailed PBPK-PD model is presented. All relevant PK and PD parameters and variables are analysed with respect to uncertainty imposed on the hypnotic depth measured by the bispectral index (BIS) and the end-tidal concentration, employed as a guideline indicator of the anaesthetic depth when no BIS measurements are available. All PK parameters are determined as a function of the individual patient, whereas the PD ones are believed to be captured best by on-line estimation proposed for propofol by Sartori *et al.* (2005). Hence, individual PD as a function of drug interaction, age dependency or pharmacogenetics is not included in this work.

3.2 Physiologically Based Patient Model

3.2.1 Pharmacokinetics

The physiologically based compartmental model for volatile anaesthetics, shown in Figure 3.1a, is based on Eger's compartmental model for volatile anaesthesia and explained in detail in Krieger *et al.* (2014), where the tissues with similar properties are lumped together resulting in three body compartments representing the vessel-rich group (VRG), the muscle group (M) and the adipose tissue (F) (Eger 1974). Each body compartment is further divided into an ideally mixed-blood and ideally mixed-tissue part. This approach is based on a model for cancer chemotherapeutic drugs first presented by Bischoff (1986). The gas, blood and tissue volumes are individually adjusted to the weight, height, gender and age of the patient.

The compartments are described assuming a flow-limited formulation. Hence, the diffusion through the capillary vessel walls is assumed to be rapid, and the mass transfer of the drug into the tissue is restricted by the perfusion of the

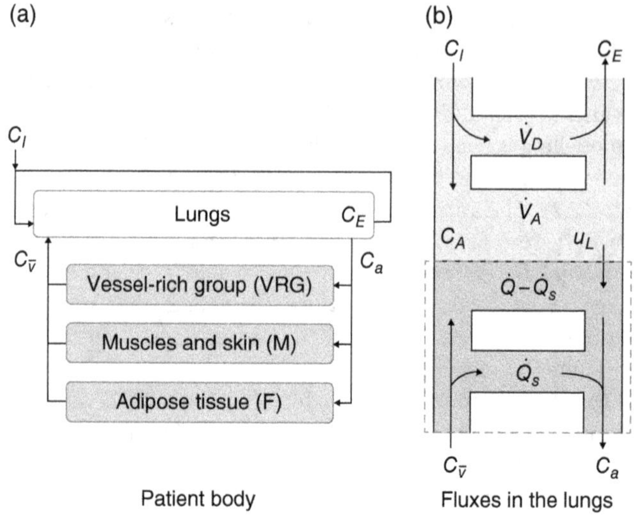

Figure 3.1 Structure of the physiologically based patient model. (a) Patient body; (b) fluxes in the lungs. *Source*: Krieger *et al.* (2014). Reproduced with permission of Elsevier.

compartment. This approximation is not fundamental to the physiological pharmacokinetic approach, but commonly used due to the lack of sufficient physiological information of, for example, membrane permeabilities, diffusion coefficients and tissue surfaces (Bischoff 1986). No inter-tissue diffusion between the compartments (e.g. from the VRG to the adipose tissue) is included (Zwart *et al.* 1972). This implies that mass exchange only occurs through the blood vessels. The transport time and the pulsatile character of the blood flow are neglected, because the equilibration times are large compared to the cardiac cycle (Zwart *et al.* 1972). All fluxes leaving a gas, blood or tissue compartment are in equilibrium with the compartment.

The uptake of the anaesthetic agent is determined by two factors, the ventilation of air and the perfusion of blood through the lungs. The ventilation is given by the product of the respiratory frequency f_R and the tidal volume V_T. Of the total minute ventilation $\dot{V}$, only the alveolar ventilation $\dot{V}_A$, which is the total ventilation less the dead space ventilation $\dot{V}_D$, is taking part in the gas exchange.

$$\dot{V}_A = \dot{V} - \dot{V}_D = f_R(V_T - V_D) \tag{3.1}$$

Here, f_R and V_T are set by the anaesthetic machine and the anaesthetist, respectively. Analogously to (3.1), the alveolar volume V_A is determined by the lung volume V_L less the dead space volume V_D.

$$V_A = V_L - V_D \tag{3.2}$$

To illustrate the applied mass balances and assumptions, the fluxes of gas and blood in the lungs are shown in Figure 3.1b.

During anaesthesia, the concentration of the inhaled volatile anaesthetic agent C_I is set by the anaesthetist. To map the respiratory cycle and changing gas concentration in the lungs, the concentration in the alveoli just after inspiration C_{A_I} is given in (3.3) analogous to the Bohr equation for carbon dioxide (Miller 2015), where the amount of inhaled anaesthetic gas is ideally mixed with the gas left in the lungs after expiration.

$$C_{A_I}\left(V_A+V_T\right)=C_I V_T+C_E V_A \tag{3.3}$$

To determine the concentration during expiration, equilibrium between the end-tidal expired concentration C_E and the mixed venous blood concentration $C_{\bar{v}}$ is assumed. Both concentrations are linked via the blood gas partition coefficient λ (Eger 1974).

$$C_E=\frac{C_{\bar{v}}}{\lambda} \tag{3.4}$$

The driving factors of the anaesthetic uptake are the perfusion of the alveoli, given by the cardiac output $\dot{Q}$ less the shunt flow $\dot{Q}_s$; the blood gas partition coefficient; and the concentration difference between the mixed venous blood and the arterial blood (Miller 2015).

$$u_L=(\dot{Q}-\dot{Q}_s)(\lambda C_{A_I}-C_{\bar{v}}) \tag{3.5}$$

Given (3.5), the concentration in the arterial blood C_a is determined by a mass balance of inlet and outlet fluxes, indicated in Figure 3.1b by the dashed line.

$$C_a\dot{Q}=C_{\bar{v}}\dot{Q}+u_L \tag{3.6}$$

The mixed venous blood concentration is given by an average of all the blood concentrations in the compartments multiplied by the perfusion of the respective body compartment. To account for the venous shunt, which is diversion of blood from the artery directly to the vein, the last term in (3.7) is added.

$$C_{\bar{v}}=\sum_i r_{\dot{Q},i}C_{b,i}+\left(1-\sum_i r_{\dot{Q},i}\right)C_a \tag{3.7}$$

The tissue compartments in Figure 3.1a are further divided into blood and tissue sub-compartments, shown in Figure 3.2.

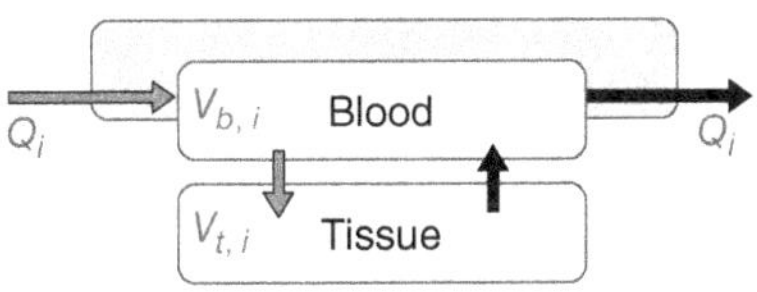

Figure 3.2 Structure of one tissue compartment.

The concentrations of the anaesthetic agent in the individual compartments are given by mass balances for each blood and tissue compartment.

$$V_{b,i}\frac{dC_{b,i}}{dt}=\dot{Q}_i(C_a-C_{b,i})-u_{t,i} \tag{3.8}$$

$$V_{t,i}\frac{dC_{t,i}}{dt}=u_{t,i} \tag{3.9}$$

The mass balance of the VRG includes an additional term for the metabolism of the anaesthetic agent in the liver m_{liv}, where $\dot{Q}_{liv}$ describes the perfusion of the liver (Saltzman 2001).

$$V_{t,VRG}\frac{dC_{t,VRG}}{dt}=u_{t,VRG}-\dot{Q}_{liv}\,C_{t,VRG}\,m_{liv} \tag{3.10}$$

The driving force of the anaesthetic uptake by the tissue $u_{t,i}$ in each compartment is the difference of the concentration in tissue at equilibrium for the given concentration in the blood $C_{b,i}$ and the actual concentration in the tissue $C_{t,i}$ (Eger 1974; Enderle and Bronzino 2011). The partition coefficients λ_i relate the concentrations in the tissue $C_{t,i}$ to the concentrations in the blood $C_{b,i}$ at equilibrium. Analogously to (3.5), the uptake of the tissue in the body compartments is described in (3.11).

$$u_{t,i}=\dot{Q}_i(\lambda_i C_{b,i}-C_{t,i}) \tag{3.11}$$

The perfusion of each compartment i is given by $\dot{Q}_i$ and the ratio of the cardiac output $r_{\dot{Q},i}$ perfusing the compartment.

$$\dot{Q}_i=r_{\dot{Q},i}\cdot\dot{Q} \tag{3.12}$$

Similarly, the parameter $r_{V_b,i}$ describes the ratio of the total blood volume V_b in compartment i and $r_{V,i}$ to the ratio of total body volume V.

$$V_{t,i}=r_{V,i}\cdot V \tag{3.13}$$

$$V_{b,i}=r_{V_b,i}\cdot V_b \tag{3.14}$$

3.2.1.1 Body Compartments

The volumes of the body compartments are given as a part of the total body volume. The mass of the adipose tissue is a function of the Body Mass Index (BMI), age and gender of the patient (Deurenberg *et al.* 1991). The percentages of the body mass of the VRG and the vessel-poor group (VPG) do not primarily depend on the BMI of the patient. Thus, they are assigned as a percentage of the ideal body weight for a person with the patient's height and a BMI = 22 [kg/m^2]

Table 3.1 Calculation of patient-specific tissue mass.

Parameter	Description	Equation	Unit	Reference
m_F	Adipose mass	$(1.2\text{BMI} - 10.8\text{gender} + 0.23\text{age} - 5.4) \cdot 0.01m$	kg	(Deurenberg *et al.* 1991)
m_{VPG}	VPG mass	$0.2 \cdot m_{ideal}$	kg	(Miller 2015)
m_{VRG}	VRG mass	$0.1 \cdot m_{ideal}$	kg	(Miller 2015)
m_M	Muscle mass	$m - m_F - m_{VPG} - m_{VRG}$	kg	

Source: Krieger *et al.* (2014). Reproduced with permission of Elsevier.

for both genders (Lemmens *et al.* 2006). The ideal body weight for a patient with height h is given in (3.15).

$$m_{ideal} = 22 \cdot h^2 \qquad \text{with BMI} = \frac{m[kg]}{(h[m])^2} \tag{3.15}$$

The body mass, which is neither allocated to the adipose tissue nor to the VPG or VRG, is assigned to the muscle group. The mass and volume of VPG are calculated to determine the volume of the muscle group and not further considered in the mathematical model, as the perfusion and anaesthetic uptake of the VPG tissue is negligible for short-term anaesthesia. The equations for the patient-specific tissue compartment mass are given in Table 3.1.

The volume of the compartments is determined by the average density of the tissue of the compartment (Heymsfield *et al.* 2005).

$$V_i = \frac{m_i}{\rho_i} \tag{3.16}$$

3.2.1.2 Blood Volume

The blood volume is adapted to the height h (in cm), weight m (in kg) and gender of the patient (published by Nadler *et al.* 1962).

rCl V_B,f = 0.3561 h^3 + 0.03308 m + 0.1833

V _ B,m = 0.3669 h^3 + 0.03219 m + 0.6041

3.2.1.3 Cardiac Output

The cardiac output $\dot{Q}$ in [L/min] as a function of the patient's BMI, age and gender is adapted from Stelfox *et al.* (2006). Gender = 1 denotes a female patient, and gender = 0 a male patient.

$$\dot{Q} = 5.84 + 0.08\,\text{BMI} - 0.03\,\text{age} - 0.62\,\text{gender} \tag{3.17}$$

Further coefficients for additional predictors such as simultaneously administered agents or the patient's health state can be found in Stelfox *et al.* (2006).

3.2.1.4 Lung Volume

The ventilated lung volume less the dead space determines the distribution volume of the inspired anaesthetic in (3.2) and (3.3). On average, men have larger lungs than women. During anaesthesia, the ventilated lung volume reduces to approximately the functional residual capacity, altered by atelectasis and anaesthetic side effects. The patient-specific functional residual capacity as a function of the BMI in litres is given by Pelosi *et al.* (1998).

$$V_L = 11.97\exp(-0.096\text{BMI}) + 0.46 \tag{3.18}$$

3.2.2 Pharmacodynamics

Pharmacodynamics describes the link between the concentration of the anaesthetic agent and the effect of the drug. An artificial effect site compartment represents the delay in the drug action:

$$\frac{dC_e}{dt} = k_{e0}\left(C_a - C_e\right) \tag{3.19}$$

where C_a denotes the concentration in the arterial blood calculated in (3.6), C_e denotes the effect site concentration and k_{e0} denotes the delay of the drug action.

The hypnotic effect is measured by the BIS, calculated as a function of the concentration in the effect site compartment C_e by the Hill equation (3.20):

$$BIS = BIS_0 + \left(BIS_{max} - BIS_0\right)\frac{C_e^{\gamma}}{C_{50}^{\gamma} + C_e^{\gamma}} \tag{3.20}$$

where C_{50} is the concentration triggering 50% of the total effect or the potency of the drug; γ is the slope of the Hill equation in (3.20); BIS_0 describes the initial effect at no anaesthetic concentration $BIS_0 = 100$; BIS_{max} describes the maximum effect $BIS_{max} = 0$; and the three PD parameters k_{e0}, C_{50} and γ are individual patient characteristics and might change during the course of anaesthesia, triggered by, for example, surgical stimulation or drug interaction.

3.2.3 Individualized Patient Variables and Parameters

In this section, the possible range of the PK and PD variables and parameters as a function of the patient's physiology (i.e. age, weight, height and gender) are calculated. Here, *variables* refer to values that might vary over time, whereas *parameters* are constant. For example, the cardiac output is referred to as a variable, because it is likely to change during the course of anaesthesia as a function of the concentration of the anaesthetic agent, other simultaneously administered drugs and/or surgical stimulation. Typically, PK parameters are the tissue volumes, which are constant during the entire course of anaesthesia. For the presented model, the individual PK variables are the cardiac output, the distribution of the cardiac output on the compartments, the lung shunt, the dead space volume and the lung volume. The PK parameters are the partition coefficients, the tissue volumes and the blood volumes.

The deviation for the blood gas partition coefficient λ and the tissue partition coefficients λ_i were summarized and published by Eger (2005). The deviation for the cardiac output $\dot{Q}$ is calculated for patients of 45–100 kg, 1.50–1.90 m, 18–90 years and both genders using (3.17). The shunt flow $\dot{Q}_s$ results from a 0% to 30% shunt of the cardiac output increased by atelectasis, often occurring during anaesthesia (Miller 2015). For the distribution of the cardiac output to the different compartments, no deviation was found in the open literature. As the baroreflex is still active during light to moderate anaesthesia and aims to provide the essential well-perfused organs with oxygen, the ratio is assumed to increase slightly, whereas the perfusion of the fat and muscle group decreases (Miller 2015). The tissue volumes are calculated for patients of 45–100 kg, 1.50–1.90 m and both genders applying the equations in Table 3.1. The blood volumes are calculated based on the assumptions that the blood volume V_b is proportional to the perfusion of the compartment and that 60% of the total blood volume is distributed on the systematic tissue.

$$V_{b,i} = 0.6 V_b r_{\dot{Q},i} \tag{3.21}$$

The deviation in the lung volumes is given by (3.18) for patients with a BMI in the range of 20 to 40 covered in the study by Pelosi *et al.* (1998). The dead space is altered from a normal value of $V_D = 150$ mL $\approx 30\%$ V_T to $V_D = 600$ mL $\approx 60\%$ V_T caused by atelectasis (Miller 2015).

The PD parameters are k_{e0}, C_{50} and γ in (3.19) and (3.20). The variation in the PD parameters were published in a study of Gentilini *et al.* (2001). The values at the boundary of the estimation problem were excluded.

All PK and PD variables and parameters, their default values and their ranges are summarized in Table 3.2.

3.3 Model Analysis

3.3.1 Uncertainty Identification via Patient Variability Analysis

In this section, the imposed uncertainty given by patient variability is identified. First, a separate simulation for the full range of PK and PD variability in Table 3.2 was performed. The course of isoflurane-based anaesthesia for

Table 3.2 Range and default values for PK and PD parameters and variables (partition coefficients at 37 °C for isoflurane).

	Symbol	Default value	Deviation	Unit	Ref.
*PK	λ	1.4	1.38–1.46	–	(Eger 2005)
	λ_F	50	43.84–55.8	–	(Eger 2005)
	λ_M	2.57	1.44–3.19	–	(Eger 2005)
	λ_{VRG}	1.65	1.45–1.86	–	(Eger 2005)
	$\dot{Q}$	5000	3520–7300	mL/min	(Stelfox *et al.* 2006)
	$\dot{Q}_s$	150	0–1500	mL/min	(Miller 2015)
	$r_{\dot{Q},F}$	0.054	0.045–0.054	–	(Eger 1974)
	$r_{\dot{Q},M}$	0.181	0.1–0.181	–	(Eger 1974)
	$r_{\dot{Q},VRG}$	0.75	0.75–0.765	–	(Eger 1974)
	V_b	4900	2875–6339	mL	(Nadler *et al.* 1962)
	$V_{b,F}$	160	69–205	mL	(Nadler *et al.* 1962; Eger 1974)
	$V_{b,M}$	410	276–688	mL	(Nadler *et al.* 1962; Eger 1974)
	$V_{b,VRG}$	1495	1293–2910	mL	(Nadler *et al.* 1962; Eger 1974)
	V_D	150	150 – 600	mL	(Miller 2015)
	V_L	2000	770–2200	mL	(Pelosi *et al.* 1998)
	$V_{t,F}$	14,500	4563–45,300	mL	(Eger 1974)
	$V_{t,M}$	33,000	20,010 – 55,789	mL	(Eger 1974)
	$V_{t,VRG}$	6000	4950–7942	mL	(Eger 1974)
*PD	C_{50}	0.7478	0.4959 – 1.094	vol%	(Gentilini *et al.* 2001)
	γ	1.534	0.7915–2.351	–	(Gentilini *et al.* 2001)
	k_{e0}	0.3853	0.0248–2.895	–	(Gentilini *et al.* 2001)

one patient with the given time-varying inputs of f_R, V_T and C_I is chosen as a case study for further analysis.

By comparison of the envelopes of uncertainty, the uncertainty introduced by PD variability is identified as more profound than the uncertainty introduced by PK variability. More specifically, the maximum deviation from the BIS for default PK values is 25% in Figure 3.3, whereas the maximum deviation of the BIS including PD variability and PK values adjusted to patient 1 is 56% in Figure 3.3.

Normally during anaesthesia, the anaesthetist modifies the inhaled concentration according to the obtained measurements in order to maintain adequate anaesthesia. The high deviation in the variables clearly supports the need of additional information about the patient in order to assure adequate hypnosis.

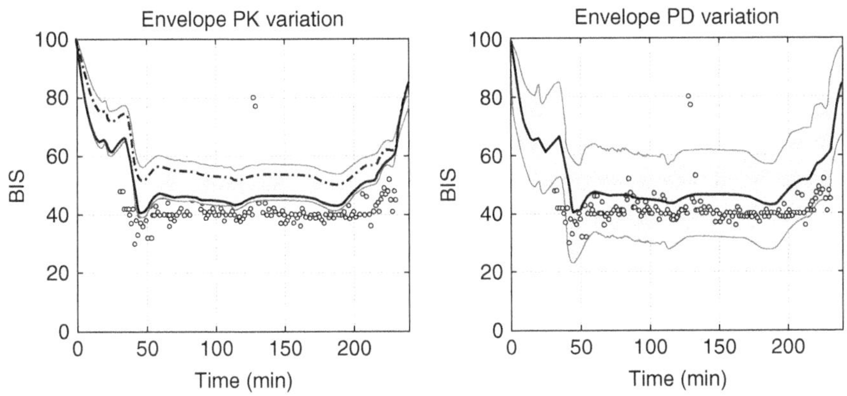

Figure 3.3 BIS for PK (left) and PD (right) variability in Table 3.2. The solid line marks the BIS for the model adjusted to patient 1. The grey dots mark the measured BIS. *Source*: Krieger *et al.* (2014). Reproduced with permission of Elsevier.

3.3.2 Global Sensitivity Analysis

In this section, the relative influence of the uncertain PK and PD parameters and variables on the measurable outputs is investigated via global sensitivity analysis. For volatile anaesthesia, the measurable outputs are the end-tidal volatile anaesthetic concentration and the BIS. The results of the global sensitivity analysis are several sensitivity indices with a value between 0 and 1, with 0 being non-influential. The sensitivity index (SI) represents the relative influence of the parameter or variable on the output of interest at the given time; the sum of all SIs for the applied Sobol method converges to 1. The SIs of the PK and PD parameters and variables presented in this section are calculated with the graphical user interface/high-dimensional model representation (GUI-HDMR) software. To perform the analysis, all PK and PD parameters and variables are varied between their bounds; the resulting output and the scaled input from 0 to 1 for a large number of sampling points are required by the GUI-HDMR software. For further details on how the SIs are derived, please see Chapter 2.

In total, four sensitivity analyses for the PK and PD variables and parameters with 26,000 sampling points were performed.

In Case 1, the influence of the PK variables and parameters on the end-tidal concentration is investigated, because the PK variables and parameters describe the distribution of the anaesthetic agent in the human body. In Case 2, the influence of the PK variables and parameters on the BIS is investigated. The PK variables and parameters affect the arterial concentration, which is linked via the effect site concentration to the BIS (3.19). In Case 3, the influence of the PD parameters, which characterize the link of the arterial blood

Table 3.3 Summary of the cases for sensitivity analysis.

	Fixed variables	Variables and parameters	Output
Case 1	PD	PK	C_E
Case 2	PD	PK	BIS
Case 3	PK	PD	BIS
Case 4	–	PK, PD	BIS

Source: Krieger *et al.* (2014). Reproduced with permission of Elsevier.

concentration to the BIS, is investigated. In the last case, Case 4, all PK and PD variables and parameters were analysed with respect to the BIS. For Case 1 and Case 2, all PD parameters were kept at their default values. In Case 3 the PK parameters were kept at their default values, while in Case 4 all PK and PD variables and parameters were varied between their lower and upper bounds given in Table 3.2. The cases are summarized in Table 3.3.

For the sensitivity analysis, the inspired concentration, respiratory frequency and tidal volume were kept constant during the entire simulation. The SIs of all PK and PD variables and parameters for Cases 1–4 are summarized in Table 3.4.

The time-varying PK and PD SIs of all cases defined in Table 3.3 are shown in Figure 3.4. The PK variables and parameters with an average SI < 0.01 were excluded for the purpose of clarity.

In Case 1, the distribution volume of the anaesthetic agent, the lung volume V_L, has the highest sensitivity index during the entire course of anaesthesia with respect to the end-tidal concentration C_E. The PK variables with the next highest SIs are the cardiac output and the shunt flow. Hence, as expected, the ventilation and perfusion are the main influencing factors for the uptake of the anaesthetic agent.

For Case 2, lung volume, cardiac output and lung shunt are identified, analogously to Case 1, as the crucial parameters with respect to the BIS, and hence the arterial blood concentration to which the BIS is directly linked through the effect site compartment via (3.19).

Case 3 shows that for short time instances, the SIs of the PD parameters γ and k_{e0} are approximately identical, while C_{50} has the highest index and hence the highest influence on the BIS. Under the assumption of a constant inspired concentration, the sensitivity of C_{50} increases to approximately 90% after 60 min. The results in Figure 3.4 for Case 3 are in accordance with the formulation of (3.19) and (3.20). The parameter γ changes the slope of the Hill equation, and k_{e0} affects the delay of the effect. Only C_{50} relates the final BIS to a given effect site concentration. Therefore, the PD parameters γ and k_{e0} only determine how fast the BIS is responding to a change in the inputs, but they are not affecting the steady-state BIS value. In Case 4, the PD parameters are identified to have

Table 3.4 Sobol's sensitivity indices using GUI-HDMR for Cases 1–4 given in Table 3.3 after 3.5 min and 20 min.

		Case 1: C_E		Case 2: BIS		Case 4: BIS	
	Variable	**5 min**	**20 min**	**5 min**	**20 min**	**5 min**	**20 min**
*PK	λ	0.0	0.0	0.0094	0.0134	0.0018	0.0031
	λ_F	0.0	0.0	0.0	0.0	0.0	0.0
	λ_M	0.0104	0.0259	0.0030	0.0127	0.0005	0.0028
	λ_{VRG}	0.0312	0.0144	0.0097	0.0101	0.0020	0.0020
	$\dot{Q}$	**0.1812**	**0.1428**	**0.0462**	**0.0919**	0.0101	0.0187
	$\dot{Q}_s$	**0.1665**	**0.1763**	**0.2362**	**0.2134**	**0.0400**	**0.0457**
	$r_{\dot{Q},F}$	0.0017	0.0034	0.0005	0.0016	0.0002	0.0005
	$r_{\dot{Q},M}$	0.0568	0.0937	0.0175	0.0489	0.0033	0.0111
	$r_{\dot{Q},VRG}$	0.0002	0.0004	0.0002	0.0006	0.0	0.0
	$V_{b,F}$	0.0024	0.0027	0.0034	0.0023	0.0	0.0
	$V_{b,M}$	0.0078	0.0085	0.0006	0.0001	0.0	0.0
	$V_{b,VRG}$	0.0132	0.0043	0.0060	0.0030	0.0010	0.0006
	V_D	0.0055	0.0054	0.0076	0.0065	0.0013	0.0013
	V_L	**0.4595**	**0.4766**	**0.6539**	**0.5663**	**0.1101**	**0.1215**
	$V_{t,F}$	0.0003	0.0003	0.0001	0.0001	0.0	0.0
	$V_{t,M}$	0.0043	0.0067	0.0267	0.0265	0.0001	0.0003
	$V_{t,VRG}$	0.0753	0.0559	0.0162	0.0369	0.0038	0.0071
*PD				*Case 3*: BIS		*Case 4*: BIS	
	C_{50}	–	–	**0.4241**	**0.7709**	**0.3124**	**0.5698**
	γ	–	–	0.2840	0.1224	**0.2813**	**0.0851**
	k_{e0}	–	–	0.2809	0.0815	**0.1947**	**0.0638**

$C_I = 1.1$ vol% is kept constant during the entire analysis.
Source: Ziehn and Tomlin (2009). Reproduced with permission of Elsevier.

the highest sensitivity towards the beginning of anaesthesia, whereas for a longer course of anaesthesia, especially the lung volume's SI increases. As stated in Case 3, γ and k_{e0} only influence how fast the BIS is following a change in the input parameters, but not the total effect.

As a conclusion of the sensitivity analysis of Case 1 and Case 2 from a physiological aspect, the cardiac output and shunt flow determine the anaesthetic uptake in the circulation, while the lung volume mainly determines the alveolar concentration in the lungs, as the driving force for the anaesthetic

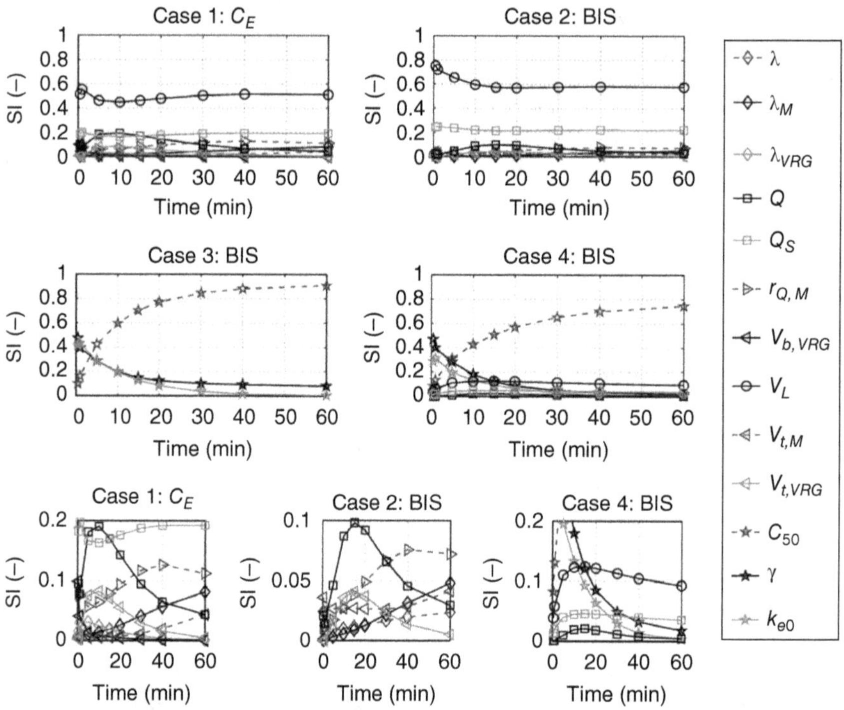

Figure 3.4 Time-varying sensitivity indices (SI) for Cases 1–4. The three bottom plots denote a zoomed-in scope for Case 1, Case 2 and Case 4. *Source*: Krieger *et al.* (2014). Reproduced with permission of Elsevier.

uptake from the ventilation site. All other PK parameters have a considerably lower SI and can be regarded as negligible compared to the lung volume, the cardiac output and the lung shunt flow. Case 3 and Case 4 illustrate that C_{50} is the most important parameter in order to obtain the correct level of anaesthesia for the individual patient.

For an in-depth understanding of the actual physical influence of the PK and PD variables and parameters, it was investigated whether an increase in the PK or PD variable or parameter increases or decreases C_E and/or BIS. The calculated outputs of C_E and BIS are compared to the outputs of C_E and BIS when changing the respective PK or PD variable or parameter one by one to the upper and the lower bounds, while keeping all other variables and parameters at their default values. The percentage of change comparing the default output of C_E and BIS to the range of change between the upper and the lower bounds is summarized in Table 3.5.

The simulations clearly confirm the results obtained by the previous sensitivity analysis.

Table 3.5 Percentage of change of C_E and BIS after 5, 20 and 60 min, compared to the output with default PK and PD variables and parameters.

		C_E			BIS		
	Variable	**5 min**	**20 min**	**60 min**	**5 min**	**20 min**	**60 min**
*PK	λ	0.0	0.0	0.0	−2.01	−3.60	−4.12
	λ_F	−0.08	−0.10	−0.16	0.01	0.04	0.08
	λ_M	−9.59	−11.27	−13.97	1.46	4.29	6.73
	λ_{VRG}	−12.24	−5.03	−0.81	1.96	2.34	0.43
	$\dot{Q}$	**28.89**	**17.34**	**8.97**	−4.11	−7.71	−4.46
	$\dot{Q}_s$	**−26.10**	**−21.25**	**−17.11**	**9.43**	**14.38**	**13.25**
	$r_{\dot{Q},F}$	−2.70	−2.86	−3.11	0.42	1.11	1.52
	$r_{\dot{Q},M}$	−18.42	−17.67	−12.64	2.90	6.75	6.19
	$r_{\dot{Q},VRG}$	−0.44	0.10	0.01	0.14	−0.04	0.00
	$V_{b,F}$	0.0	0.0	0.00	0.0	0.0	0.0
	$V_{b,M}$	−0.26	−0.12	−0.10	0.06	0.05	0.05
	$V_{b,VRG}$	−8.26	−3.06	−0.47	1.50	1.44	0.25
	V_D	5.39	4.09	3.14	−1.92	−2.56	−2.24
	V_L	**−77.81**	**−50.87**	**−36.32**	**25.79**	**27.63**	**23.06**
	$V_{t,F}$	−0.06	−0.24	−0.88	0.01	0.08	0.41
	$V_{t,M}$	−1.07	−4.02	−11.35	0.12	1.39	5.41
	$V_{t,VRG}$	−19.24	−10.59	−1.61	2.59	4.84	0.87
*PD	C_{50}		–	–	**28.06**	**49.25**	**56.19**
	γ	–		–	**27.71**	**14.34**	4.55
	k_{e0}	–	–	–	**−37.69**	**−52.55**	**−23.98**

$C_I = 1.1$ vol%, $f_R = 12$ and $V_T = 500$ mL performed with gPROMS.
Source: Krieger *et al.* (2014). Reproduced with permission of Elsevier.

3.3.3 Correlation Analysis and Parameter Estimation

The envelope of BIS uncertainty by PK variability is significantly smaller than the envelope of uncertainty by PD variability (Figure 3.3). This motivates an attempt to estimate the PD parameters in order to capture the uncertainty as a consequence of PK variability. First, all three PD parameters and the PK variable with the highest sensitivity, V_L, are included in the parameter estimation problem.

The parameter estimation problem is evaluated by the correlation matrix C of the estimated parameters. An entry in the off-diagonal elements of the correlation matrix C close to one ($|C_{ij}| \approx 1$) indicates a high correlation of the corresponding parameters i and j, whereas an entry of zero ($|C_{ij}| \approx 0$) indicates no correlation. The entries of the correlation matrix are calculated based on the variance–covariance matrix V, the variance of a parameter is given on the diagonal elements (V_{ii}) and the covariance of two parameters i and j on the off-diagonal elements (V_{ij}). Further details can be found in the gPROMS user guide (PSE 2015).

$$C_{ij} = \frac{V_{ij}}{\sqrt{V_{ii}V_{jj}}},\ i \neq j \qquad C_{ii} = 1 \tag{3.22}$$

During the following analysis, the upper bound of the envelope shown in Figure 3.3 is referred to as PK_u, while the lower bound is referred to as PK_l. The evaluation of the quality of the estimates is performed for both cases. The correlation matrix of V_L and the three PD parameters obtained using gPROMS (PSE 2015) is given in Table 3.6.

The results show that V_L and the PD parameters are highly correlated, in particular C_{50} and V_L, where $C_{C_{50},V_L} \approx -0.99$. As a consequence, C_{50} and V_L cannot be estimated independently, or, for this case, the uncertainty imposed by variability in the PK variables and parameters can be captured, and a sufficiently accurate BIS can be reproduced, by the adjustment of the PD parameters only. This statement is investigated for PK_l and PK_u, this time only estimating the PD parameters. The correlation matrix of the PD parameters for PK_l and PK_u and the final values for all PD parameters are given in Table 3.7.

In the second parameter estimation problem, the PD parameters γ and C_{50} show minor correlation originated by the nature of (3.20). All values of the PD parameters are obtained by solving a maximum likelihood parameter estimation problem with gPROMS and lie within their respective bounds (Table 3.7; see also Bard 1974; PSE 2015). The results of the fit are shown in Figure 3.5. Here, PK_l and PK_u denote the upper and lower bounds of the PK

Table 3.6 Correlation matrix C of C_{50}, γ, k_{e0} and V_L for the parameter estimation problem PK_u and PK_u; PK_u above diagonal and PK_l below diagonal.

	C_{50}	γ	k_{e0}	V_L
C_{50}	1	0.644	−0.691	**−0.993**
γ	0.662	1	−0.312	−0.713
k_{e0}	−0.753	−0.368	1	0.697
V_L	**−0.992**	−0.663	0.791	1

Source: Krieger *et al.* (2014). Reproduced with permission of Elsevier.

Table 3.7 Correlation matrix of the PD parameters, entries for PK_u above diagonal and PK_l below diagonal, and estimated PD parameters for PK_l and PK_u.

	C_{50}	γ	k_{e0}		PK_l	PK_u
C_{50}	1	−0.759	−0.0033	C_{50}	0.6177	0.8962
γ	0.0563	1	0.383	γ	1.4458	1.6369
k_{e0}	0.431	0.348	1	k_{e0}	0.4308	0.2978

Source: Krieger *et al.* (2014). Reproduced with permission of Elsevier.

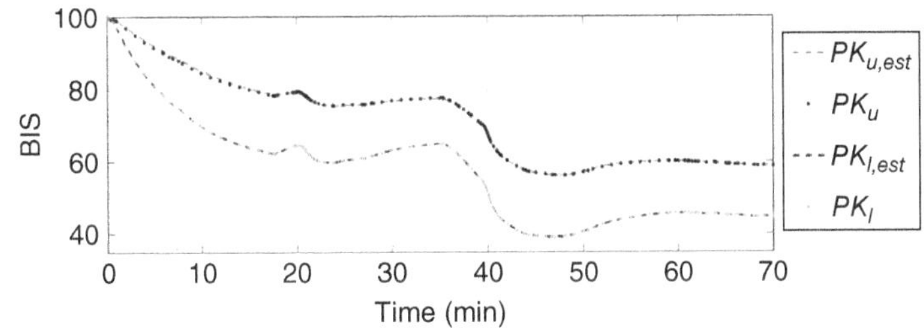

Figure 3.5 BIS output for estimated PD parameters in Table 3.7 capturing PK variability
Source: Krieger *et al.* (2014). Reproduced with permission of Elsevier.

uncertainty envelope, and $PK_{l,est}$ and $PK_{u,est}$ the output for values of the PD parameters given in Table 3.7.

This analysis shows that, via PD parameter estimation, it is possible to capture the uncertainty introduced by potential PK variability and that the PD parameters k_{e0}, C_{50} and γ are sufficient to predict the correct BIS under uncertainty in the PK and PD variables and parameters. This statement is further investigated for a set of measured patient data in Section 3.3.4.

3.3.4 Simulation Results

The anonymized data used for simulation purposes were provided by the Department of Medical Informatics in Anesthesiology and Intensive Care Medicine of the University of Gießen in Germany. The data of isoflurane concentration were measured with an anaesthesia ventilator (Primus, Draeger Medical), and BIS was measured by patient monitoring (IntelliVue MP70, Phillips). The data were recorded on-line with an anaesthesia information management system (NarkoData, IMESO GmbH). The individual PD parameters for each of three patients were obtained by a parameter estimation problem as described in Section 3.3.3. For comparison, the expected BIS for the default PD parameters in Table 3.2 is added. All results are shown in Figure 3.6,

Table 3.8 Patients' characteristics, calculated values of the lung volume and cardiac output and estimated PD parameters.

	Patient 1	Patient 2	Patient 3	Units
Age	61	65	66	Years
BMI	31.5	14.5	26.0	kg/m^2
Height	1.69	1.7	1.63	m
Mass	90	42	69	kg
Gender	m	m	f	m/f
V_L	1041	2200†	1449	mL
$\dot{Q}$	6530	5052	5317	mL/min
C_{50}	0.3989	0.4107	0.5661	vol%
γ	0.4945	0.5939	1.9974	–
k_{e0}	1.7577	0.0248	0.374	–

$^{\dagger}V_L$ at the upper bound.
Source: Krieger *et al.* (2014). Reproduced with permission of Elsevier.

where the measured data points are denoted with *(meas)*, the BIS for individually estimated PD parameters is denoted with *(est)* and the expected BIS for the PD default values is denoted with *(def)*. The characteristics of the three patients and the values of the estimated PD parameters are given in Table 3.8.

As seen in Figure 3.6, the simulation results of the end-tidal concentration C_E for Patient 1 and Patient 3 are in good accordance with the measurements C_E (meas). Hence, the PK model shows a good fit of the data. However, for Patient 2 the model is not predicting the measured end-tidal concentration. This might be related to the underweight of the patient, BMI = 14.5, for which the PK parameters have to be modified with additional knowledge about the patient's health state. The expected BIS for PD default variables, BIS (def), published by Gentilini *et al.* (2001) showed the best match with the measured BIS for Patient 1. Especially for Patient 3, a considerable offset between the measurement and the predicted BIS is observed. For individually estimated PD parameters, the predicted BIS is in good accordance with the measurement for all three patients.

This study shows the high inter-patient variability of the PD parameters; already in the study for three patients in Table 3.8. a very wide range for the estimated PD parameters can be observed. This shows the need for on-line adjustment of the PD parameters. The strategy of choice to assure sufficient and correct prediction of the anaesthetic state through the mathematical model is an on-line parameter estimation (Parker and Doyle 2001).

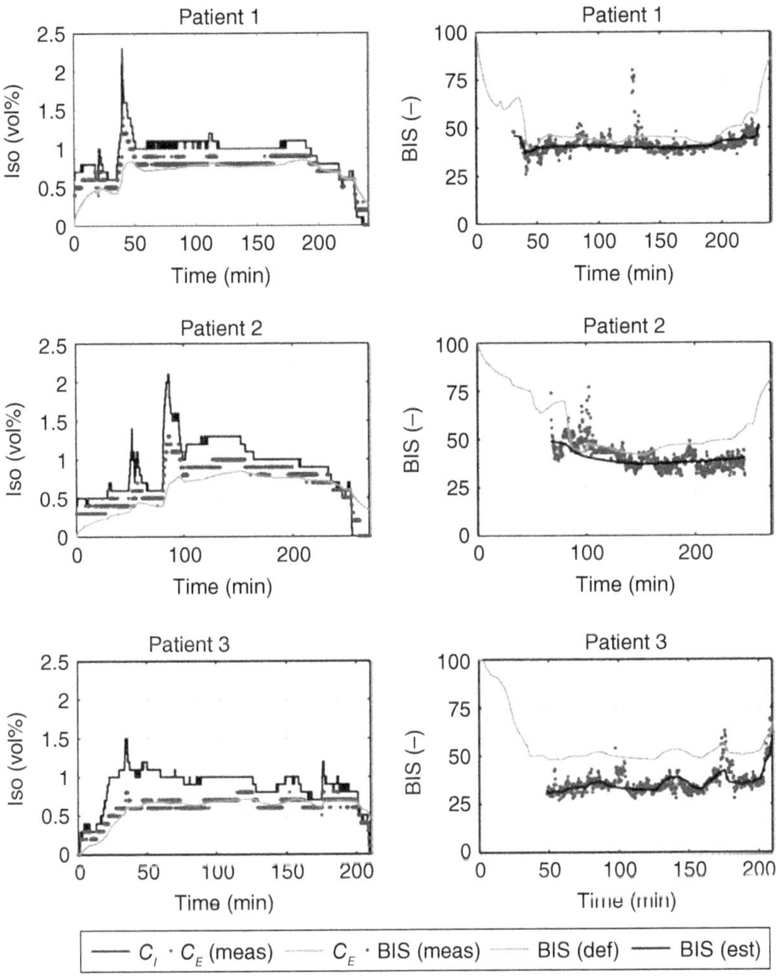

Figure 3.6 Inspired and expired isoflurane concentrations and BIS for patient characteristics and parameters given in Table 3.8.

Motivated by the high PK and PD patient variability (Mertens and Vuyk 1998) and our interest in applying the derived model as the underlying model for model predictive control, this work provided an in-depth analysis of an individualized physiologically based model for volatile anaesthesia and the influence of the individual patient variables and parameters on the outputs of interest, which are the BIS, as a measure of the hypnotic level, and the end-tidal concentration of the volatile anaesthetic agent.

First, the PK variables and parameters were individually adjusted to the specific patient characteristics, and their influence was investigated by a global sensitivity analysis and by changing the parameters one at a time within their bounds. Here, the parameters primarily determining the ventilation and perfusion of the lungs, the lung volume, the shunt flow and the cardiac output were found to be the crucial factors.

Second, motivated by the high variability in the PD parameters of up to 400% (Mertens and Vuyk 1998), the PD parameters were included in the analysis and indeed proofed to be the crucial parameters with respect to the BIS, where C_{50}, the effect site concentration at BIS = 50, was found to be the parameter with the highest sensitivity.

As a final remark, when applying the presented model as the underlying model for model predictive control, to assure safe control action and safe anaesthesia as well as account for the high inter- and intra-patient variability, the PD parameters are recommended to be adjusted during the course of anaesthesia based on the BIS measurements and updated accordingly (Ionescu *et al.* 2008). The adjustment of the PD parameters only was found to be sufficient, due to the correlation of the PK and PD variables and parameters with respect to the BIS. As a consequence, even for a non-correct prediction of the PK variable V_L with the highest sensitivity, an estimate of the PD parameters can capture the introduced uncertainty. This strategy, of an on-line parameter estimation of the PD parameters to update the underlying PD model, is believed to be suitable to capture the imposed model uncertainty by inter- and intra-patient variability.

3.4 Control Design for Volatile Anaesthesia

In this section, the design and testing of the closed-loop control strategy is presented. A schematic of the closed-loop control structure is depicted in Figure 3.7.

The objective is to achieve a fast onset and stable maintenance of the desired depth of hypnosis measured by the BIS. In order to achieve this, the MPC manipulates the control input, the inspired concentration C_I. The state feedback MPC calculates the optimal control strategy as a function of the states of the system.

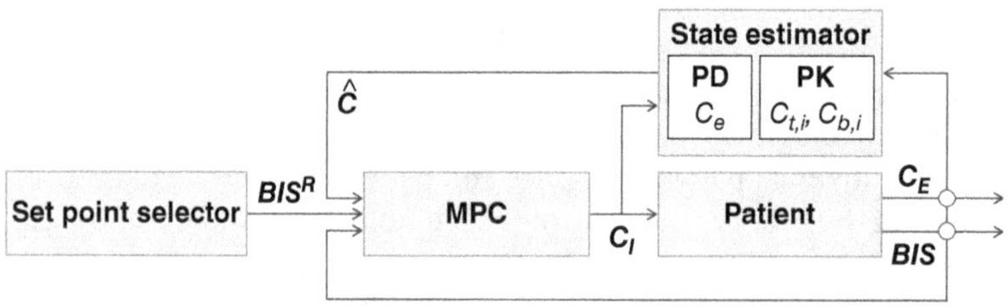

Figure 3.7 Closed-loop control design for volatile anaesthesia.

The available measurements are the BIS and the end-tidal concentration C_E. Given these measurements and the control input C_I, the state estimator obtains the predicted states $\hat{C}$ of the system that are not measurable.

For closed-loop control validation, the patient model is simulated with other PK and PD variables and parameters compared to the nominal values.

3.4.1 State Estimation

For the applied state feedback MPC design, the optimal control law is obtained as a function of the system's states. This is indicated by the state estimator block in the control design in Figure 3.7. For the control of volatile anaesthesia, the two measurable outputs are the end-tidal concentration C_E and the BIS. Due to the identified high uncertainty in the PD parameters, the strategy of choice is to estimate the states, which are the concentrations in the respective compartments of the PK model and the effect site concentration C_e of the PD model, based on the measurement of the end-tidal concentration C_E. Due to the formulation of (3.19), C_e is not observable from the measurement of C_E.

The observability matrix

$$O = \begin{bmatrix} C \\ CA \\ \vdots \\ CA^{n-1} \end{bmatrix} \tag{3.23}$$

for the system has low rank:

$$\text{rank}(O) = 6 < n = 7$$

Hence, the strategy of choice is to estimate only the states of the PK model:

$$\begin{aligned} \hat{x}_{k+1} &= A\hat{x}_k + Bu_k + w_k \\ \hat{y}_k &= C\hat{x}_k + v_k \end{aligned} \tag{3.24}$$

where the state vector corresponds to the concentrations in the blood and tissue compartments of the model. Additionally, the effect site concentration C_e of the PD model is obtained as a function of the arterial concentration C_a according to (3.19) given as follows:

$$C_a = \left(1 - \frac{\dot{Q}_S}{\dot{Q}}\right)\left(\frac{\lambda\left(C_I V_T + \hat{C}_E V_A\right)}{V_A + V_T}\right) - \frac{\dot{Q}_S}{\dot{Q}}\hat{C}_E\lambda \tag{3.25}$$

where $\hat{C}_E = \hat{y}_{C_E}$ refers to the estimated end-tidal concentration.

An estimation of $\hat{C}_E$ based on the BIS measurement and the PD parameters of the Hill equation is not performed due to the high uncertainty in the PD parameters by inter- and intra-patient variability and the tendency towards an inaccurate estimation.

3.4.1.1 Model Linearization

Given the assumption of constant inputs of respiratory frequency fR and VT and constant PK variables, the model equations presented here result in a linear system with the seven states:

$$x = [C_e C_{b;V\,RG} C_{t;V\,RG} C_{b;M} C_{t;M} C_{b;F} C_{t;F}]'$$

The only non-linearity is introduced by the Hill equation (3.21), which relates the linear PK model to the effect measured by the BIS. In this work, we only consider linear mp-MPC algorithms. For this specific case, as the Hill equation is an algebraic equation and no ODE, two options to compensate for the non-linearity are considered:

I) Algebraic inverse of the Hill equation
II) Linearized Hill equation
  a) Linearization at BIS reference point
  b) Set of piecewise affine functions.

Both options and their advantages and disadvantages are discussed in this section.

I. Algebraic Hill Equation

One option is to calculate the reference effect site concentration *CRe* by the inverse of the Hill equation for the reference BIS^R (Ionescu *et al.* 2008; Naşcu *et al.* 2012):

$$C_e^R = C_{50}\left(\frac{BIS^R - BIS_0}{BIS_{\max} - BIS}\right)^{\frac{1}{\gamma}} \tag{3.26}$$

The control design consisting of the mp-MPC controller, a state estimator, the patient and the inverse Hill equation is depicted in Figure 3.8.

This design requires robustification against the uncertainty in the PD parameters C_{50} and γ, which are parameters in the Hill equation (3.21) and (3.26). For the proposed design of compensating the non-linearity by the inverse of the Hill equation, these parameters are not included in the control design where they can be compensated by the disturbance rejection formulation. Hence, this design can only compensate uncertainty in the PD parameter k_{e0}.

The advantages and disadvantages of algebraic inverse of the Hill equation are summarized as follows:

✓ Exact approximation of the Hill equation
× Robustification strategy for inter-patient variability in C_{50} and γ.

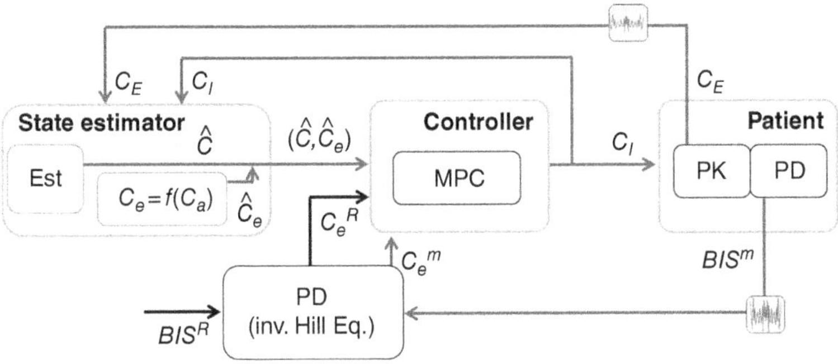

Figure 3.8 Control design for algebraic Hill equation.

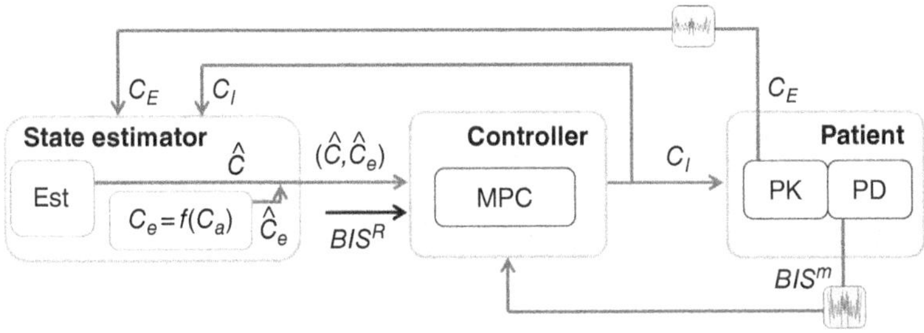

Figure 3.9 Control design for linearized Hill equation.

II. Linearized Hill Equation

The second design is a linearization of the Hill equation at the desired reference point. The given control design is depicted in Figure 3.9.

The linearized Hill equation is given by:

$$BIS\big|_{BIS_{lin}} = \alpha_{BIS_{lin}} C_e + b_{BIS_{lin}} \tag{3.27}$$

where *lin* denotes the linearization point; that is, for linearization at BIS = 50, the linearization constants are given by:

$$\alpha_{BIS_{50}} = \left(BIS_{\max} - BIS_0\right)\left(\frac{\gamma}{4C_{50}}\right) \tag{3.28}$$

$$b_{BIS_{50}} = BIS_0 + \left(\frac{BIS_{\max} + BIS_0}{2}\right) - \alpha_{BIS_{50}} C_{50} \tag{3.29}$$

II.i Linearization at Reference Point The linearized Hill equation at the operating point of BIS = 50 was applied by Gentilini *et al.* (2001) and Yelneedi *et al.* (2009), resulting in the visualized linearization in Figure 3.10 for the nominal PD isourane parameters.

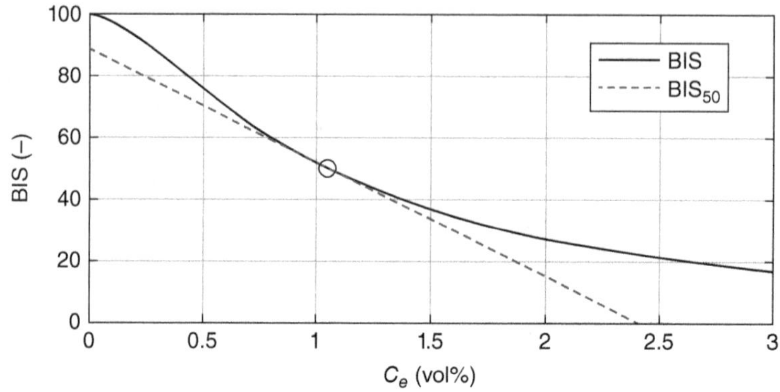

Figure 3.10 Linearized Hill equation at BIS = 50. The dot marks the linearization point.

However, this might not be accurate when the anaesthetist decides another operating point (e.g. BIS = 40 or BIS = 25), as in the case study for desflurane. Furthermore, the intersection of the linearized Hill equation and the *y*-axis does not coincide with the initial condition of the patient during induction, where BIS = 100. Hence, this strategy results in a large offset during induction of anaesthesia.

The advantages and disadvantages of linearization at reference point only are summarized as follows:

✓ Good approximation at reference point
✓ Straightforward implementation
× Linearization error out of linearization region.

II.ii Set of Piecewise Affine Functions Hence, a safer and more accurate linearization procedure to achieve a smooth transition of the non-linearity for the full Hill equation is a set of linear approximations, where the Hill equation is linearized at BIS = 60 and BIS = 30, and the controller is switching at the intersection points. The linearization for induction is obtained by a line through the points (BIS = 100, $C_e = 0$) and (BIS = 60, $C_e = C_{e}$; BIS = 60) (Figure 3.11).

The advantages and disadvantages of the piecewise affine linearization of the Hill equation are summarized as follows:

✓ Linearization of the full parameter space
✓ Compensation of uncertainty in C_{50} and γ
× Implementation of controller switching to guarantee stability.

3.4.2 On-Line Parameter Estimation

The strategy of on-line parameter estimation for anaesthesia control was performed for propofol by Sartori *et al.* (2005) and Robayo *et al.* (2010). In Robayo *et al.* (2010), the authors estimate the slope of the linearized Hill

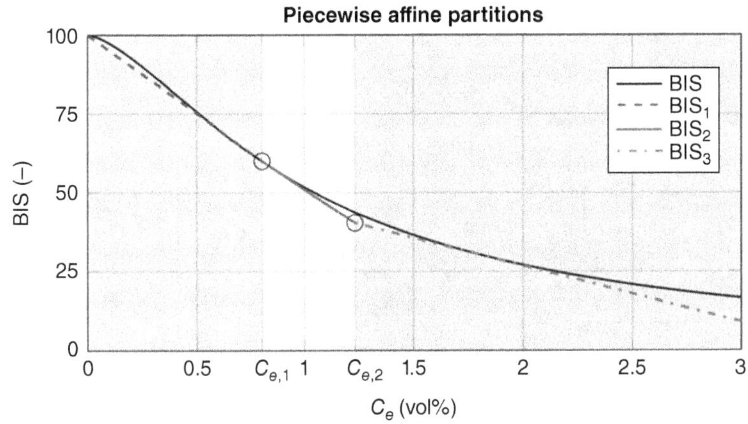

Figure 3.11 Piecewise linearization of the Hill equation. The dots mark the intersection of the linearization functions and the switching points of the controllers, respectively.

equation at BIS = 50 as a function of the cross correlation between measurement in the intensive care unit and prediction of the BIS. Sartori *et al.* (2005) formulate the non-linear PK-PD system by adding the parameters C_{50} and k_{e0} as states, the system is linearized at every step, and the states and parameters are estimated by an implementation of the Kalman filter. In Sreenivas *et al.* (2009), the authors mention an improved prediction of the BIS for an estimation of C_{50} for isourane-based anaesthesia, given the measurement during induction. However, no method for the estimation of C_{50} is described.

3.4.2.1 Control and Algorithm Design

For the proposed control design, an additional block for the on-line estimation of C_{50} is added to the control structure depicted in Figure 3.12 (Krieger and Pistikopoulos 2014). The non-linearity of the Hill equation is compensated by its inverse (3.20) analogously to the design in Figure 3.13. Hence, the reference point on the effect site concentration C_e^R is calculated as a function of the reference point on the hypnotic depth given by BIS^R and the PD parameters C_{50} and γ.

The process of the on-line estimation block in Figure 3.12 is depicted in Figure 3.13. It can manually be switched on or off. If active, C_{50} is estimated and updated under the conditions depicted in the flowchart presented in Figure 3.13.

The first estimation and update of C_{50} occur at least 5 min after induction of anaesthesia ($t > 5$ min). If the on-line parameter estimation is switched on, the parameter estimation block becomes active, when an error between measurement BIS^m and the predicted $\widehat{BIS}$ by the Hill equation with the current parameters

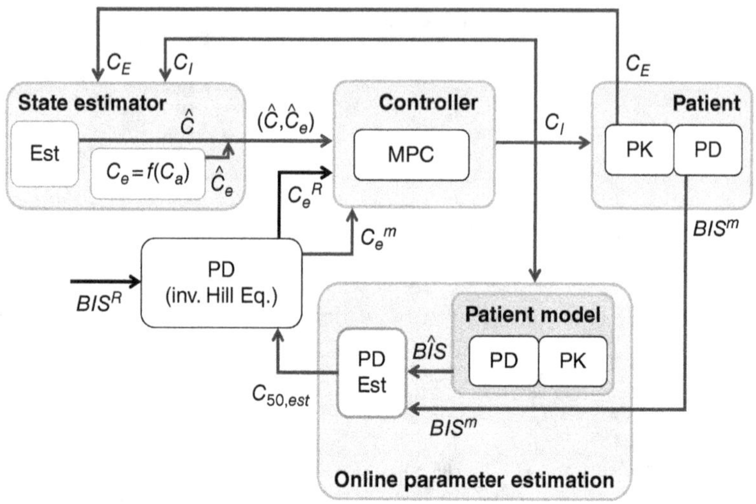

Figure 3.12 Closed-loop control design for on-line parameter estimation of C_{50}.

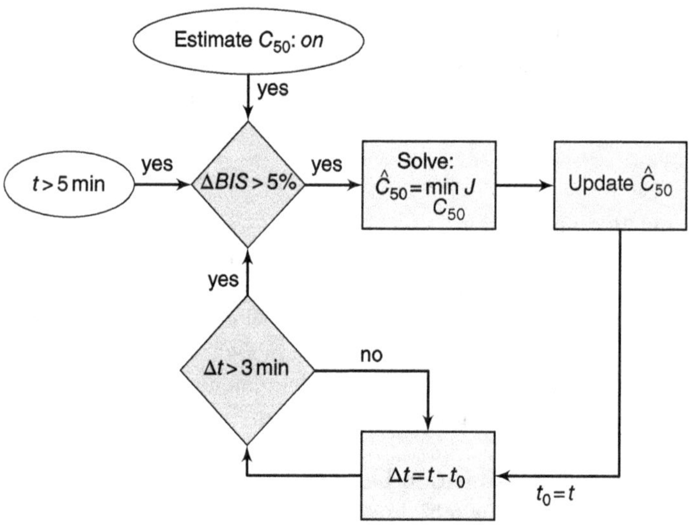

Figure 3.13 Decision process of the on-line parameter estimation bloc. *Source*: Krieger *et al.* (2014). Reproduced with permission of Elsevier.

is detected. A *mismatch* is defined as a deviation of prediction and measurement of more than 5% in the last 3 min, $\Delta BIS > 0{:}05$:

$$\Delta BIS = \frac{1}{n}\sum_{i=t_0}^{t}\left(\left(\frac{BIS_i^m - \widehat{BIS}_i}{BIS_i^m}\right)^2\right)^{\frac{1}{2}} \tag{3.26}$$

This triggers the on-line estimation by solving a constrained non-linear least-squares problem. The solution of C_{50} is minimizing the error between $\widehat{BIS}$ and BIS^m:

$$\begin{aligned} &\min_{C_{50,t}} J = \sum_{i=t_0}^{t}\left(BIS_i^m - \widehat{BIS}_i\right)^2 \\ &s.t. 0.8C_{50,t-1} \le C_{50,t} \le 1.2C_{50,t-1} \\ &\quad C_{50,min} \le C_{50,t} \le C_{50,max} \end{aligned} \tag{3.27}$$

Constraints on the change of the estimated value of $C_{50,t}$ aim for a smooth transition of the parameter to the real value and secure stability against short-term disturbances and/or measurement errors. Before anaesthesia, $C_{50,t-1}$ is set to the nominal value. Throughout the simulation, $C_{50,t}$ is initialized with its previous value $C_{50,t-1}$. The solution of the estimation problem is constrained by ±20% of its previous value $C_{50,t-1}$ and a lower and upper bound given in Table 3.2. Given a feasible and optimal value of $C_{50,t}$ of the estimation problem, the inverse Hill equation (3.20) is updated accordingly after each on-line estimation step. The least-squares estimation problem is solved using GAMS and the global solver BARON (GAMS 2013). The estimated parameter is sent to MATLAB via GDXMRW. An additional measure to enhance a smooth transition of the parameter to the real value is a minimum interval of 3 min between each on-line estimation step (Figure 3.13).

3.4.2.2 Testing of the On-Line Estimation Algorithm

This strategy is now tested for patient 3 as patient 3 is generally showing the highest offset for all control strategies (Krieger 2014). The mp-MPC is designed using a 'Perfect' Observer for state estimation (CD_1). The state estimator and mp-MPC are both based on the derived PK-PD model with nominal patient values, whereas the patient model is based on individualized variables and parameters. The described control strategy ($\widehat{CD}$) is tested for a constant reference point of $BIS^R = 40$ for 100 min. The simulation results are shown in Figures 10.3–10.6.

During the initial 5 min, both controllers give an identical input, while after 5 min the input of $\widehat{CD}$ is adjusted according to the update of C_{50} in Figure 3.14. This update triggers a reference point change on the effect site concentration C_e^R as a result of the updated value of C_{50} in the inverse Hill equation and better knowledge of the patient's individual parameters. Figure 3.15 shows the

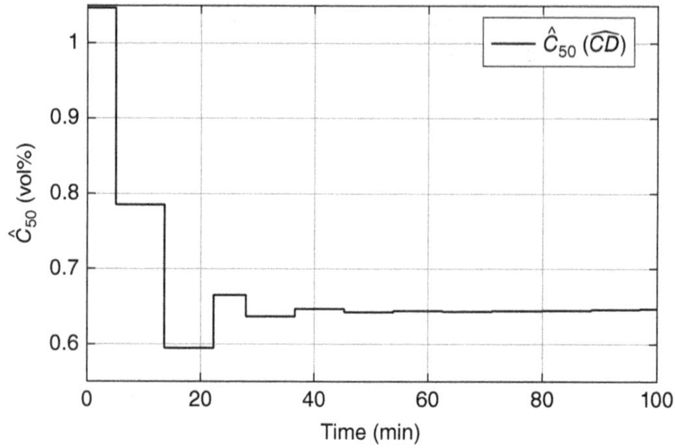

Figure 3.14 Estimated $\widehat{C_{50}}$ of $\widehat{CD}$ for Patient 3.

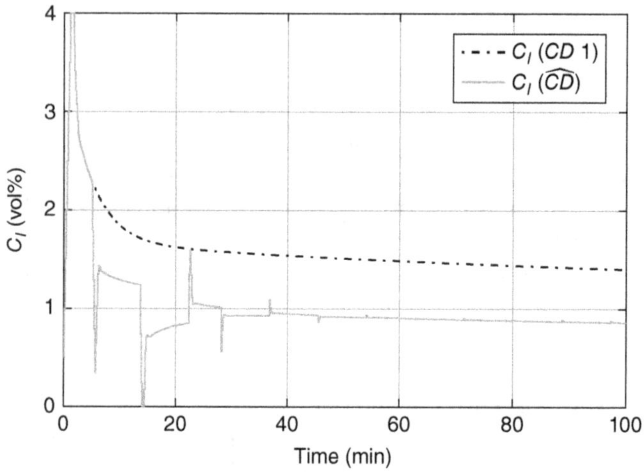

Figure 3.15 Control input for Patient 3 of $\mathbf{CD_1}$ and $\widehat{\mathbf{CD}}$.

changing inlet concentration of the controller as a consequence of this reference point change of C_e^R depicted in Figure 3.16. The reference concentration C_e^R is updated with every new estimate of C_{50}.

Figure 3.16 also shows the large offset for control design CD_1 and emphasizes the need of an offset free control design. This offset in the effect site concentration originated from different

PK and PD variables and parameters of the nominal patient, and patient 3 causes the further offset in the BIS shown in Figure 3.17.

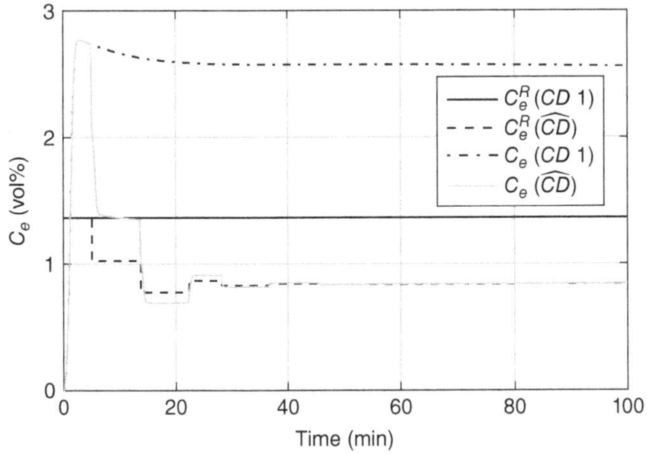

Figure 3.16 $\mathbf{C_e^R}$ and actual $\mathbf{C_e}$ for Patient 3 of $\mathbf{CD_1}$ and $\widehat{\mathbf{CD}}$.

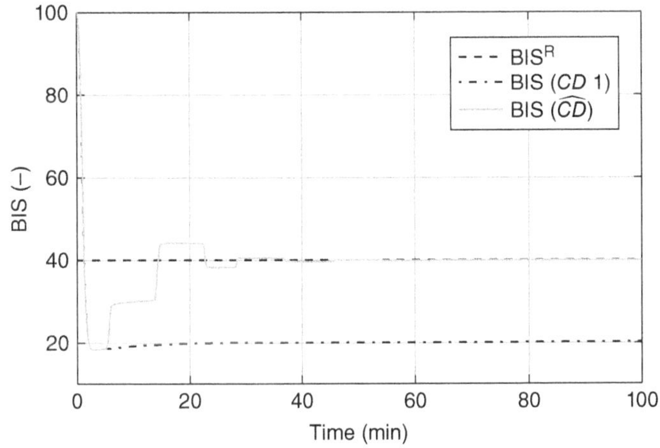

Figure 3.17 $\mathbf{BIS^R}$ and actual BIS for Patient 3 of $\mathbf{CD_1}$ and $\widehat{\mathbf{CD}}$.

However, by the estimation of C_{50}, $\widehat{CD}$ converges to the reference point BIS = 40 (Figure 3.17). Likewise, the estimated value of C_{50} converges to a final value of C_{50} = 0.647 (vol%).

To confirm and validate this result, the least-squares parameter estimation with GAMS (2013), a maximum likelihood parameter estimation for nominal values of all the PK and PD variables and initialized with the nominal value of C_{50} was performed with gPROMS (PSE 2015). The obtained estimated value was C_{50} = 0.612 (vol%). This result is reasonably close to the result obtained by the solution of the least-squares problem and confirms the accuracy of the PD estimation result with GAMS (2013).

3.4.3 Case Study: Controller Testing for Isourane-Based Anaesthesia

The control design of *CD* is now tested for all three patients undergoing isourane-based anaesthesia for a reference point change from BIS = 40 to BIS = 60 after 60 min of anaesthesia. The simulation results of all three patients are shown in Figures 3.18–3.21.

During the initial 5 min, the MPC computes an identical input for all three patients. However, after this short induction time, C_{50} of each patient is estimated individually based on the obtained measurements of the BIS during the last 3 min depicted in Figure 3.20. The estimation of C_{50} results in an update of the reference effect site concentration C_e^R in Figure 3.19. The measured BIS of all three patients converges to the reference point BIS^{R} in Figure 3.20.

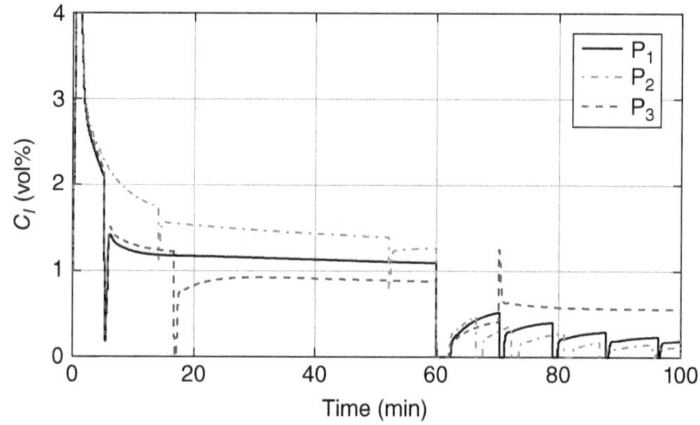

Figure 3.18 Control input for Patients 1–3 of $\widehat{CD}$.

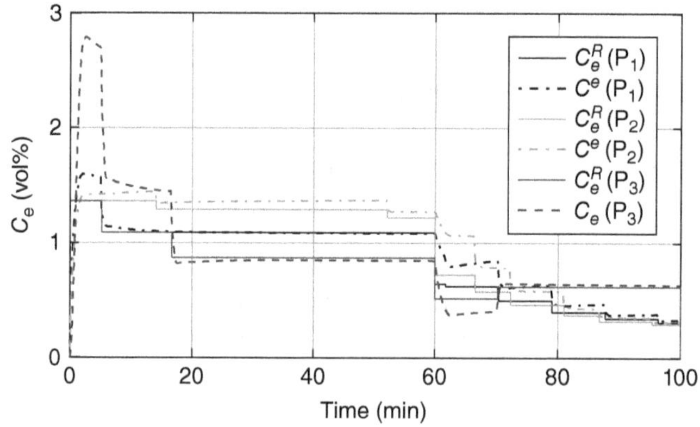

Figure 3.19 C_e^R and actual C_e for Patients 1–3 of $\widehat{CD}$.

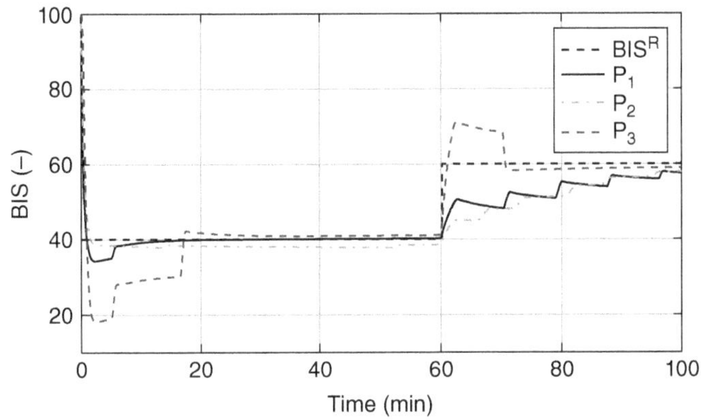

Figure 3.20 $\mathbf{BIS^R}$ and actual BIS for Patients 1–3 of $\widehat{CD}$.

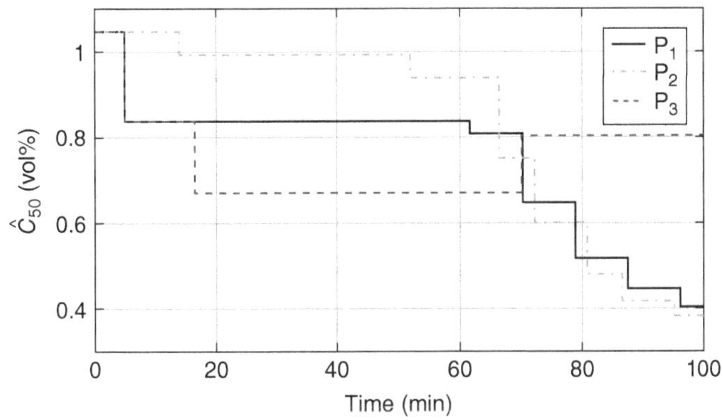

Figure 3.21 Estimated $\widehat{C_{50}}$ of CD for Patients 1–3.

The required target effect site concentration to obtain BIS^R varies significantly between patients due to large inter-individual variability. Figure 3.18 shows the individual control inputs obtained correctly through the individualized parameter estimation of C_{50} shown in Figure 3.21.

The estimated values of C_{50} converge to a constant value in less than 20 min of anaesthesia. Due to the change in BIS^R, the estimation of C_{50} is triggered repeatedly (Figure 3.21). Here the updated parameter, C_{50}, for patient 3 converges faster to a steady value. C_{50} of patient 1 and patient 2 is repeatedly updated every 3 min, and at the constraint at the lower bound $C_{50,t} \geq 0.8C_{50,t-1}$ is active. This is originated from the different slope of the individual Hill equations and the distinct deviation from the nominal value at BIS = 40 and BIS = 60.

The on-line estimation of C_{50} shows promising results towards an individualized control strategy of anaesthesia. This strategy allows to adjust the controller to the individual sensitivity of the patient towards the anaesthetic agent. Furthermore, the anaesthetist gains understanding of the patient's sensitivity, which could be advantageous for future surgeries of the same patient. The presented strategy is believed to be safe for the patient, ensured by constraints in the controller configuration and in the formulation of the parameter estimation problem.

The tuning parameters of this strategy are (a) the permitted deviation from the initial value for C_{50} in the parameter estimation problem, which was set to ±20% in this study; (b) the percentage of deviation from the measured BIS, which triggers an estimation of C_{50}, set to 5% in this study; and (c) Δt the sampling time between each parameter estimation, which was set to 3 min.

Conclusions

The framework presented in this thesis provided a valuable guideline for model development and analysis when aiming for a robust control strategy and the design of a safe drug delivery system for anaesthesia (Figure 3.22).

Closing the loop of the anaesthetic system implies automatic drug infusion based on the model predictions and the feedback through the measured patient variables. Apart from hypnotic depth, anaesthesia is defined by amnesia, analgesia, muscle relaxation and the maintenance of the vital functions. A multiple-input/multiple-output (MIMO) controller could regulate all of these variables (Biro 2013).

The presented control strategy combines mp-MPC and online parameter estimation of C_{50} to address control of anesthesia under uncertainty. This strategy showed a good performance during induction of anesthesia and adapted the controller's dynamics to the individual patient's sensitivity. The safety of the patient is assured by constraints in (a) the online parameter estimation problem

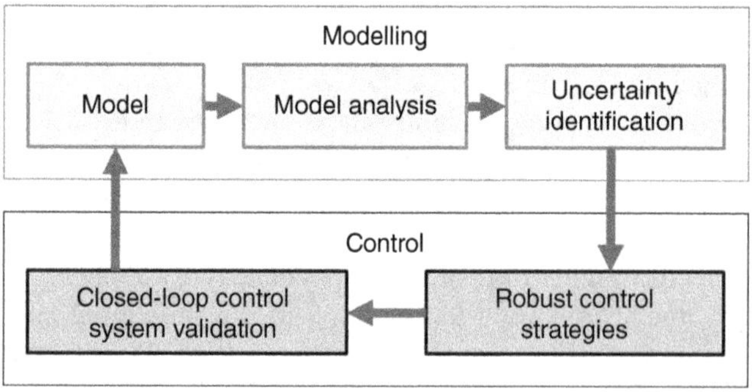

Figure 3.22 Framework presented in this thesis for volatile anaesthesia.

and (b) the mp-MPC specification. This set-up allows extensive advance testing of the control performance. The control strategy was further evaluated for disturbance rejection of commonly occurring disturbances during the course of surgery. Here, the online estimation of C_{50} showed promising results for slowly varying disturbances. However, further investigation is needed to guarantee safe and robust control also during fast-acting disturbances. The online estimation contributes to further understanding of the patient's sensitivity, which could be advantageous for future surgeries of the same patient. In Chapter 4, we will focus on the control of intravenous anaesthesia.

Appendix

Variable Lists

Table 3A.1 Denotation of variables and parameters of the PBPK/PD model.

Symbol	Denotation	Units	Symbol	Denotation	Units
BIS	Bispectral index	–	$\dot{Q}$	Blood flow	$\frac{\text{mL}}{\text{min}}$
C	Concentration volatile anaesthetic	vol%	$\dot{Q}_s$	Shunt flow	$\frac{\text{mL}}{\text{min}}$
C_{50}	Drug concentration at 50% effect	vol%	r	Ratio	–
f_R	Respiratory frequency	$\frac{1}{\text{min}}$	u	Uptake	$\frac{\text{mL}}{\text{min}}$
γ	Slope Hill equation	–	V	Volume	mL
λ	Partition coefficient	–	V_D	Dead space	mL
l_s	Lung shunt	–	$\dot{V}$	Air flow	$\frac{\text{mL}}{\text{min}}$
m_{liv}	Metabolism liver	$\frac{\text{mL}}{\text{min}}$	V_T	Tidal volume	mL

Table 3A.2 Denotation of subscripts in the PBPK/PD model.

Subscript	Denotation	Subscript	Denotation
A	Alveoli	i	Compartment
a	Arterial	I	Inhaled
b	Blood	L	Lungs
e	Effect compartment	t	Tissue
E	Exhaled	$\bar{v}$	Mixed venous return

References

(PSE) P. S. E. L. (2015). *gPROMS Model Developer Guide*. Release version 4.1.

Bard, Y. (1974). *Nonlinear Parameter Estimation*. Academic Press, London.

Bischoff, K. B. (1986). "Physiological pharmacokinetics." *Bulletin of Mathematical Biology* **48**(3–4): 309–322.

Brunner, M. D., P. Braithwaite, R. Jhaveri, A. I. McEwan, D. I. Goodman, L. R. Smith and P. S. A. Glass (1994). "MAC reduction of isoflurane by sufentanil." *British Journal of Anaesthesia* **72**(1): 42–46.

Deurenberg, P., J. A. Weststrate and J. C. Seidell (1991). "Body mass index as a measure of body fatness: Age- and sex-specific prediction formulas." *British Journal of Nutrition* **65**(2): 105–114.

Dua, P., V. Dua and E. N. Pistikopoulos (2010). "Modelling and multi-parametric control for delivery of anaesthetic agents." *Medical and Biological Engineering and Computing* **48**(6): 543–553.

Eger, E. I. (1974). *Anesthetic Uptake and Action*. Williams & Wilkins, Baltimore.

Eger, E. I. (2005). "The pharmacology of inhaled anesthetics." *Seminars in Anesthesia, Perioperative Medicine and Pain* **24**(2): 89–100.

Enderle, J. D. and J. D. Bronzino (2011). *Introduction to Biomedical Engineering*. Elsevier, New York.

Fiserova-Bergerova, V. (1992). "Inhalation anesthesia using physiologically based pharmacokinetic models." *Drug Metabolism Reviews* **24**(4): X2–557.

Gentilini, A., M. Rossoni-Gerosa, C. W. Frei, R. Wymann, M. Morari, A. M. Zbinden and T. W. Schnider (2001). "Modeling and closed-loop control of hypnosis by means of bispectral index (BIS) with isoflurane." *IEEE Transactions on Biomedical Engineering* **48**(8): 874–889.

Glass, P. S. A., T. J. Gan, S. Howell and B. Ginsberg (1997). "Drug interactions: Volatile anesthetics and opioids." *Journal of Clinical Anesthesia* **9**(Suppl. 6): 18S–22S.

Haggard, H. W. (1924). "The absorption, distribution, and elimination of ethyl ether: II. Analysis of the mechanism of absorption and elimination of such a gas or vapor as ethyl ether." *Journal of Biological Chemistry* **59**: 753–770.

Hall, C., A. M. Lueshen and A. A. Linninger (2012). "Pharmacokinetics, pharmacodynamics and drug metabolism interspecies scaling in pharmacokinetics: A novel whole-body physiologically based modeling framework to discover drug biodistribution mechanisms in vivo." *Pharmaceutical Sciences* **101**(3): 1221–1224.

Heymsfield, S., T. Lohman, Z. M. Wang and S. Going (2005). *Human Body Composition*. Human Kinetics Press, Champaign, IL.

Ionescu, C. M., R. D. Keyser, B. C. Torrico, T. D. Smet, M. M. R. F. Struys and J. E. Normey-Rico (2008). "Robust predictive control strategy applied for propofol dosing using BIS as a controlled variable during anesthesia." *IEEE Transactions on Biomedical Engineering* **55**(9): 2161–2170.

Kety, S. S. (1951). "The theory and applications of the exchange of inert gas at the lungs and tissues." *Pharmacological Reviews* **3**(1): 1–41.

Krieger, A. (2014). *Modelling, Optimisation and Explicit Model Predictive Control of Anaesthesia Drug Delivery Systems*. PhD thesis, Imperial College London.

Krieger, A., N. Panoskaltsis, A. Mantalaris, M. C. Georgiadis and E. N. Pistikopoulos (2014). "Modeling and analysis of individualized pharmacokinetics and pharmacodynamics for volatile anesthesia." *IEEE Transactions on Biomedical Engineering* **61**(1): 25–34.

Krieger, A. and E. N. Pistikopoulos (2014). "Model predictive control of anesthesia under uncertainty." *Computers and Chemical Engineering* **71**: 699–707.

Lemmens, H. J. M., D. P. Bernstein and J. B. Brodsky (2006). "Estimating blood volume in obese and morbidly obese patients." *Obesity Surgery* **16**(6): 773–776.

Lerou, J. G. C., R. Dirksen, H. H. Beneken Kolmer and L. H. D. J. Booij (1991a). "A system model for closed-circuit inhalation anesthesia: I. Computer study." *Anesthesiology* **75**(2): 345–355.

Lerou, J. G. C., R. Dirksen, H. H. B. Kolmer, L. H. D. J. Booij and G. F. Borm (1991b). "A system model for closed-circuit inhalation anesthesia. II. Clinical validation." *Anesthesiology* **75**(2): 230–237.

Mapleson, W. W. (1963). "An electric analogue for uptake and exchange of inert gases and other agents." *Journal of Applied Physiology* **18**(2): 197–204.

Mapleson, W. W. (1996). "Effect of age on MAC in humans: A meta-analysis." *British Journal of Anaesthesia* **76**(2): 179–185.

Mertens, M. J. and J. Vuyk (1998). "Interactions between intravenous anaesthetic agents." *Bailliere's Clinical Anaesthesiology* **12**(2): 247–261.

Miller, R. D. (2015). *Miller's Anesthesia*. Elsevier Health, New York.

Nadler, S. B., J. U. Hidalgo and T. Bloch (1962). "Prediction of blood volume in normal human adults." *Surgery* **51**(2): 224–232.

Nascu, I., I. Nascu, C. M. Ionescu and R. De Keyser (2012). "Adaptive EPSAC predictive control of the hypnotic component in anesthesia." In *IEEE International Conference on Automation, Quality and Testing, Robotics, Proceedings*.

Parker, R. S. and F. J. Doyle (2001). "Control-relevant modeling in drug delivery." *Advanced Drug Delivery Reviews* **48**(2–3): 211–228.

Pelosi, P., M. Croci, I. Ravagnan, S. Tredici, A. Pedoto, A. Lissoni and L. Gattinoni (1998). "The effects of body mass on lung volumes, respiratory mechanics, and gas exchange during general anesthesia." *Anesthesia and Analgesia* **87**(3): 654–660.

Robayo, F., D. Sendoya, R. Hodrea, C. Ionescu and R. De Keyser (2010). "Estimating the time-delay for predictive control in general anesthesia." Paper presented at the 2010 Chinese Control and Decision Conference, CCDC 2010.

Rosow, C. E. (1997). "Anesthetic drug interaction: An overview." *Journal of Clinical Anesthesia* **9**(Suppl. 6): 27S–32S.

Saltzman, W. M. (2001). *Drug Delivery Engineering Principles for Drug Therapy*. Oxford University Press, Oxford.

Sartori, V., P. M. Schumacher, T. Bouillon, M. Luginbuehl and M. Morari (2005). "On-line estimation of propofol pharmacodynamic parameters." In *Annual International Conference of the IEEE Engineering in Medicine and Biology – Proceedings.*

Searle, R. and P. M. Hopkins (2009). "Pharmacogenomic variability and anaesthesia." *British Journal of Anaesthesia* **103**(1): 14–25.

Stelfox, H. T., S. B. Ahmed, R. A. Ribeiro, E. M. Gettings, E. Pomerantsev and U. Schmidt (2006). "Hemodynamic monitoring in obese patients: The impact of body mass index on cardiac output and stroke volume." *Critical Care Medicine* **34**(4): 1243–1246.

Teorell, T. (1937a). "Kinetics of distribution of substances administered to the body, I." *Archives internationales de pharmacodynamie et de therapie* **57**: 205–225.

Teorell, T. (1937b). "Kinetics of distribution of substances administered to the body, II." *Archives internationales de pharmacodynamie et de therapie* **57**: 226–240.

Yelneedi, S., L. Samavedham and G. P. Rangaiah (2009). "Advanced control strategies for the regulation of hypnosis with propofol." *Industrial & Engineering Chemistry Research* **48**(8): 3880–3897.

Ziehn, T. and A. S. Tomlin (2009). "GUI-HDMR: A software tool for global sensitivity analysis of complex models." *Environmental Modelling and Software* **24**(7): 775–785.

Zwart, A., N. Ty Smith and J. E. W. Beneken (1972). "Multiple model approach to uptake and distribution of halothane: The use of an analog computer." *Computers and Biomedical Research* **5**(3): 228–238.

4

Intravenous Anaesthesia

Ioana Naşcu[1], Alexandra Krieger[2], Romain Lambert[3], and Efstratios N. Pistikopoulos[4]

[1] *Artie McFerrin Department of Chemical Engineering, Texas A&M University, College Station, USA*
[2] *Jacobs Consultancy, Kreisfreie Stadt Aachen Area, Germany*
[3] *Department of Chemical Engineering, Imperial College London, UK*
[4] *Texas A&M Energy Institute, Artie McFerrin Department of Chemical Engineering, Texas A&M University, USA*

4.1 A Multiparametric Model-based Approach to Intravenous Anaesthesia

4.1.1 Introduction

Anaesthesia can be defined as a reversible pharmacological state where the patient muscle relaxation, analgesia and hypnosis are guaranteed. It is characterized by unconsciousness through the action of anaesthetics, but also by loss of the ability to perceive pain through the action of analgesics. Analgesics block the sensation of pain; the hypnotics produce unconsciousness, while the muscle relaxants prevent unwanted movement of muscle tone.

The concept of intravenous (IV) anaesthesia compared to volatile anaesthesia is simpler: it requires an IV line, and everything needed for general anaesthesia is supplied through this line. This will eliminate the need for sophisticated gas delivery systems or time-consuming procedures such as establishing regional blocks or neuraxial blocks (Eikaas and Raeder 2009). The drugs used in IV anaesthesia are usually less toxic than inhalational agents, with less risk of malignant hyperthermia and no pollution of environmental air or the atmosphere. IV anaesthesia usually implies giving dedicated component therapy with different drugs for different effects, in general one drug for the hypnotic agent (propofol, ketamine, methohexital or midazolam) and another drug for analgesia and anti-nociception (remifentanil, other opioids or ketamine).

The role of the anaesthetist has become more and more complex and indispensable to maintain the patients' vital functions before, during and after surgery. In order to do this, averaged population models are used to estimate the drug effect

Modelling Optimization and Control of Biomedical Systems, First Edition.
Edited by Efstratios N. Pistikopoulos, Ioana Naşcu, and Eirini G. Velliou.

in the patient's body and calculate the corresponding drug infusion rates. These strategies do not take into account any measured variable in a feedback control scheme, and even if they reach the desired level of sedation fast, they can result in minimal values (undershoot) that are not safe for the patient. But if the control of *depth of anaesthesia* (DOA) is done automatically, anaesthetists will have more time to focus on critical issues that can threaten the safety of the patient.

Anaesthesia control deals with many challenges: inter- and intra-patient variability, multivariable characteristics, variable time delays, dynamics dependent on hypnotic agent and stability issues. Hitherto, many proportional–integral–derivative (PID) tuning techniques have been elaborated during recent decades, but since these classical controllers cannot anticipate the response of the patient and have no prior knowledge of the drug metabolism, their performance was sub-optimal. Therefore, model-based strategies using fuzzy, predictive and adaptive control algorithms have been elaborated and used in clinical trials.

The purpose of computer-controlled closed-loop systems is to formalize the process of observation and intervention to provide better and more accurate control. Such systems use a near continuous signal of drug effect, calculate the error between the observed value and the set-point value (selected by the user), and use this error in an algorithm to make frequent and regular adjustments to drug administration rates. Moreover, some computer-controlled systems try to predict the future drug effect so that better adjustments will be done in advance

Drug administration is an asymmetrical process: we can actively infuse but cannot actively remove the drug. Because the relationship between dose and plasma concentration is so complex, target-controlled infusion (TCI) systems are a logical choice of control actuator, so that the control input is a target concentration rather than an infusion rate. Many assumptions underpin the pharmacokinetic models used in TCI systems, and some are obviously incorrect (such as that of instant mixing of an administered drug within the central compartment). Not surprisingly, the predictive accuracy of current models is imperfect, and the choice of model for propofol is controversial.

4.1.2 Patient Model

For IV administration, the infused drug enters the circulatory system, where under the action of the heart it is mixed and evenly distributed. The drug must diffuse out of the circulatory system into extracellular volumes before it reaches the target organ or cells. Because it acts on the target, the controlling drug may also be subject to excretion by the kidneys and intestines, as well as biotransformation and inactivation by organs such as the liver, the renal epithelium, and the intestinal mucosa.

The model used for prediction should not be too complex, in order not to take too much computational time. On the other hand, it should capture very well the dynamics of the patient in response to an applied propofol signal.

The relationship between the infusion rate of propofol and its effect can be described with pharmacokinetic (PK) and pharmacodynamic (PD) models. A PK model describes the distribution of propofol in the body, and a PD model describes the relationship between propofol blood concentration and its clinical effect. A compartmental model is used to describe the PK–PD blocks representing the distribution of drugs in the body (i.e. mass balance). In each compartment, the drug concentration is assumed to be uniform, as perfect and instantaneous mixing is assumed. The structure of the compartmental model is depicted in Figure 4.1 (Schnider *et al.* 1998; Struys *et al.* 2004).

The PK–PD models most commonly used for propofol are the fourth-order compartmental models described by Schnider *et al.* (1998, 1999) and Minto *et al.* (1997a, 1997b), respectively.

The PK model and the first term of the PD model are considered linear and are represented by the following equations:

$$\begin{aligned}\dot{C}_1(t) &= -[k_{10}+k_{12}+k_{13}]\cdot C_1(t)+k_{21}\cdot C_2(t)+k_{31}\cdot C_3(t)+u(t)/V_1\\ \dot{C}_2(t) &= k_{12}\cdot C_1(t)-k_{21}\cdot C_2(t)\\ \dot{C}_3(t) &= k_{13}\cdot C_1(t)-k_{31}\cdot C_3(t)\end{aligned} \tag{4.1}$$

where C_1 represents the drug concentration in the central compartment [mg/l]. The peripheral compartments 2 (muscle) and 3 (fat) model the drug exchange of the blood with well and poorly diffused body tissues. The concentrations of

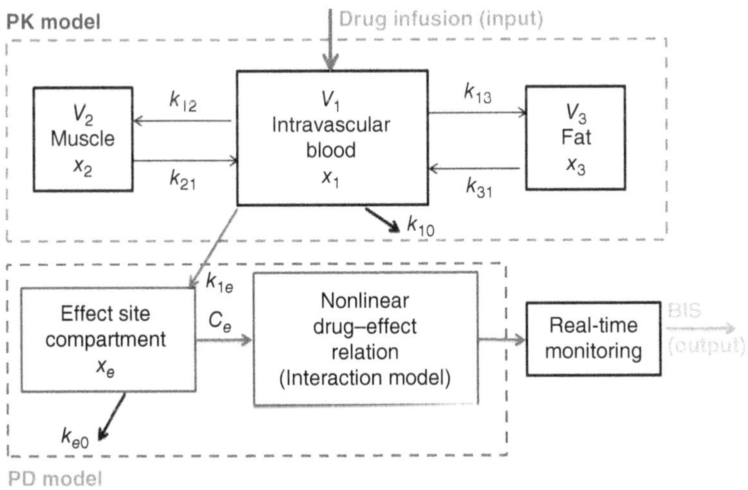

Figure 4.1 Compartmental model of the patient. PK = the pharmacokinetic model; PD = the pharmacodynamic model. *Source:* Naşcu *et al.* (2014a). Reproduced with permission of IEEE Transactions on Biomedical Engineering.

drug in the fast and slow-equilibrating peripheral compartments are denoted by C_2 and C_3, respectively. The parameters k_{ij} for $i = 1:3, i \neq j$, denote the drug transfer frequency from the ith to the jth compartment; k_{10} is the frequency of drug removal from the central compartment; and $u(t)$ [mg/min] is the infusion rate of the anaesthetic or analgesic drug into the central compartment. The parameters k_{ij} of the PK models depend on age, weight, height and gender and can be calculated for propofol (Naşcu *et al.* 2014a) as follows:

$$\begin{aligned}
&V_1 = 4.27[l], V_2 = 18.9 - 0.391\cdot(age - 53)[l], V_3 = 2.38[l] \\
&C_{l1} = 1.89 + 0.456(weight - 77) - 0.0681(lbm - 59) \\
&\qquad + 0.264(height - 177)\left[l/\text{min}\right] \\
&C_{l2} = 1.29 - 0.024(age - 53)\left[l/\text{min}\right], C_{13} = 0.836\left[l/\text{min}\right] \\
&k_{10} = \frac{C_{l1}}{V_1}\left[\text{min}^{-1}\right], k_{12} = \frac{C_{l2}}{V_1}\left[\text{min}^{-1}\right], k_{13} = \frac{C_{l3}}{V_1}\left[\text{min}^{-1}\right], \\
&k_{21} = \frac{C_{l2}}{V_2}\left[\text{min}^{-1}\right], k_{31} = \frac{C_{l3}}{V_3}\left[\text{min}^{-1}\right] \\
&k_{e0} = 0.456\left[\text{min}^{-1}\right]
\end{aligned} \tag{4.2}$$

where C_{l1} is the rate at which the drug is cleared from the body, and C_{l2} and C_{l3} are the rates at which the drug is removed from the central compartment to the other two compartments by distribution.

The lean body mass (*lbm*) for men (m) and women (f) are calculated by:

$$lbm\_m = 1.1\cdot weight - 128\frac{weight^2}{height^2} \tag{4.3}$$

$$lbm\_f = 1.07\cdot weight - 148\frac{weight^2}{height^2}$$

The pharmacodynamics describe the link of concentration of the anaesthetic agent to the effect of the drug. The PD mathematical model is presented as follows:

$$\dot{C}_e(t) = k_{e0}\left(C_e(t) - x_1(t)\right) \tag{4.4}$$

$$BIS(t) = E_0 - E_{\max}\cdot\frac{C_e(t)^\gamma}{C_e(t)^\gamma + EC_{50}^\gamma} \tag{4.5}$$

An additional hypothetical effect compartment is added to represent the lag between plasma drug concentration and drug response. The drug concentration in this compartment is represented by x_e, called the *effect-site compartment concentration*. The effect compartment receives drug from the central compartment by a first-order process, and it is considered as a virtual additional compartment. Therefore, the drug transfer frequency for propofol from the central compartment to the effect-site compartment is considered in clinical practice to be equal to the frequency of drug removal from the effect-site compartment $k_{e0} = k_{1e} = 0.456[\min^{-1}]$ (Schnider *et al.* 1998, 1999; Nunes *et al.* 2009).

When considering the drug effect observed on the patient, the Bispectral Index (BIS) variable can be related to the effect drug concentration C_e by the empirical static nonlinear relationship (4.5) (Schnider *et al.* 1998, 1999; Struys *et al.* 2003; Ionescu *et al.* 2008), also called the *Hill curve*, which corresponds to the second part of the PD model. E_0 denotes the baseline value (awake state = without drug), which by convention is typically assigned a value of 100; E_{max} denotes the maximum effect achieved by the drug infusion; EC_{50} is the drug concentration at 50% of the maximal effect and represents the patient sensitivity to the drug; and γ determines the steepness of the curve.

The inverse of the Hill curve can also be defined as follows:

$$C_e(t) = EC_{50}\left(\frac{E_0 - BIS(t)}{E_{\max} - E_0 + BIS(t)}\right)^{\frac{1}{\gamma}} \tag{4.6}$$

For the automatic regulation of DOA in Figure 4.2, the anaesthetic agent (i.e. propofol) is the input and the BIS is the output of the system. Propofol is a powerful anaesthetic, for which the pharmacologic properties have been well described and studied on different types of patients. Because of its pharmacological profile, propofol is applicable for both induction and maintenance of hypnosis during anaesthesia and intensive care sedation (Ionescu *et al.* 2015).

Equations (4.1)–(4.4) complete the PK–PD patient model for IV anaesthesia.

The BIS is a signal that is derived from the electro-encephalogram (EEG) used to assess the level of consciousness during anaesthesia. A BIS value of 0 equals EEG silence, while a BIS value of 100 is the expected value of a fully awake and conscious adult patient; the 60–70 and 40–60 ranges represent light and moderate hypnotic conditions, respectively. The target value during

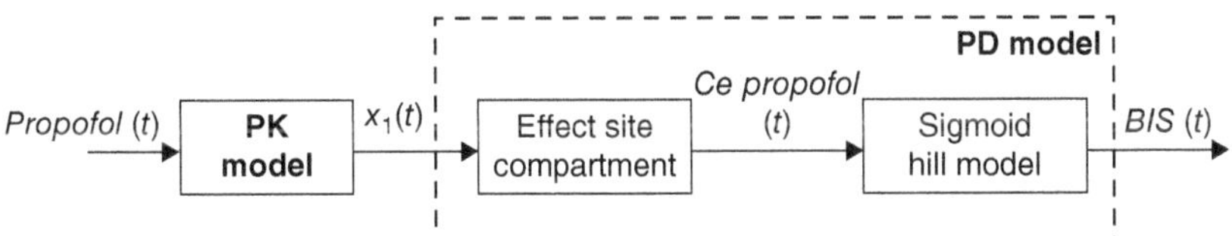

Figure 4.2 Schematic representation of the nonlinear SISO patient model for intravenous anaesthesia. *Source*: Naşcu *et al.* (2014a). Reproduced with permission of IEEE Transactions on Biomedical Engineering.

surgery is 50, giving us a gap between 40 and 60 to guarantee adequate sedation (Haddad *et al.* 2003; Bailey and Haddad 2005; Absalom *et al.* 2011).

4.1.3 Sensitivity Analysis

In this section, the methodologies presented in Chapter 2 are tested on the IV anaesthesia process. Based on the mathematical model presented there, global sensitivity analysis is first applied, in particular Sobol's method of sensitivity indices, the high-dimensional model representation (HDMR) approach and group method of data handling (GMDH)-HDMR, to determine the relative influence of the PK–PD parameters and variables. GMDH-HDMR relies on the direct construction of the HDMR expansion through GMDH inductive modelling. By analysing the anaesthesia model, it can be observed that the dynamics of the linear part is influenced by the age, height, weight and gender parameters. The characteristic of the nonlinear part is influenced by EC_{50}, E_0 and γ parameters. The relative influence of the uncertain PK and PD parameters and the variables on the measurable outputs is investigated. The sensitivity index (SI) represents the relative influence of the parameter or variable on the output at a given time. To perform the analysis using the mathematical models, an anaesthesia experiment was simulated. A step of 50 [mg/min] was applied on the anaesthetic drug infusion rate that represents the input of the model. The evolution of the output was investigated for an interval of 100 min, until a steady-state regime is reached. During simulations, all parameters and variables were varied between their bounds. S1–S7 represent the sensitivity indices for age, height, weight, EC_{50}, E_0, γ and gender, respectively.

Figure 4.3 presents the evolution of the first-order Sobol SIs at different sample points starting from $t = 2$ min to $t = 100$ min. At the beginning of anaesthesia E_0, the baseline value (the awake state, without drug) has the highest SI with respect to the BIS, but it converges asymptotically to 0 with time as the patient enters the deep anaesthesia state. Analysing the SIs from Figure 4.3, it is observed that the most important parameter is EC_{50} (SI: S4), the drug concentration at 50% of the maximal effect, representing the patient sensitivity to the drug. Note that EC_{50} (S4) increases exponentially and stabilizes when the BIS reaches its steady-state regime. Also note that because the nonlinearity is represented by a sigmoid, the parameter γ (S6), which determines the steepness of the curve, has more influence in the beginning of anaesthesia due to the high nonlinearity of this zone. Note also that the SIs of the parameters of the linear part (PK) increase with the nonlinearity slope, but the corresponding values are less important than the ones for the parameters of the nonlinear part.

Table 4.1 represents a comparison of the results obtained with the various techniques: (a) Sobol's sensitivity analysis, (b) GMDH-HDMR and (c) HDMR reported for four of the parameters. These parameters are weight (S3), EC_{50} (S4), E_0 (S5) and γ (S6), and they were chosen as they have the highest SI at

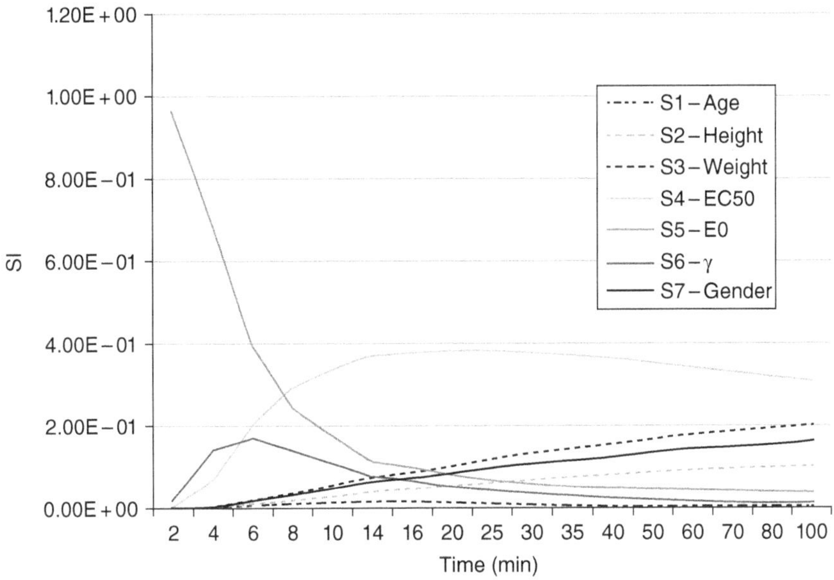

Figure 4.3 Evolution of the first-order sensitivity indices.

Table 4.1 Comparison of GMDH and HDMR at $t = 14$.

	Sobol	GMDH-HDMR (40)	HDMR (40)
S4	0.37122289	0.44268	0.47
S5	0.11501718	0.111865	0.8305
S3	0.07972202	0.08903	0
S6	0.07690754	0.0715	0.99

$t = 14$ min when using Sobol's sensitivity analysis as presented in Figure 4.3. The GMDH-HDMR approach only computes the indices for the parameter it selects as important; therefore, S1 (age), S2 (height) and S7 (gender) are not calculated.

Sensitivity analysis using GMDH-HDMR can be performed on a limited number of sample points, and as a result, the most important individual contributions are detected. HDMR is able to do this with 256 Sobol sample points, but it will fail to operate properly for only 40 data points. For example, as depicted in Figure 4.4, the individual contribution of the third parameter (S3) is completely ignored, and the interaction is not detected either. The advantage of GMDH-HDMR is that most of the time, it gives more accurate results than HDMR with a small number of data samples. HDMR becomes unreliable with

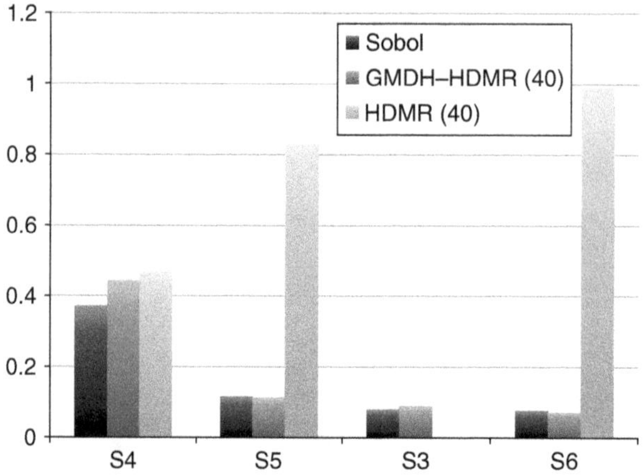

Figure 4.4 Comparison GMDH-HDMR for small data samples ($N=40$), $t=14$.

limited sets of samples as it can miss important first-order contributions. Moreover, the strength of GMDH-HDMR is the ability to be economical in the number of simulations required. This can be useful in the case of computationally expensive high-fidelity models. GMDH inductively selects the most important parameters, performing as a sparse method for calculating SIs, with scarce recourse to model simulations. HDMR, in contrast, relies on the calculation of the SIs for all parameters and potential interactions. It captures nonlinear system outputs as a summation of variable correlations in a hierarchical order. In a full model expansion, it considers all possible variable interactions and their contribution to the original function. The first-order term describes the average value of the fitness landscape near a reference point called the *cut-centre.* Second-order terms express the independent effects of each variable if decision variables are deviated from the cut-centre. Higher order terms denote the residual group effect of variables. Therefore, the terms are independent or orthogonal to each other to form a mean convergent series. From Table 4.1, it can be observed that γ (S6) has the highest SI with respect to the BIS. Since this method exhibits better results for this type of model, γ (S6) shows which patient is most sensitive to the drug.

4.1.4 Advanced Model-based Control Strategies

Model predictive control (MPC) is a control methodology based on two main principles: explicit online use of a *process model* to *predict* the process output at future time instants, and the computation of an optimal control action by minimizing a cost function, including *constraints* on the process variables.

The main differences between the different types of MPC algorithms are: (a) the *type of model* used to represent the process and its *disturbances*; (b) the *cost function(s)* to be minimized, with or without *constraints*; and (c) the type of optimization performed. Details of the method formulation are presented in Chapter 2.

4.1.4.1 Extended Predictive Self-adaptive Control (EPSAC) Strategy

EPSAC is one of the various MPC design methods reported in the literature (Su *et al.* 2016), and it adopts the approach of iterative optimization based on a predefined input trajectory (De Keyser and Van Cauwenberghe 1985; De Keyser 2003). A potential drawback of previous EPSAC methods is the incorporation of a convolution model in the formulation of the control algorithms. Since model parameters are obtained by introducing a step change to the current input value specified by the base input trajectory, the predicted outputs at sampling instants further away from the current sampling instant become less accurate due to process nonlinearity, leading to inevitable modelling error that degrades the achievable closed-loop performance. Another potential downside of this method is that the optimization problem has to be solved online. This issue is addressed in this chapter.

For the EPSAC approach (see De Keyser 2003), the controller output is obtained by minimizing the cost function:

$$\sum_{k=N_1}^{N_2}\left[r\left(t+k/t\right)-y\left(t+k/t\right)\right]^2+\lambda\sum_{k=0}^{N_u-1}\left[\Delta u\left(t+k/t\right)\right]^2 \tag{4.7}$$

where the design parameters are: N_1 = the minimum costing horizon; N_2 = the maximum costing horizon; N_2-N_1 = the prediction horizon; N_u = control horizon; λ = weight parameter; and r = reference trajectory.

In our case, the process input is represented by the propofol infusion rate applied to the patient. The process output is the BIS and is predicted at time instant t over the prediction horizon N_2-N_1, based on the measurements available at that moment and the future outputs of the control signal. The cost function is an extended EPSAC cost function that penalizes the control movements using the weight parameter λ.

4.1.4.2 Multiparametric Strategy

For the mp-MPC (see Chapter 2), the generic optimization problem solved is:

$$\begin{aligned}
&\min_u J=\sum_{k=1}^{N}\left(BIS_k-BIS_k^R\right)^T QR_k\left(BIS_k-BIS_k^R\right)+\sum_{k=0}^{N_u-1}\Delta u_k^T R_k \Delta u_k\\
&s.t.\ \mathrm{x}_{t+1}=Ax_t+Bu_t\\
&\quad \mathrm{y}_t=Cx_t\\
&\quad \mathrm{BIS}_{\min}\le y\le BIS_{\max}\\
&\quad \Delta\mathrm{u}_{\min}\le \Delta u\le \Delta u_{\max}\\
&\quad x_t\in X\subseteq \Re^p, u_t\in U\subseteq \Re
\end{aligned} \tag{4.8}$$

where x = the states; y = the outputs; and u = the controls; all are (discrete) time-dependent vectors, with the tracked output variables having time-dependent set points y^R. Δu corresponds to changes in control variables, $\Delta u(k) = u(k) - u(k-1)$; N = the prediction horizon; N_u = the control horizon; and X, U are the sets of the state and input constraints that contain the origin in their interior. The weight matrix for manipulated variables, R, is a positive definite diagonal matrix ($R \succ 0$); QR is the weight for tracked outputs; and $R1$ is a weight matrix for the control action changes (Δu). The control problem is posed as a quadratic convex optimization problem for which an explicit solution can be obtained. The key idea is to derive the optimal control inputs as a set of explicit affine functions of the current state of the system:

$$u = f(x) = \begin{cases} K_1 x + c_1 \text{ if } x \in CR^1 \\ \dots \\ K_s x + c_s \text{ if } x \in CR^s \end{cases} \tag{4.9}$$

where s is the number of critical regions.

4.1.5 Control Design

The presence of the Hill nonlinearity complicates the use of linear controller synthesis. Two methods to overcome this problem have been proposed: (a) exact and (b) local linearization.

The local linearization is based on the linearized patient model (4.1), (4.4) and (4.5) using the parameter values of the nominal patient, known *a priori*, for a BIS value of 50 obtained using gPROMS (PSE 2015). The controller is designed using the linearized patient model as presented in Figure 4.5.

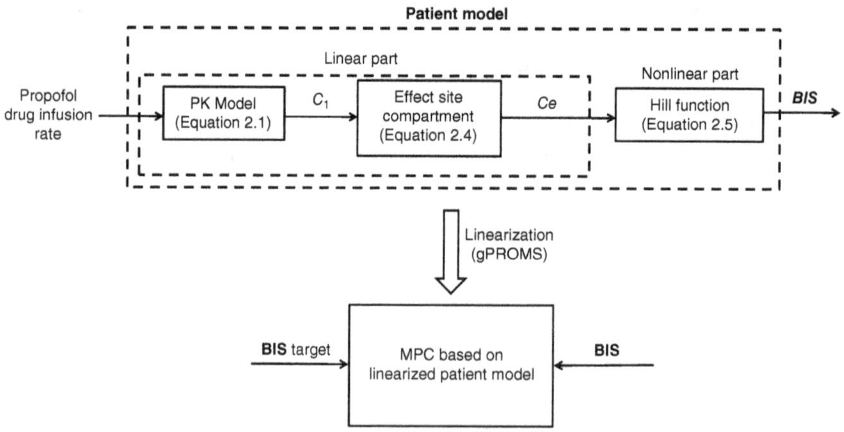

Figure 4.5 Controller design using local linearization.

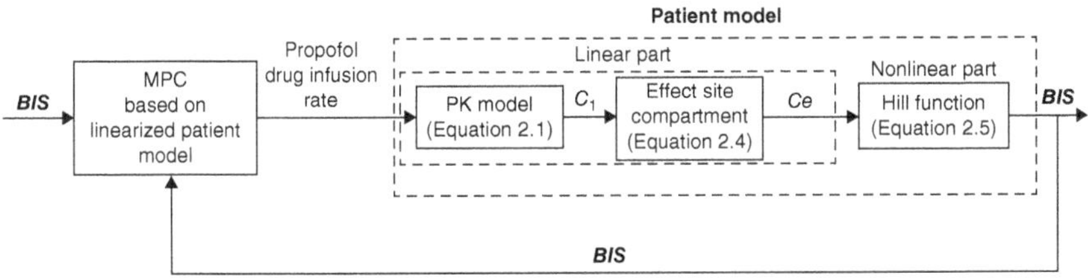

Figure 4.6 Controller scheme using local linearization.

Figure 4.6 depicts the control design scheme where the designed controller minimizes the error between the *BIS target value* and the *measured BIS*, giving the patient the optimal *propofol drug infusion rate* that will derive the patient to the desired set point value.

Exact linearization is based on the compensation of the nonlinearity introduced by the Hill curve (4.5) in the PD model (4.4) and (Figure 4.7). Since the Hill nonlinearity is a monotonic function of the normalized effect site concentration, it has an inverse presented in Equation (4.6). Using a parameter scheduling technique, the inverse Hill function could be implemented in the control design scheme, as illustrated by the block diagram in Figure 4.7. In the *patient model*, the *Hill curve* uses the nonlinear parameters of the real patient (E_0, E_{max}, EC_{50}, γ), while the *inverse Hill function* is using the nonlinear parameter corresponding to the nominal patient known *a priori* (${E_0}^{mean}$, ${E_{max}}^{mean}$, ${EC_{50}}^{mean}$, γ^{mean}).

For this method, the controller is designed using the *linear part* (4.1) and (4.4) of the patient model (4.1), (4.4) and (4.5), with the linear parameters of the nominal patient (*age, height, weight* and *gender*) as presented in Figure 4.8. Note that the *BIS target* here is transformed in the C_e *target* using the inverse Hill function since the controlled variable is the estimated drug concentration $\hat{C}_e$.

An exact linearization takes place only for the case where the patient model is identical to the nominal model; in this case, the nonlinearity is cancelled out (i.e. $\hat{C}_e = C_e$).

An important challenge of depth of anaesthesia (DOA) control is the high inter- and intra-patient variability. This results in different dynamics in the PK model, and changes in the parameters of the Hill function for each patient model. Four control strategies, a model predictive controller, EPSAC and three

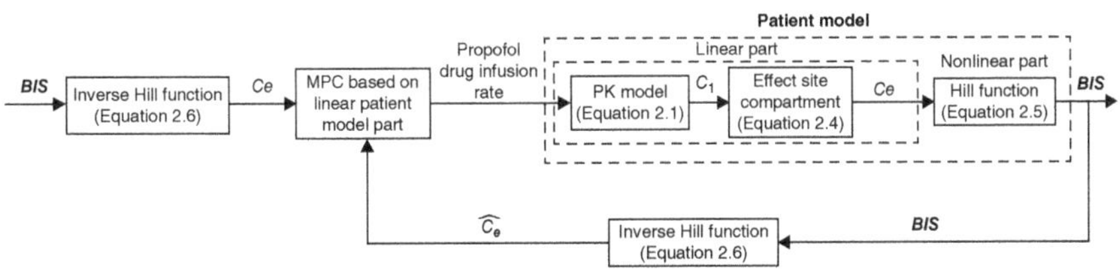

Figure 4.7 Controller scheme using exact linearization.

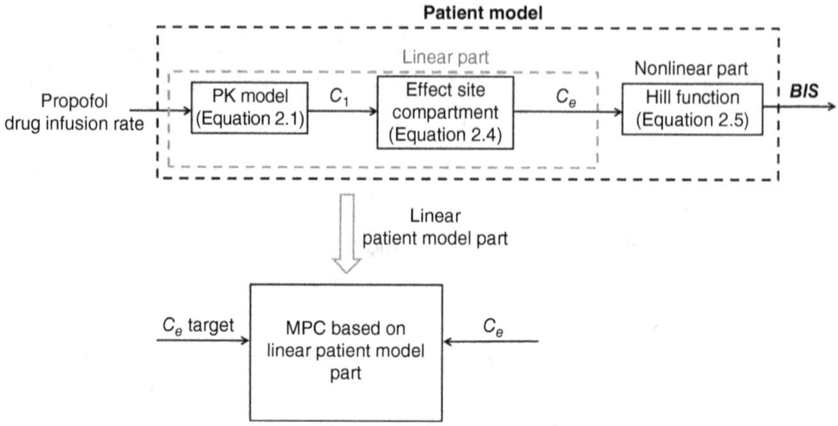

Figure 4.8 Controller scheme using exact linearization.

different mp-MPCs are designed and evaluated. The framework for the different ways of designing the controllers is presented in Figure 4.9.

The patient response is simulated using the *patient model* block, composed of the PK–PD linear part (4.1), the effect site concentration (4.4) and the

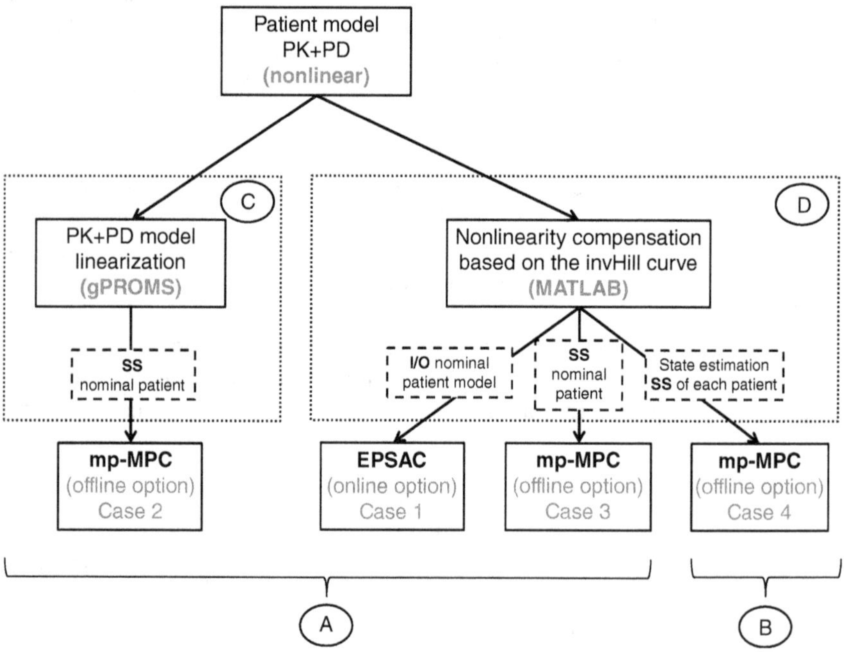

Figure 4.9 Control scheme development flowchart. *Source*: Naşcu *et al.* (2014a). Reproduced with permission of IEEE Transactions on Biomedical Engineering.

nonlinear PD part composed of the Hill nonlinearity (4.5). BIS can be measured; however, the states cannot and have to be estimated, either by using the drug rate input and the nominal state-space patient model or by using the drug rate input and measured output (BIS) of the process, the state-space nominal model and a correction estimator based on the output changes.

To analyse the influence of the changes in the dynamics of the PK model on the control performances, two types of control schemes are implemented. One uses the states given by the nominal model (A), and the other uses an estimator to adjust the states based on the dynamics of each patient (B).

The influence on the changes of parameters of the Hill curve on the control performances is analysed by two types of control schemes; the first uses local linearization of the complete PK–PD patient model (C), and the second is based on the exact linearization (D). The following design parameters are used: the weight matrix for tracked outputs (y), $QR = 1000$, the weight matrix for manipulated variables (u), $R = 1$, the control horizon $N_u = 1$ and the prediction horizon $N = 20$ in both mp-MPC and EPSAC. The EPSAC has an extra weighting factor λ from Equation (4.7), for which its default value $\lambda = 0$ was used. The states used in the design of the controllers are C_1, C_2, C_3, C_e, as described in Equation (4.1). The clinically recommended sampling time is 5 s (Ionescu *et al.* 2008). N_1, N_2, and N_u are chosen based on the characteristics of the process and the desired performances. Based on Clarke *et al.* (1987) and Mohtadi *et al.* (1987), N should be large: at least $2n-1$ but not larger than the rise time of the process. For anaesthesia, due to medical procedures, we are constrained to use a small sampling time, which leads to a choice of a greater value for N. Also, the dead time is not considered since it is very small and does not affect the process; therefore, $N_1 = 1$. In choosing N_u for processes with no unstable or underdamped poles, like anaesthesia, $N_u = 1$ is generally satisfactory. A choice of the Q, R and QR is given by Bryson's rule (Franklin *et al.* 2001). Usually the pump for the propofol drug infusion is limited to 200 mg/min (3.3 mg/s), but it can be observed from simulations as well as literature that the controller gives maximum values of 70 mg/min for the drug infusion. To test the ability of the controller in dealing with severe constraints and avoid giving the patient unnecessary amounts of drug infusion (leading to longer recovery time), in this work the drug infusion is limited to 50 mg/min.

4.1.5.1 Case 1: EPSAC

In this section, we apply a particular case of online MPC, the EPSAC. For the EPSAC approach, described in detail in De Keyser (2003), the controller output is obtained by minimizing the cost function (4.7)

In our case, the process input is represented by the propofol infusion rate applied to the patient. The process output is predicted at time instant t over the

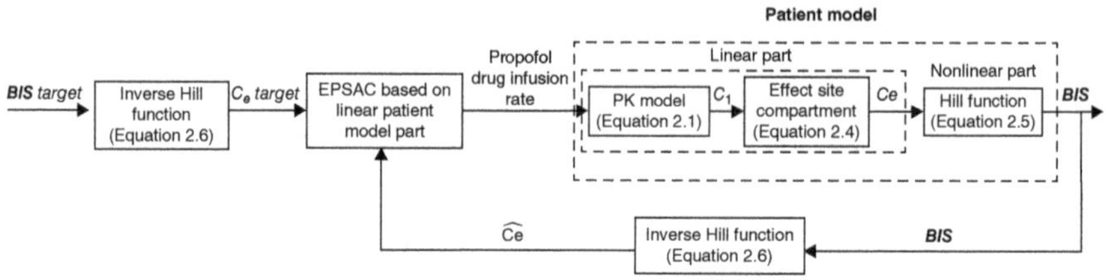

Figure 4.10 Case 1: EPSAC control scheme.

prediction horizon $N_2 - N_1$, based on the measurements available at that moment and the future outputs of the control signal. The cost function is an extended EPSAC cost function that penalizes the control movements using the weight parameter λ. The structure of the control system proposed in this section is shown in Figure 4.10.

The patient model block is composed of the *PK model* (4.1), the *effect-site compartment* (4.4) and the Hill function (4.5). The control strategy is based on the exact linearization (see Figure 4.7) and the input–output (I/O) linear nominal part of the patient model (see Figure 4.7). The controller output is obtained by minimizing the cost function (4.7) with the design parameters in Section 4.1.4. The control algorithm uses for prediction a transfer function derived from the linear part (4.1) and (4.4) of the patient model. The inverse of the Hill curve (4.6) is used to compensate the nonlinearity as presented in Figure 4.8. The patient model uses the parameter values of the real patient, while the inverse of the Hill curve and the linear model used for the controller use the parameter values of the nominal patient from Table 4.2.

4.1.5.2 Case 2: mp-MPC Without Nonlinearity Compensation

The structure of the control scheme is presented in Figure 4.11. This approach uses the explicit/multiparametric MPC strategy described in Chapter 2 based on local linearization of the patient model for the parameter values of the nominal patient from Table 4.2, as presented in Figure 4.5. The controller uses the error between the measured BIS and the BIS target value as well as the state-space model of the nominal patient to give the patient the optimal propofol drug infusion rate.

To obtain the linearized patient model, the PK and PD model for the nominal patient is implemented in gPROMS (PSE 2015), and the state space of the linearized nominal patient model at BIS = 50 is determined. Using these matrices, the mp-QP optimization problem (4.8) is solved to obtain the control laws using a MATLAB implementation of a multiparametric quadratic programming algorithm (Pistikopoulos *et al.* 1999; ParOs 2004) and to determine the mp-MPC control laws that will be used to calculate the optimal control action.

Table 4.2 Biometric values of the virtual patients.

Patient	Age	Height (cm)	Weight (kg)	Gender	EC_{50}	E_0	γ
1	40	163	54	M	6.33	98.8	2.24
2	36	163	50	M	6.76	98.6	4.29
3	28	164	52	M	8.44	91.2	4.1
4	50	163	83	M	6.44	95.9	2.18
5	28	164	60	F	4.93	94.7	2.46
6	43	163	59	M	12.0	90.2	2.42
7	37	187	75	F	8.02	92.0	2.1
8	38	174	80	M	6.56	95.5	4.12
9	41	170	70	M	6.15	89.2	6.89
10	37	167	58	M	13.7	83.1	1.65
11	42	179	78	F	4.82	91.8	1.85
12	34	172	58	M	4.95	96.2	1.84
Mean	38	169	65	M	7.42	93.1	3

Source: Naşcu *et al.* (2014a). Reproduced with permission of IEEE Transactions on Biomedical Engineering.

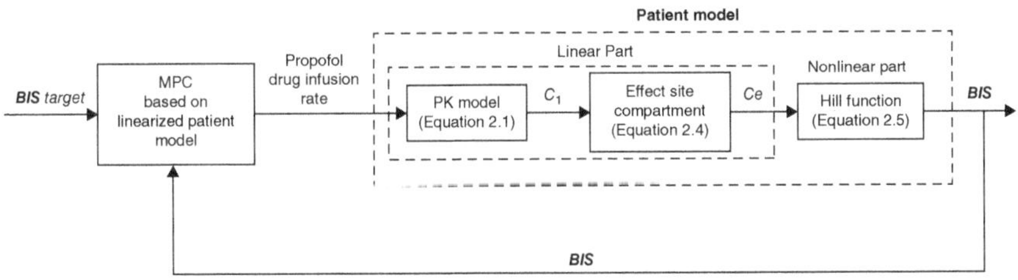

Figure 4.11 Case 2:mp-MPC without nonlinearity compensation – control scheme.

4.1.5.3 Case 3: mp-MPC With Nonlinear Compensation

This approach uses explicit/multiparametric MPC based on exact linearization, as presented in Figure 4.7. The controller is designed on the linear part of the patient model using the values of the nominal patient from Table 4.2. The optimization problem (4.8) is solved offline using POP (ParOs 2004) to obtain the explicit control laws. Since the controller is now designed only on the linear part, it will use the error between the C_e target and the $\hat{C}_e$ given by the inverse Hill function as well as the state space of the nominal patient values to give the optimal propofol drug infusion rate, as depicted in Figure 4.12. Note that here the inverse Hill function uses the values of the nominal patient, while the patient model uses the real values of the real simulated patient.

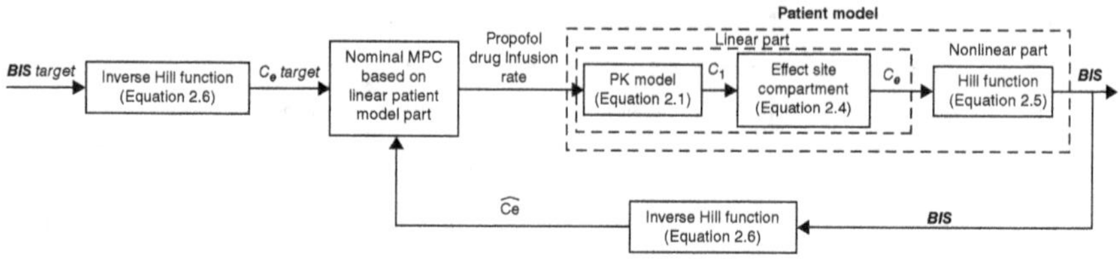

Figure 4.12 Case 3:mp-MPC with nonlinearity compensation – control scheme.

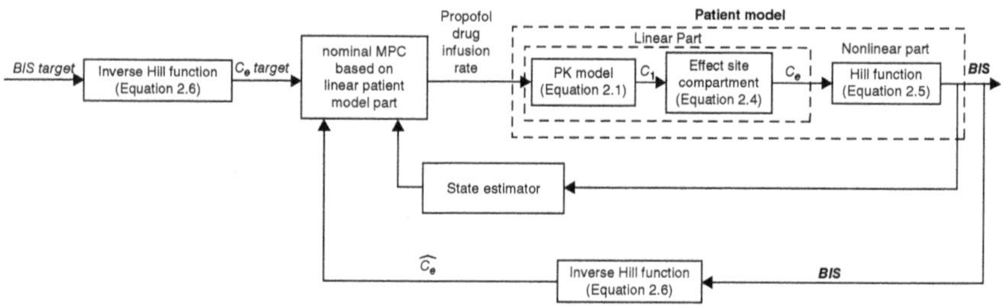

Figure 4.13 Case 4: mp-MPC with nonlinearity compensation and estimator – control scheme.

4.1.5.4 Case 4: mp-MPC With Nonlinearity Compensation and Estimation

This approach, like the one presented in Section 4.1.5.3, uses the explicit/multiparametric MPC strategy and has a similar control scheme. In Figure 4.13, the design of the controller is based on exact linearization (Figure 4.7).

The difference is the use of a state estimator that will give the controller the states of the real patient. The real patient states are estimated using a Kalman filter (Welch and Bishop 2001) based on the state-space model of the nominal patient, the online BIS measurement and the drug rate.

The Kalman filter is a standard method for unconstrained state estimations and follows a two-step procedure to calculate the maximum *a posteriori* Bayesian estimate (Welch and Bishop 2001). The first step is the time update, which uses the nominal patient model to predict the current state of the system based on the last estimate. The second step is the measurement update. The prediction from the previous step is updated by using the sensor information. Therefore, we can say that the Kalman filter is a predictor-corrector type estimator that is optimal in the sense that it minimizes the estimated error covariance (Welch and Bishop 2001).

4.1.6 Results

In this section, the results of a simulation study to evaluate the four control strategies for the administration of propofol are presented. The controllers are based on model-based predictive control algorithms for automatic induction and control of DOA. DOA is monitored using the BIS during the induction and maintenance phases of general anaesthesia. The closed-loop control tests are performed on a set of 12 patients (Ionescu *et al.* 2011) plus an extra patient representing the nominal values of all 12 patients (PaN: patient nominal). The parameter values of these patients are given in Table 4.2 and are also used to calculate the parameters of the patient model. For a particular patient, E_0 can be measured in the awake state and E_{max} is considered to have the same value, $E_{max} = E_0$. These parameters are considered known *a priori* in the simulations.

All of the designed controllers are simulated first for the whole set of data presented in Table 4.2 in order to have a better understanding of their behaviour with the different types of patients, and also to be able to analyse the inter- and intra-patient variability. Next, the four controllers will be tested against each other and simulated for different patients to compare their performances by means of the BIS and the corresponding propofol infusion rates. The performances of the four controllers are evaluated, and the results are analysed comparatively. The target value during surgery is 50, giving us a range between 40 and 60 to guarantee adequate sedation, resulting in an overshoot/undershoot lower than 10%.

4.1.6.1 Induction Phase

Ideally, the induction phase of the patient in an operational DOA is performed as fast as possible, such that little time is lost before the surgeon can start operating. It is therefore desirable that the patient reaches the BIS = 50 target and remains within the target value without much undershoot or overshoot (i.e. values below BIS = 40 and above BIS = 60 should be avoided). In common practice, the operation procedure does not start until the patient reaches an adequate DOA, usually taking up to 15 min. Thus, a rise time between 5 and 7 min gives good performances.

In Figures 4.14, 4.16, 4.18 and 4.19, we have the simulations of the four controllers for all 12 patients and the nominal one in the induction phase. Figure 4.15 presents the map of the critical regions (CRs; a CR is the region in the space of the parameters where the objective and optimization variables obtained as a function of the varying parameters are valid) for the controller using local linearization (Case 2). And in Figure 4.17, we have the map of the CRs for the controllers designed using exact linearization, by using the inverse of the Hill curve (Cases 3 and 4).

Simulations of some patients show very small oscillations around the steady-state values. The average settling time for EPSAC is approximately 7 min, and for the mp-MPC controllers is approximately 5 min. As mentioned, adequate DOA takes up to 15 min, and a rise time between 5 and 7 min is preferable.

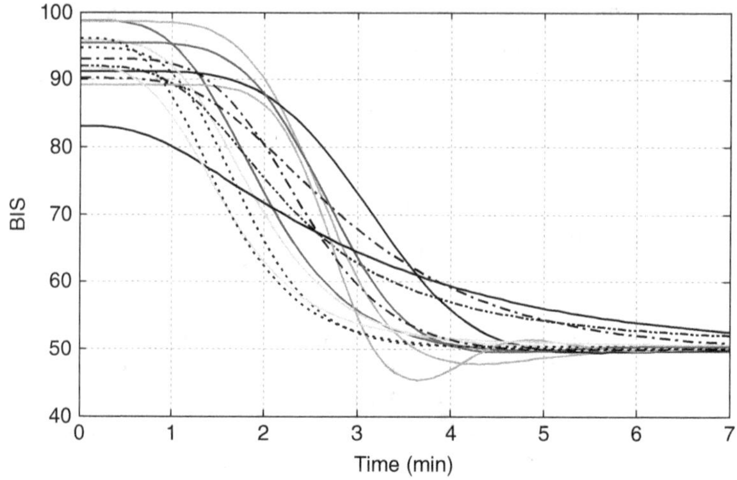

Figure 4.14 BIS output for all 13 patients for Case 1. *Source*: Naşcu *et al.* (2014a). Reproduced with permission of IEEE Transactions on Biomedical Engineering.

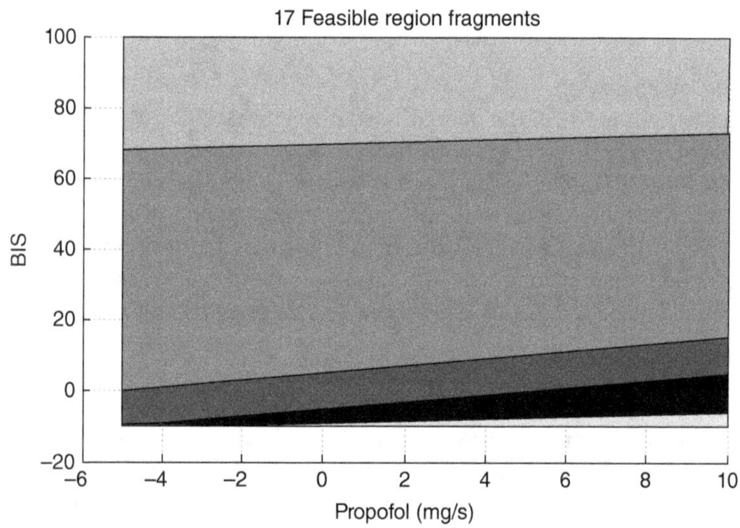

Figure 4.15 Map of critical regions, Case 2. *Source*: Naşcu *et al.* (2014a). Reproduced with permission of IEEE Transactions on Biomedical Engineering.

The best performances are obtained for Case 2. It seems that the local linearization is able to deal with inter- and intra-patient variability. Also, the process was linearized at BIS = 50, which is the value of the controller set point. The EPSAC controller is more influenced by inter-patient variability, and for some patients the settling time has greater values.

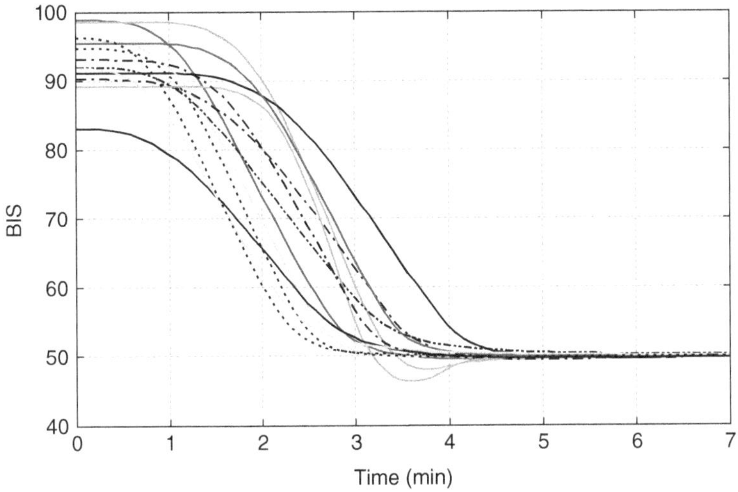

Figure 4.16 BIS output for all 13 patients for Case 2. *Source*: Naşcu *et al.* (2014a). Reproduced with permission of IEEE Transactions on Biomedical Engineering.

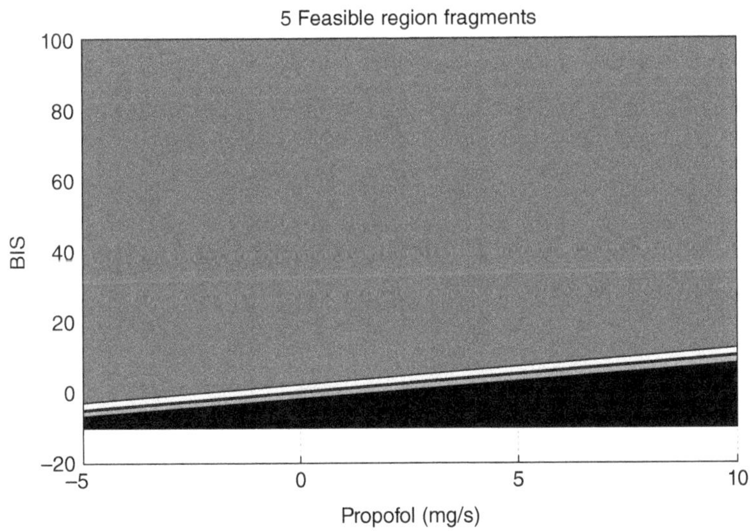

Figure 4.17 Map of critical regions, Case 3 and Case 4. *Source*: Naşcu *et al.* (2014a). Reproduced with permission of IEEE Transactions on Biomedical Engineering.

For the nominal patient PaN, the four controllers (EPSAC and the mp-MPC controllers) are simulated; the results are compared and presented in Figure 4.20. For patient 9, the most sensitive patient, this simulation is presented in Figure 4.22. In Figure 4.21 and Figure 4.23, we have the corresponding propofol infusion rates for the two patients. We can observe

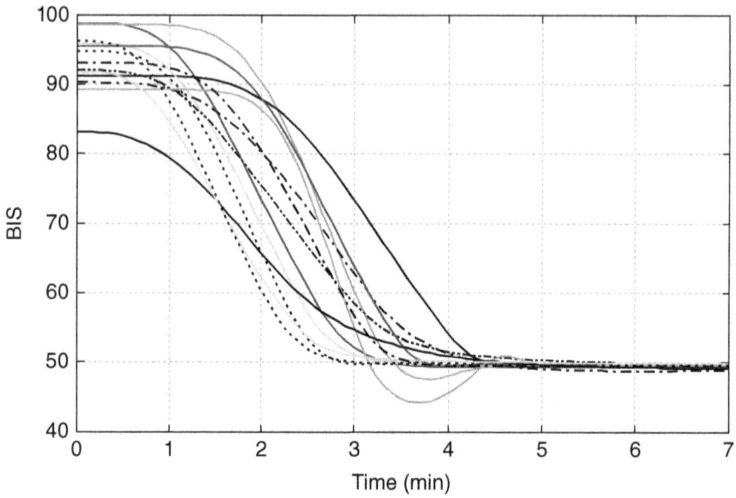

Figure 4.18 BIS output for all 13 patients for Case 3. *Source*: Naşcu *et al.* (2014a). Reproduced with permission of IEEE Transactions on Biomedical Engineering.

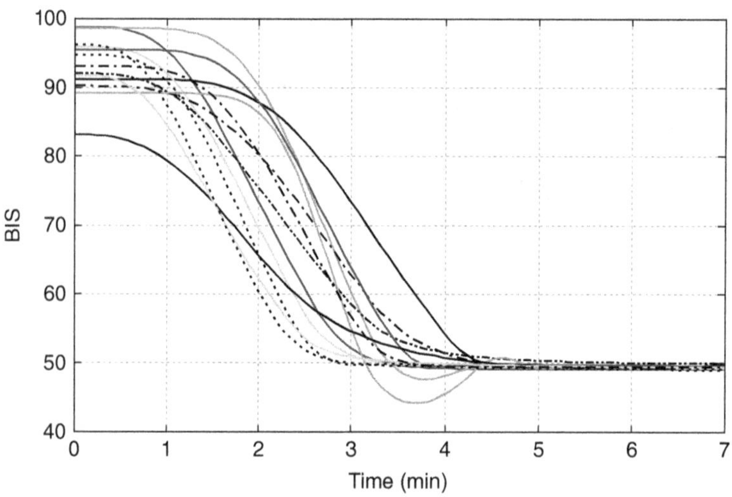

Figure 4.19 BIS output for all 13 patients for Case 4. *Source*: Naşcu *et al.* (2014a). Reproduced with permission of IEEE Transactions on Biomedical Engineering.

that due to the less aggressive behaviour of the EPSAC controller, the output evolution will be smoother. In all cases, the propofol infusion rates are limited to 50 (mg/min). The same conclusions as for Figures 4.20–4.23 are valid here. For both simulated patients, the EPSAC controller has a slower response.

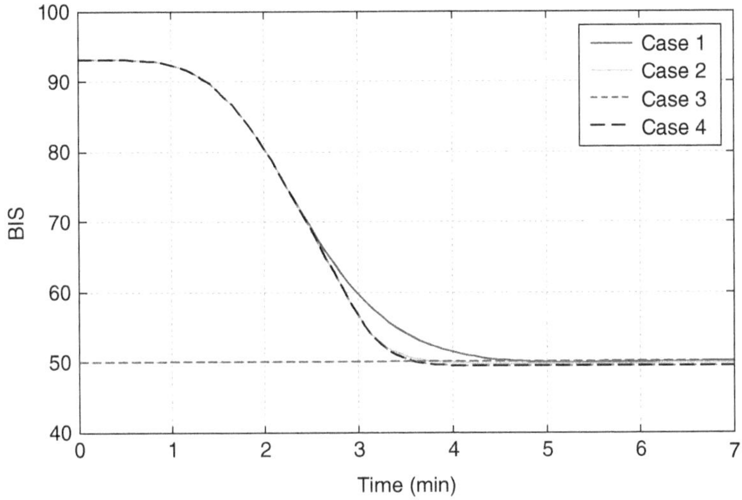

Figure 4.20 BIS response for the four controllers for PaN. *Source*: Naşcu *et al.* (2014a). Reproduced with permission of IEEE Transactions on Biomedical Engineering.

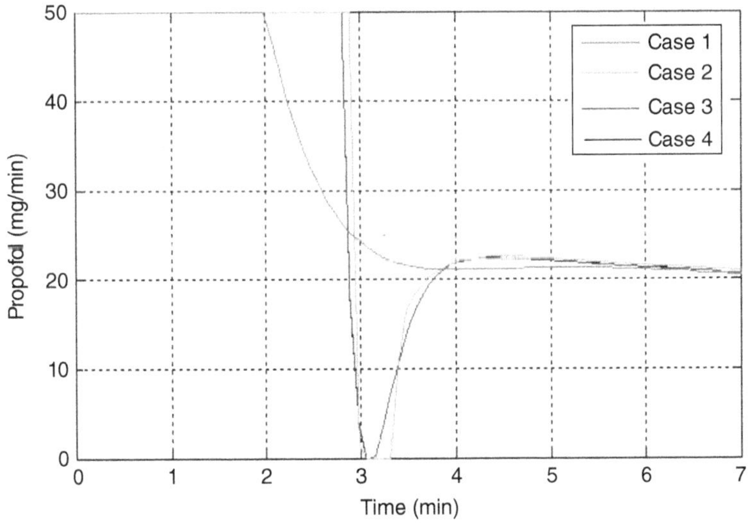

Figure 4.21 Output for the four controllers for PaN. *Source*: Naşcu *et al.* (2014a). Reproduced with permission of IEEE Transactions on Biomedical Engineering.

4.1.6.2 Maintenance Phase

During the maintenance phase, it is important that the controller rejects the disturbances occurring during surgery as fast as possible and bring the patient to the BIS target value. In this phase, typical disturbances can be applied additively to the output of the process to check the controller's ability to reject

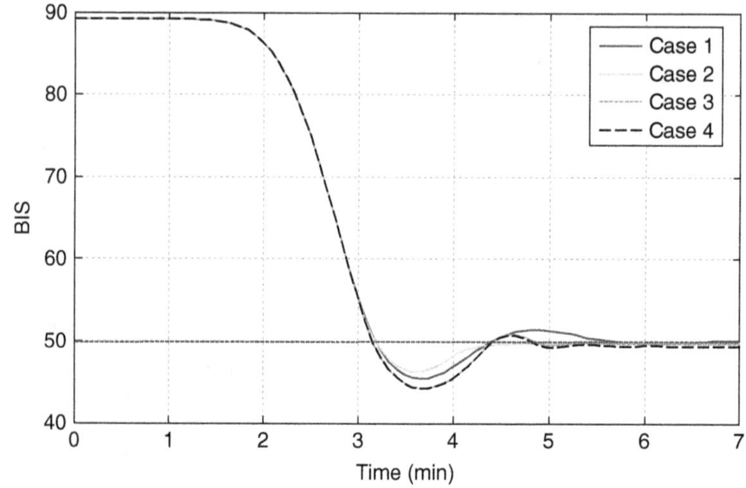

Figure 4.22 BIS response for the four controllers for patient 9. *Source*: Naşcu *et al.* (2014a). Reproduced with permission of IEEE Transactions on Biomedical Engineering.

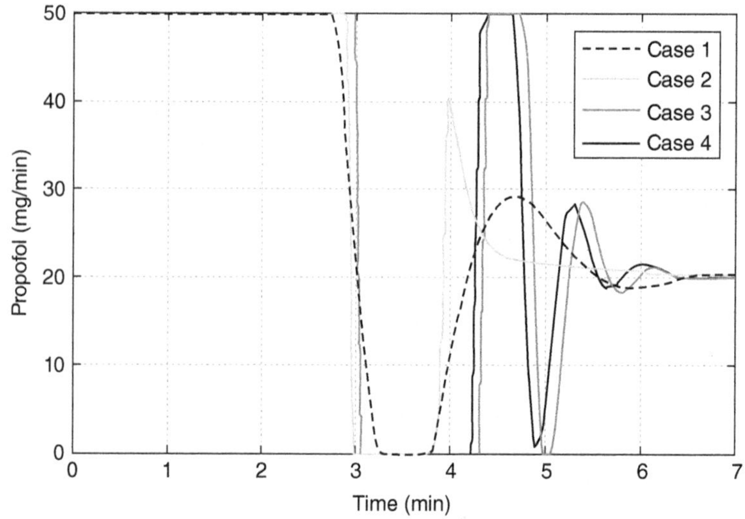

Figure 4.23 Output for the four controllers for patient 9. *Source*: Naşcu *et al.* (2014a). Reproduced with permission of IEEE Transactions on Biomedical Engineering.

them (West *et al.* 2013). A standard stimulus profile is defined and is presented in Figure 4.24. Each interval denotes a specific event in the operation theatre. Stimulus A represents a response to intubation; B is a surgical incision that is followed by a period of no surgical stimulation (i.e. waiting for pathology result); C mimics an abrupt stimulus after a period of low-level

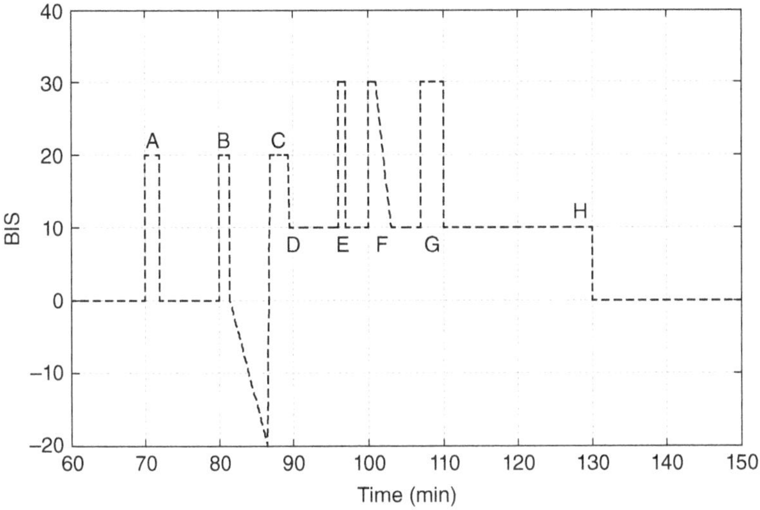

Figure 4.24 The artificially generated disturbance signal. *Source*: Naşcu *et al.* (2014a). Reproduced with permission of IEEE Transactions on Biomedical Engineering.

stimulation; D is the onset of a continuous normal surgical stimulation; E, F and G simulate short-lasting, larger stimuli within the surgical period; and H represents the withdrawal of stimulation during the closing period (Yelneedi *et al.* 2009).

In Figure 4.25 and Figure 4.27, the performance of disturbance rejection of the four controllers for PaN and a more sensitive patient (patient 9) are shown. The figures present the most challenging part of the disturbance rejection test, namely figure sections B–E. In Figure 4.26 and Figure 4.27, we have the corresponding propofol infusion rate for PaN and patient 9, limited between 0 and 50 mg/min. The simulations are performed for the maintenance phase using the disturbance signal (see Figure 4.24) between 60 and 140 min. The simulations show only small differences between the controllers and, thus, comparable performances of all four controllers. For the first control scheme, the behaviour of the controller is less aggressive; the response is slower, but it also has the smallest values of the undershoot.

4.1.6.3 Discussion

The aim of this study is to evaluate the performance of a model-based predictive control algorithm and model predictive multiparametric control for automatic induction and control of DOA during the induction and maintenance phases. In order to implement the control strategies, an accurate model is needed. For many control techniques used in the control of DOA, compartmental models are used to represent the drug distribution in the body for patients undergoing anaesthesia.

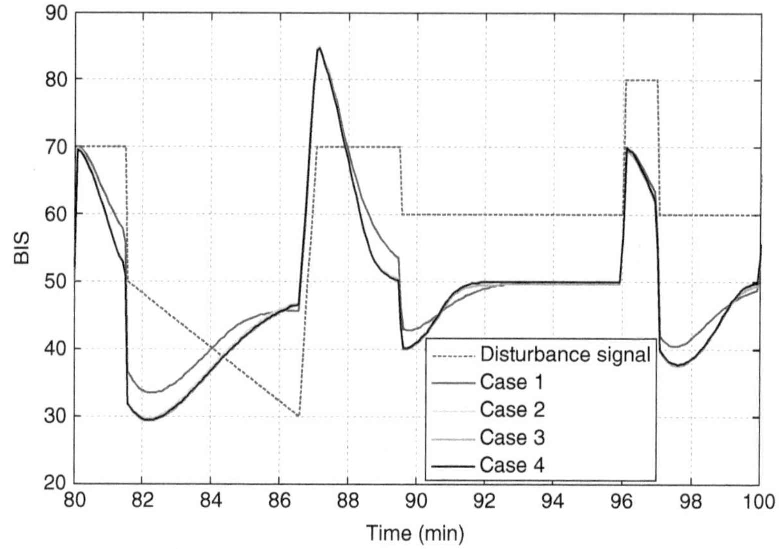

Figure 4.25 BIS response for the four controllers for PaN with disturbance. *Source*: Naşcu *et al.* (2014a). Reproduced with permission of IEEE Transactions on Biomedical Engineering.

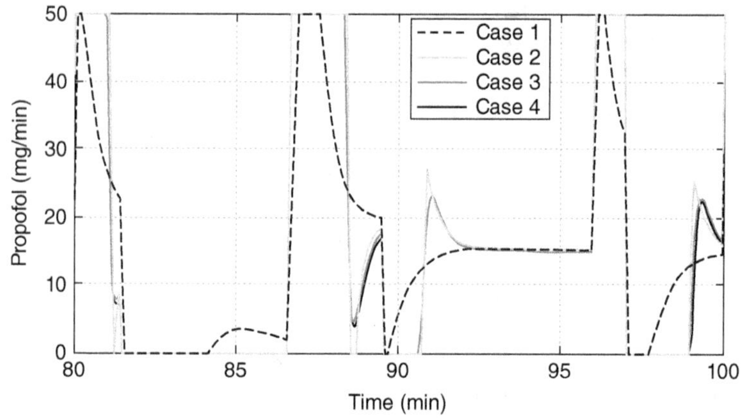

Figure 4.26 Output for the four controllers for PaN with disturbance. *Source*: Naşcu *et al.* (2014a). Reproduced with permission of IEEE Transactions on Biomedical Engineering.

Some of the most important aspects of this application are the high inter- and intra-patient variability, variable time delays, dynamics dependent on the hypnotic agent and model analysis variability. These are just some of the issues that are dealt with when trying to control the DOA.

The hypnotic agent propofol is given as input, and the output is described by the BIS, resulting in a single-input–single-output system (SISO). SISO patient

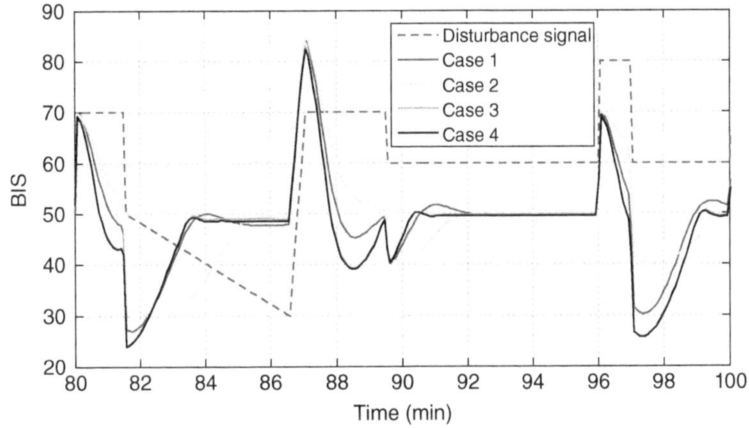

Figure 4.27 BIS response for the four controllers for patient 9 with disturbance. *Source*: Naşcu *et al.* (2014a). Reproduced with permission of IEEE Transactions on Biomedical Engineering.

models for control of most anaesthetic drugs already exist in the literature, and their parameters are estimated based on age, weight, gender and height.

Four different types of controllers are designed and tested. The first controller is based on the online optimization EPSAC MPC technique. The other three controllers are based on the offline optimization mp-MPC: one uses the linearized patient model, and the other two use the compensation of the nonlinear part of the patient model. The difference between the two control strategies using nonlinearity compensation is that for one of them, the states are computed using the nominal patient model, whereas the other one uses an online estimator.

In order to address the issue of inter- and intra-patient variability, each of the four controllers are first tested for the whole set of patients presented in Table 4.2 for the induction and the maintenance phases. The maps of the critical regions for the mp-MPC are presented in Figure 4.15 and Figure 4.17. One can observe that for the controllers using the nonlinearity compensation (exact linearization), there are less critical regions than for the controller using local linearization. This will make the controllers from Cases 3 and 4 easier to implement on embedded devices.

For the induction phase, the aim is to reach the target value as fast as possible and with as little undershoot as possible, trying to avoid BIS values under 40. For Case 1, representing the online EPSAC controller, we have an average settling time of 390 s. The undershoot of the most sensitive patient is 4.6%. As can be observed from Figures 4.14, 4.16, 4.18 and 4.19, representing the BIS response of the mp-MPC controllers, the three cases have very similar settling times, lower than for the EPSAC strategy, an average of 270 s. For the undershoot evaluation, we will consider the worst-case scenario, meaning the most sensitive patient. We obtain for the first controller (Case 2) an undershoot of

3.7%; for Case 3, an undershoot of 5.8%; and, for Case 4, 5.78%. All undershoots are below 10%, which represents the maximum limit. For the induction phase, it can be said that all four controllers perform well, each of them having their own advantages (e.g. lower settling time or smaller undershoot).

The controllers are tested in the maintenance phase in order to see how well they can deal with disturbance rejection. In this phase, it is important for the controller to reject the disturbance as fast as possible and with as little undershoot/overshoot as possible. In Figure 4.25 and Figure 4.27, we can observe the four controllers' response to a disturbance signal that mimics the events that occur in an operation theatre for PaN and for patient 9.

All four controllers are tested against each other for the induction and maintenance phases for two different patients. This will allow a comparison of their performances. The first patient is PaN, and the second patient used for comparison, patient 9, represents the most sensitive patient. It is worth mentioning that the controllers are designed using the values of the nominal patient, which means that for this patient we will have the best behaviour of the controllers. As can be observed from Figures 4.20, 4.21, 4.25 and 4.26, the BIS response and the output for PaN in the induction phase and the maintenance phase, respectively, the three offline controllers have a very similar behaviour. All the controllers present no undershoot and a fast settling time. The EPSAC controller has a less aggressive behaviour, and hence a longer settling time compared to the mp-MPC controllers; but, as can be observed in the maintenance phase, it will have less undershoot. In Figures 4.22, 4.23, 4.27 and 4.28, we have the BIS response and the output for patient 9 in the induction phase and the maintenance phase. This patient represents the worst-case scenario since it is the most sensitive patient. We can observe from the figures that all four controllers

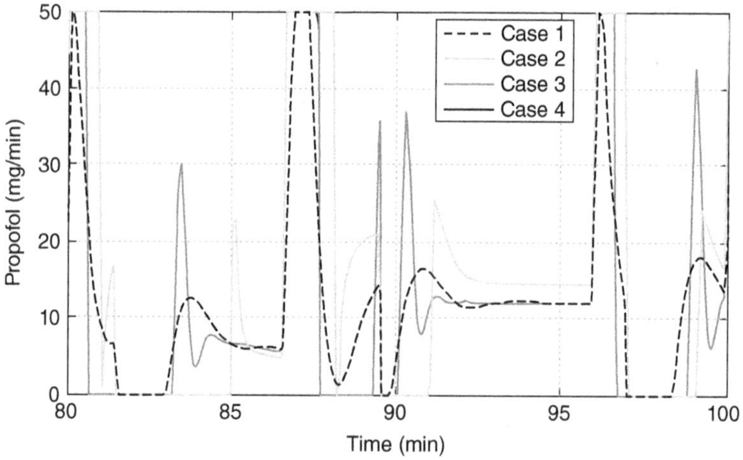

Figure 4.28 Output for the four controllers for patient 9 with disturbance.

have good performances and their responses are very close to each other. However, the controller from Case 2 gives the best performances for this patient in the induction phase, particularly; lower undershoot, 3.7%; and faster settling time, 300 s. This shows that the combination between the linearization method based on gPROMS and optimization methods based on mp-MPC gives good results, even without the nonlinearity compensation.

It is important to state that the mp-MPC controller designed using the linearized patient model is the simplest version of the four controllers, since it doesn't use an estimator and it avoids using the nonlinearity compensation which introduces additional complexity in the DOA control. Moreover, it obtains the best performances, which can be explained through the fact that the nonlinearity of the Hill curve is more intense at extreme values of the BIS and weaker around the BIS value of 50, where the model was linearized and where the BIS target is set. If the induction phase and the maintenance phase are kept around the value of 50%, Case 2 will give very good performances. But if the disturbances take the process out of the 50% area, we can observe that the performances are not as good as in the case of nonlinearity compensation. Also, this is the reason why during the maintenance phase, the controller from Case 2 does not provide good performances if the disturbances are substantial. Due to the Hill nonlinearity, the real patient model has smaller gains at the extreme values of the control variable. In the case of substantial disturbances, the control variable goes to the extreme values, and the controller has a slower response but also a lower undershoot/overshoot.

Using nonlinearity compensation is a good alternative in this case. Moreover, the computations required for the nonlinearity compensation are rather straightforward (the inverse of the Hill curve), and there are no recursive computations that might lead to accumulation of errors.

The estimator used for the mp-MPC with nonlinearity compensation can also be applied for the mp-MPC using local linearization. It was not used for this study because, as can be observed from the simulations, a case with nonlinearity compensation is more meaningful in the presence of disturbance.

The aim of the studies on control of anaesthesia is to be able to implement the controllers on embedded devices. These types of devices do not have the same computational power as the computers where simulations are performed in real time. This would make classical MPC more difficult to implement since matrix operations are harder to program on embedded devices. The mp programming algorithms derive the explicit mapping of the optimal control actions as a function of the current states, resulting in the implementation of a simple lookup table and simple function evaluations. This makes the mp-MPC controllers much easier to implement for the control of DOA.

For each patient, there will be a variable dose–response relationship. For the same reference value, the controller sends different drug rates, and the blood and effect-site concentration levels are different for each patient. The safety limits for propofol blood concentration and effect-site concentration are fulfilled by maintaining the drug infusion rate below 50 mg/min. It can be observed from Figures 4.21, 4.23, 4.26 and 4.28 that the drug infusion rates are maintained below this limit.

Note that the robustness of the performance is analysed by having the controllers designed on a nominal model (Ionescu *et al.* 2008) and then tested on a wide set of patient model parameters, where the impacts of parameter uncertainties were analysed. Formal robust criteria can also be included (Kouramas *et al.* 2011), and this represents a topic of our ongoing research.

4.2 Simultaneous Estimation and Advanced Control

4.2.1 Introduction

In this chapter, we have designed and studied different controllers for the regulation of DOA during the induction and maintenance phases. The implementation of these controllers is based on the assumption that the state values are readily available from the system measurements and that we have a clear measurable output with not much noise influence. However, in reality, the measured output may be noisy and the system measurements may not produce this information directly – instead, the state information needs to be inferred from the available output measurements. This can be done by developing state estimators.

The use of estimation techniques will enable researchers to: (a) estimate the state of each individual patient, and adjust them based on his or her corresponding dynamics; (2) overcome the noisy output measurements; and (c) deal with the system constraints (in conjunction with an MPC structure).

While in Section 4.1.5.4 for Case 4, a Kalman filter was implemented simultaneously with mp-MPC, it was observed that the performance of the controller was not significantly improved due to its limitations, such as handling system constraints and noise. As a result, in this section, the online and offline moving horizon estimator (MHE) will be investigated and implemented simultaneously with mp-MPC. For comparison purposes, a Kalman filter also will be tested. All resulting controllers with the corresponding estimation techniques are tested for both the induction and maintenance phases for a set of 12 patients' data.

4.2.2 Multiparametric Moving Horizon Estimation (mp-MHE)

The idea of MHE is to estimate the state using a moving and fixed-size window of data. Once a new measurement becomes available, the oldest measurement is discarded and the new measurement is added. The concept is to penalize deviations between measurement data and predicted outputs. In addition – for theoretical reasons – a regularization term on the initial state estimate is added to the objective function. There are two main characteristics that distinguish MHE from other estimation strategies, such as the Kalman filter: (a) prior information in the form of constraints on the states, disturbances and parameters can be included; and (b) since MHE is optimization based, it is able to handle explicitly nonlinear system dynamics through the use of approximate nonlinear optimization algorithms. Here, MHE is first formulated as a multiparametric problem and then combined with an explicit mp-MPC strategy.

Based on the MHE theory presented in Chapter 2, the resulting multiparametric MHE formulation can be obtained:

$$\left[\hat{x}_{T-N+1},\{\hat{w}\}_{T-N+1}^{T-1}\right]=\min_{x_{T-N+1},\{w\}_{T-N+1}^{T-N}} \sum_{k=T-N+1}^{T-N}\left(w_k-\overline{w}\right)^T\cdot Q^{-1}\cdot\left(w_k-\overline{w}\right)+\sum_{k=T-N+1}^{T}\left(v_k^T\cdot R^{-1}\cdot v_k\right)$$
$$+\left(x_{T-N+1}-\overline{x}_{T-N+1/T-N}\right)^T\cdot P_{SS}^{-1}\cdot\left(x_{T-N+1}-\overline{x}_{T-N+1/T-N}\right) \tag{4.10}$$

subject to:

$$x_{k+1}=A\cdot x_k+B\cdot u_k+G\cdot w_k \qquad \text{discrete state – space formulation}$$
$$y_k=C\cdot x_k+v_k \qquad \rightarrow v_k=y_k-C\cdot x_k$$
$$BIS_{\min}\le y\le BIS_{\max}$$
$$\Delta u_{\min}\le\Delta u\le\Delta u_{\max}$$
$$\begin{bmatrix}1&0&0&0\\-1&0&0&0\\0&1&0&0\\0&-1&0&0\\0&0&1&0\\0&0&-1&0\\0&0&0&1\\0&0&0&-1\end{bmatrix}\cdot\begin{bmatrix}x_1\\x_2\\x_3\\x_e\end{bmatrix}\le\begin{bmatrix}100\\0\\100\\0\\100\\0\\100\\0\end{bmatrix} \qquad \text{path constraints on state variables}$$
$$\begin{bmatrix}1\\-1\end{bmatrix}\cdot w\le\begin{bmatrix}1\\-1\end{bmatrix} \qquad \text{path constraints on the noise w}$$
$$\begin{bmatrix}1\\-1\end{bmatrix}\cdot v\le\begin{bmatrix}1\\-1\end{bmatrix} \qquad \text{path constraints on the noise v}$$
$$\overline{x}_{T-N+1/T-N}=A\cdot\check{x}_{T-N/T-N}+B\cdot u_k+G\cdot\overline{w} \qquad \text{update of the cost to arrive}$$

where N is the length of the horizon; T is the current point in time; Q and R are positive definite diagonal weighting matrices on the noises; P_{SS} is the steady-state solution for the Kalman filter; $\{\hat{w}\}_{T-N+1}^{T-N}$ and $\{v\}_{T-N+1}^{T}$ are sequences of independent, normally distributed random numbers with mean values $\overline{w}$ for $\{w\}$ and zero-mean for $\{v\}$; $\overline{x}_{T-N+1/T-N}$ is the arrival cost, which captures the previous measurements that are not considered anymore; and $\hat{x}_{T-N/T-N}$ is the solution of the MHE at the previous time step.

4.2.3 Simultaneous Estimation and mp-MPC Strategy

In this section, developed estimation strategies are implemented simultaneously with mp-MPC for the control of IV anaesthesia. The proposed control design scheme for simultaneous estimation and mp-MPC is presented in Figure 4.29.

The *patient* is simulated using the mathematical model of the patient composed of the PK part (linear) and the PD part, including the Hill curve (nonlinear) as described in Section 4.1.2. The inverse Hill function block, designed based on the nominal patient model, is used to compensate for the nonlinearity introduced by the Hill curve in the PD model. This block uses the measured BIS and provides the corresponding $\hat{C}_e$ to the *mp-MPC block*. The BIS target can be set by the user, but for general anaesthesia it is set to the value of 50. Since the mp-MPC block uses the corresponding $\hat{C}_e$ of the measured BIS, the BIS target will be transformed in C_e target by using the *inverse of the Hill curve block*. The *estimator block* is used to estimate the state of each individual patient. The commonly used technique to generate the states for the controller in the absence of the real state measurements is to compute them by using the control action for the simulated patient and the nominal state-space model. By estimating the states of every simulated patient, the controller (instead of using the states computed using the nominal patient model) will use the estimated values of states corresponding to each individual patient. Finally, the mp-MPC block, using the individual patient states given by the estimator block, calculates the error between the $\hat{C}_e$ corresponding to the measured BIS from the patient and the target C_e, and provides the optimal *drug rate u* to the patient block in order to drive it to the desired target value.

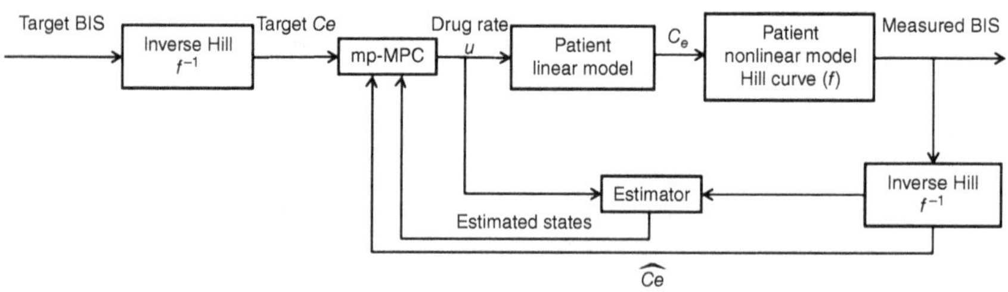

Figure 4.29 Schematic of simultaneous mp-MHE and mp-MPC for intravenous anaesthesia. *Source*: Naşcu and Pistikopoulos (2017). Reproduced with permission of Elsevier.

The Hill curve (4.5) introduces nonlinearities in the system which complicate the use of linear MPC controllers' synthesis. To compensate for the nonlinearity, a parameter scheduling technique, as presented in Section 4.1.4, the inverse of the Hill curve (4.6), is implemented in the controller with the nominal patient model parameters as shown in Figure 4.29; *f* is using the nonlinearity parameter of the real patient (E_0, E_{max}, EC_{50}, γ), while f^{-1} is using the parameter assumed by the controller (the nominal patient nonlinear parameters known *a priori*: ${E_0}^{mean}$, ${E_{max}}^{mean}$, ${EC_{50}}^{mean}$, γ^{mean}). The controller then aims at controlling the estimated drug concentration $\hat{C}_e$ using a linear controller.

Using the explicit/multiparametric MPC formulation described in Chapter 2, the control strategy is based on the nonlinearity compensation and the state-space model of the PK–PD linear part for the nominal patient model. The following mp-QP optimization problem, Equation (4.11), is solved to obtain the control laws using the POP toolbox (Pistikopoulos *et al.* 1999) and determine the controller.

$$
\begin{aligned}
\min_u J = {} & \hat{x}_N^T P \hat{x}_N + \sum_{k=1}^{N-1} x_k Q_k x_k + \sum_{k=1}^{N-1} \left(y_k - y_k^R\right)^T QR_k \left(y_k - y_k^R\right) \\
& + \sum_{k=0}^{N_u-1} \left(u_k - u^R\right)^T R_k \left(u_k - u^R\right) + \sum_{k=0}^{N_u-1} \Delta u_k R1_k \Delta u_k \\
s.t. \quad & x_{t+1} = Ax_t + Bu_t + Gw_t \\
& y_t = Cx_t + v_t \\
& BIS_{\min} \le y \le BIS_{\max} \\
& u_{\min} \le u \le u_{\max} \\
& w_{\min} \le w \le w_{\max} \\
& v_{\min} \le v \le v_{\max} \\
& x_t \in X \subseteq \Re^p, u_t \in U \subseteq \Re^s, w_t \in W, v_t \in V
\end{aligned}
\tag{4.11}
$$

where $\hat{x}$ are the estimated states given by the state estimator; y = outputs; u = controls; w = process disturbances; and v is the measurement noise of all (discrete) time-dependent vectors. The subsets of output variables that get tracked have time-dependent set points y^R. Finally, Δu are changes in control variables, $\Delta u(k) = u(k) - u(k-1)$. The prediction horizon is denoted by N, and the control horizon by N_u. X, U are the sets of the state and input constraints that contain the origin in their interior. Both $Q \succ 0$, the objective coefficient for the states, and $P \succ 0$, the terminal weight matrix for the states, are positive definite diagonal matrices. The weight matrix for manipulated variables $R \succ 0$ is a positive definite diagonal matrix, QR is the weight matrix for tracked outputs and $R1$ is a weight matrix for the control action changes (Δu).

For the design of the controller, the following tuning parameters are used: the objective coefficients for states (x), $Q = 0$ when we have no state estimation and

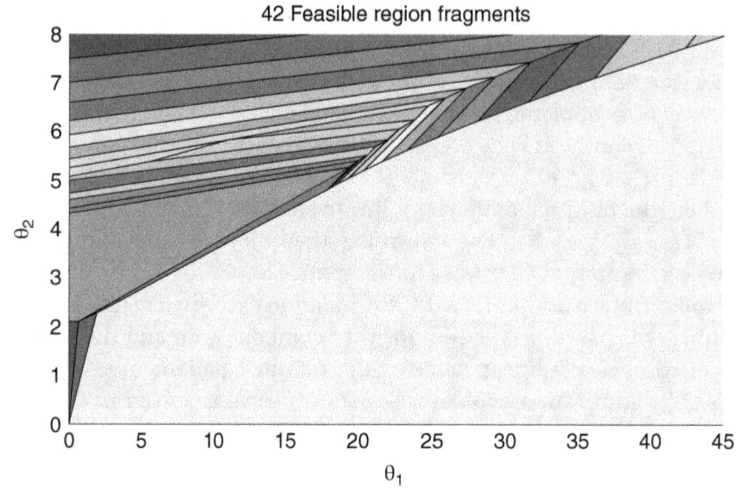

Figure 4.30 Map of critical regions – mp-MPC. *Source*: Naşcu and Pistikopoulos (2017). Reproduced with permission of Elsevier.

Q = 1 in the case with state estimation; the weight matrix for tracked outputs (y), $QR = 1000$; the weight matrix for manipulated variables (u), $R = 1$; the control horizon, $N_u = 1$; and the prediction horizon, $N = 20$. The states used in the design of the controllers are C_1, C_2, C_3, C_e as described in Equations (4.1) and (4.4). The clinically recommended sampling time is 5 s (Ionescu *et al.* 2008). N and N_u are chosen based on the characteristics of the process and the desired performances. Based on Clarke *et al.* (1987) and Mohtadi *et al.* (1987), N should be large, at least $2n - 1$ but no larger than the rise time of the process. For anaesthesia, due to medical procedures, we are constrained to use a small sampling time that will lead to a choice of a greater value for N. In choosing N_u, for processes with no unstable/underdamped poles, like anaesthesia, $Nu = 1$ is generally satisfactory. A choice of the Q, R and QR is given by Bryson's rule (Franklin *et al.* 2001).

Figure 4.30 presents a typical solution of the multiparametric programming problem in the form of two-dimensional projection of the critical space. The parametric vector θ consists of: the estimated states, the current time output and the output reference ($BIS = 50$). Here θ_1 and θ_2 represent the concentration of the effect-site compartment C_e and BIS.

4.2.4 Results

In this section, the results of a simulation study to evaluate the three controllers for the administration of propofol are presented. The three controllers are: (a) the nominal controller that uses no state estimation, (b) the simultaneous mp-MPC and Kalman filter and (c) the simultaneous mp-MPC and mp-MHE.

DOA is monitored using BIS during the induction and maintenance phases of general anaesthesia. The closed-loop control sets are performed on a set of 12 patients plus an extra patient (Table 4.2) representing the nominal values of all 12 patients. The designed controllers are tested both under the assumption that the output of the system is not influenced by noise and with the output corrupted by noise and disturbances.

The simulations are performed first for the set of data presented in Table 4.2, to have a better understanding of their behaviour on different types of patients, and analyse the inter- and intra-patient variability. Next, the three controllers are tested against each other and simulated for different patients to compare their performances by means of BIS and the corresponding propofol infusion rates.

4.2.4.1 Induction Phase

Ideally, the induction of the patient is performed as fast as possible, such that little time is lost before the surgeon can start operating. It is therefore desirable that the patient reaches the $BIS = 50$ target and remains within the target value without much undershoot or overshoot, (i.e. values below BIS = 40 and over BIS = 60 should be avoided).

Figures 4.31–4.36 present the simulations of the nominal controller and the two simultaneous mp-MPC and estimation in the induction phase. Simulations of some patients show very small oscillations around the steady-state values. The average settling time for the designed controller is: (a) 280 s for the nominal controller, (b) 240 s for the simultaneous mp-MPC and Kalman filter and (c) 225 s for the simultaneous mp-MPC and mp-MHE. The best performance is obtained for the simultaneous mp-MPC and mp-MHE. The mp-MHE is able to deal better with the inter- and intra-patient variability.

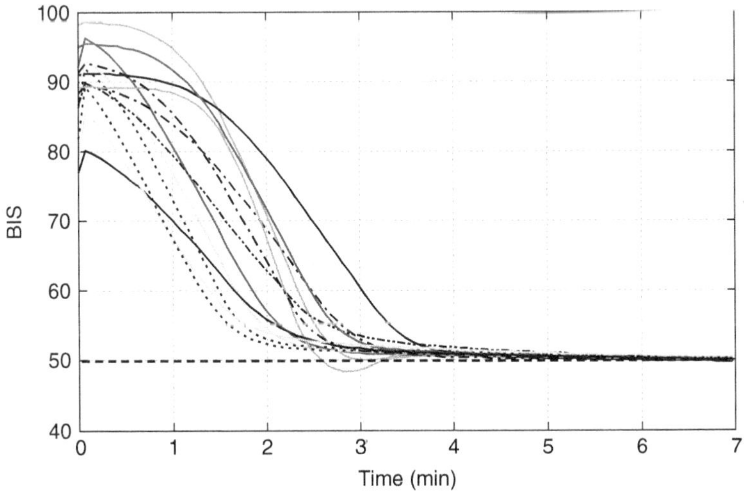

Figure 4.31 BIS response for all 13 patients in the induction phase – nominal mp-MPC.

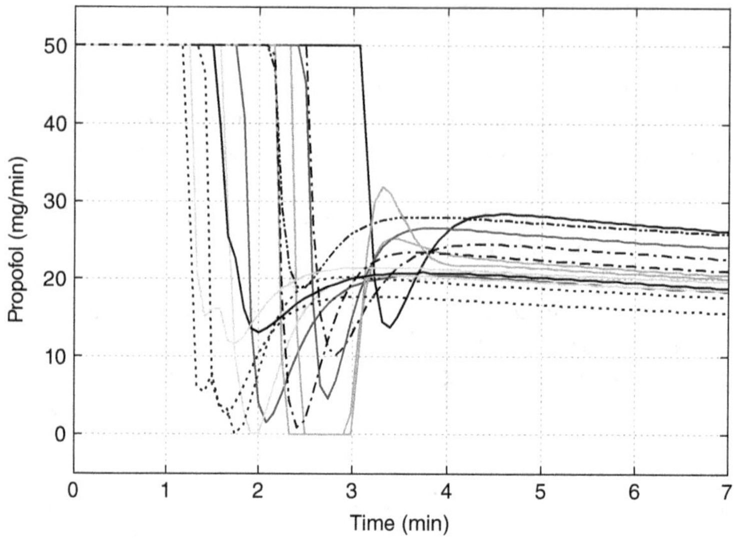

Figure 4.32 Propofol infusion rate for all 13 patients in the induction phase – nominal mp-MPC.

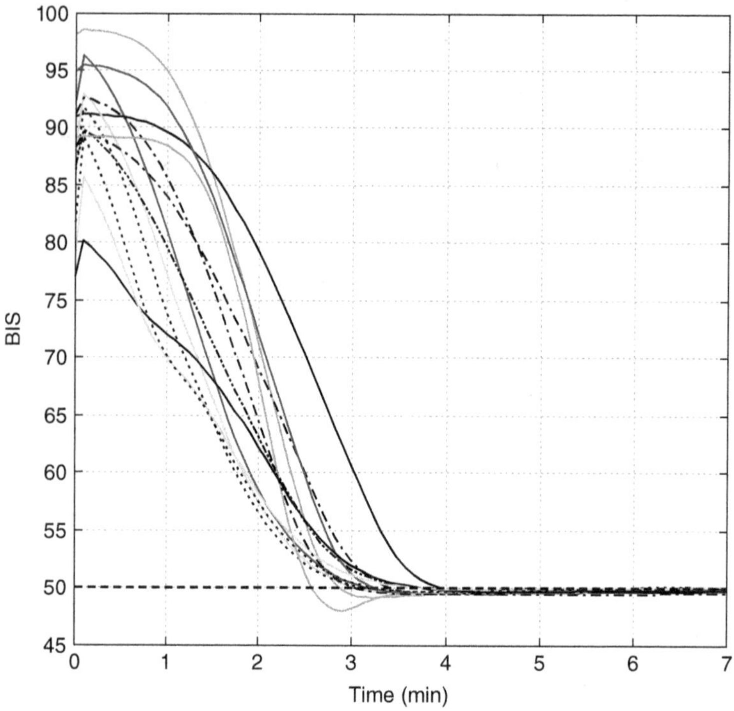

Figure 4.33 BIS response for all 13 patients in the induction phase – simultaneous mp-MPC and Kalman filter.

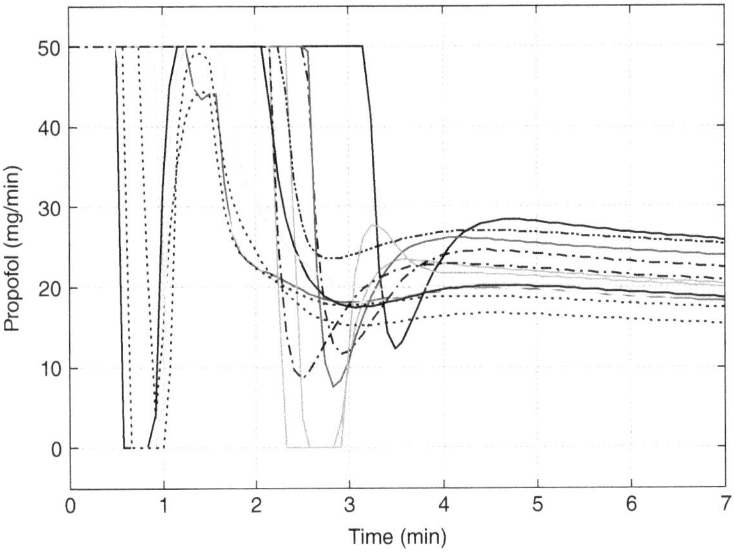

Figure 4.34 Propofol infusion rate for all 13 patients in the induction phase – simultaneous mp-MPC and Kalman filter.

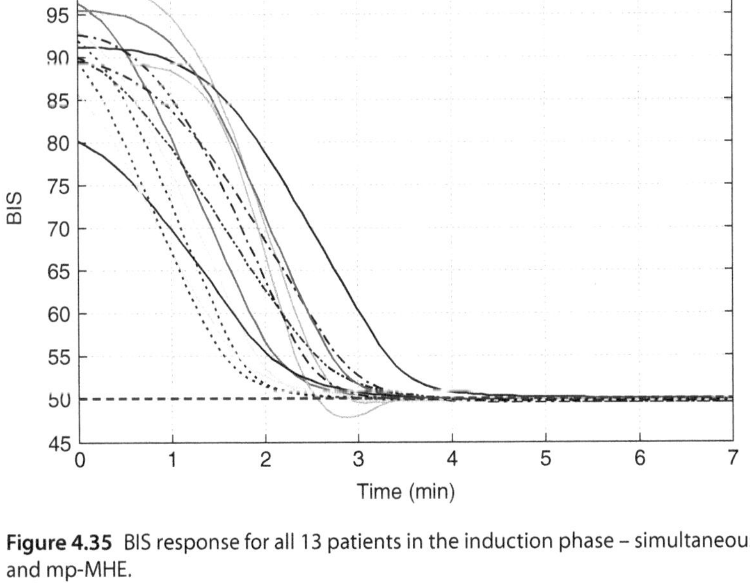

Figure 4.35 BIS response for all 13 patients in the induction phase – simultaneous mp-MPC and mp-MHE.

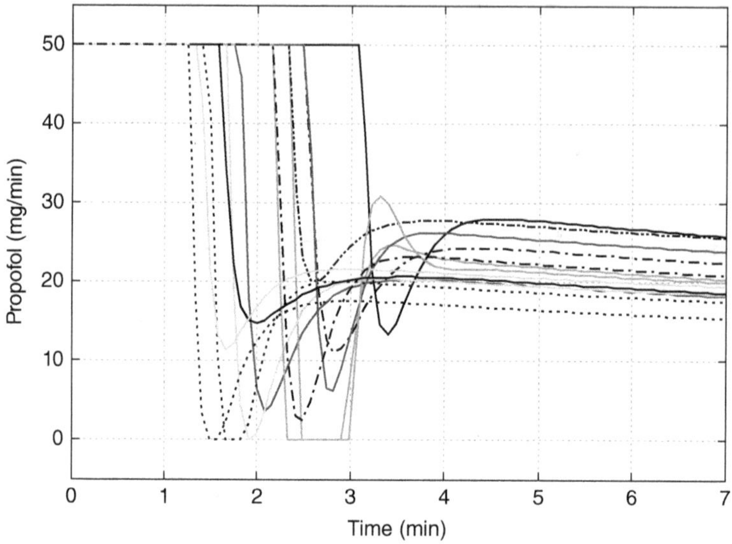

Figure 4.36 Propofol infusion rate for all 13 patients in the induction phase – simultaneous mp-MHE and mp-MPC.

The designed controllers are tested for the most sensitive patient, patient 9. Figure 4.37 and Figure 4.38 represent the BIS response of the patients, while Figure 4.39 shows the corresponding control action (drug infusion). It can be observed that all three controllers exhibit similar performances with a settling time of 200 s. Note that for all simulations, we have: (a) with blue, the mp-MPC without state estimation; (b) with black, the simultaneous mp-MPC and Kalman filter; (c) with magenta, the simultaneous mp-MPC and mp-MHE; and (d) with red, the set point for the BIS index. From Figure 4.38, where we zoom in on the BIS response of the three controllers, we can observe that the undershoot of the most sensitive patient is 2.2%. For the undershoot, the worst case is considered to mean the most sensitive patient (patient 9). Also, the simultaneous Kalman filter and mp-MPC as well as mp-MHE and mp-MPC have similar performances for this case, where we have no noise corrupting the output. However, the simultaneous mp-MHE and mp-MPC combination has slightly better performance for both patients, since it is able to give better estimates of the patient states.

4.2.4.2 Maintenance Phase

During the maintenance phase, it is important that the controller rejects the disturbances occurring during surgery as fast as possible and brings the patient to the BIS target value. In this phase, typical disturbances can be applied additively to the output of the process to check the controller's ability to reject them (West *et al.* 2013). A standard stimulus profile is defined and presented in Figure 4.24. Each interval denotes a specific event in the operation theatre, as presented in Section 4.1.5.2.

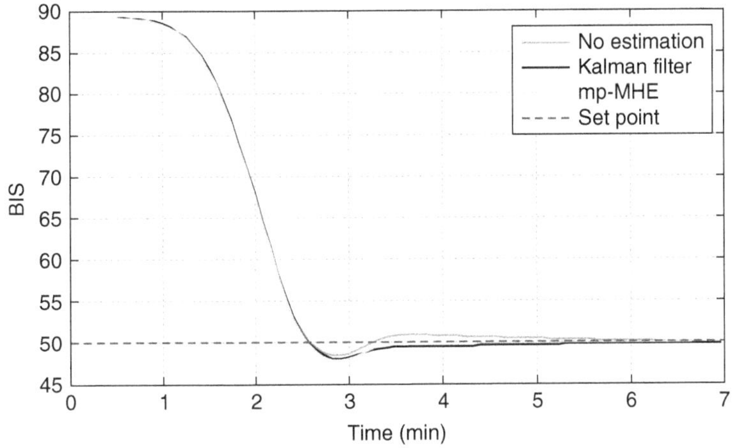

Figure 4.37 BIS response of the three controllers for patient 9 in the induction phase without noise. *Source*: Naşcu and Pistikopoulos (2017). Reproduced with permission of Elsevier.

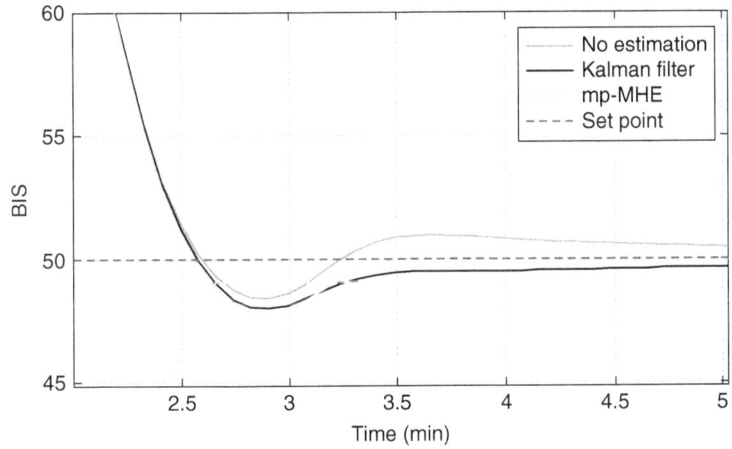

Figure 4.38 BIS response of the three controllers for patient 9 in the induction phase without noise – zoomed in. *Source*: Naşcu and Pistikopoulos (2017). Reproduced with permission of Elsevier.

In Figure 4.40 and Figure 4.41, the performance of the disturbance rejection for the most sensitive patient (patient 9) is shown. Figure 4.42 presents the BIS response for the most challenging part of the disturbance rejection test, namely parts B–E, while in Figure 4.43 we have the corresponding propofol infusion rate for patient 9. The simulations are performed for the maintenance phase using the disturbance signal (see Figure 4.24). The simulations show only small differences between the controllers. The controller using mp-MHE has a less aggressive behaviour and thus a smaller value for the undershoot.

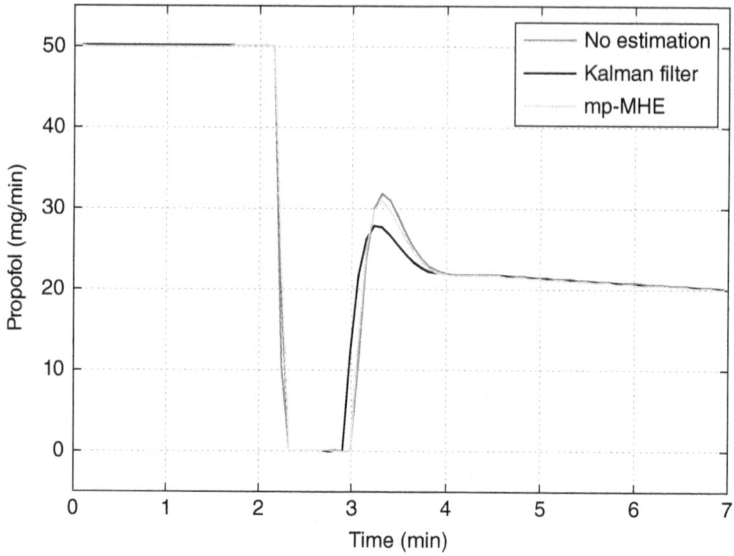

Figure 4.39 Propofol infusion rate of the three controllers for patient 9 in the induction phase without noise. *Source*: Naşcu and Pistikopoulos (2017). Reproduced with permission of Elsevier.

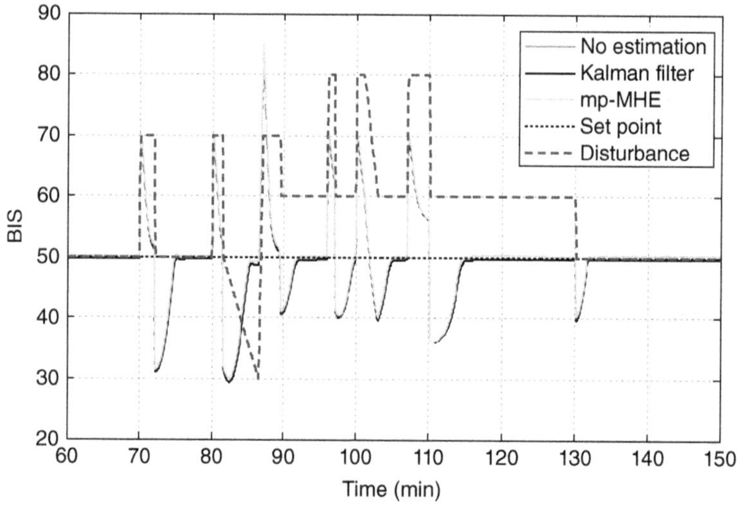

Figure 4.40 BIS response of the three controllers for patient 9 in the maintenance phase without noise.

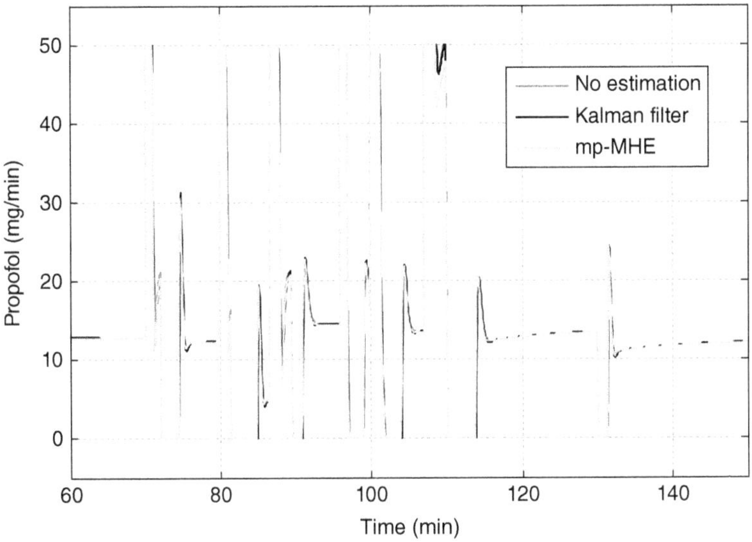

Figure 4.41 Propofol infusion rate of the three controllers for patient 9 in the maintenance phase without noise.

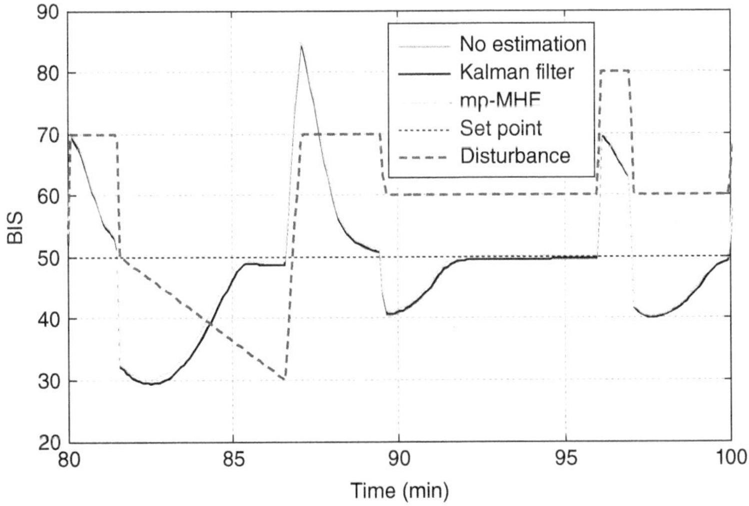

Figure 4.42 BIS response of the three controllers for patient 9 in the maintenance phase – the B–C–D–E interval – without noise.

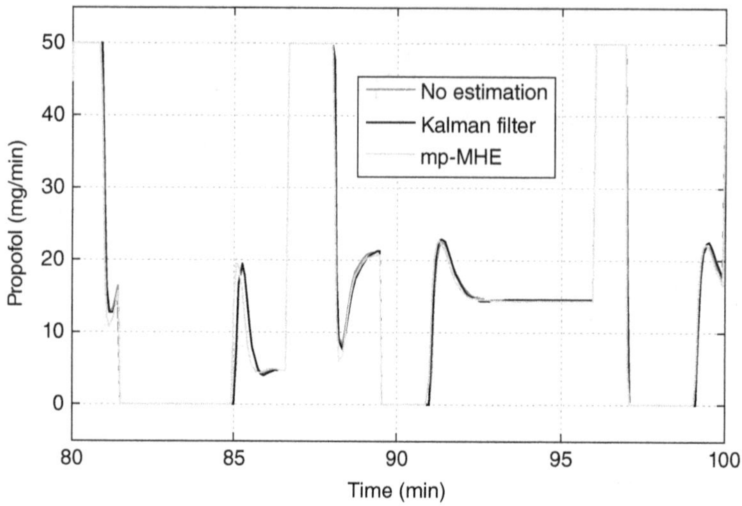

Figure 4.43 Propofol infusion rate of the three controllers for patient 9 in the maintenance phase – the B–C–D–E interval – without noise.

4.3 Hybrid Model Predictive Control Strategies

4.3.1 Introduction

In most drug delivery systems, such as controlling the DOA, the nonlinearities are typically present in the PD model of the system and are described by the Hill curve representing the relation between the concentration of the drug and the effect observed on the patient. For the case of infusion of anaesthetic agents, the nonlinear Hill curve approximation has been used in both volatile (Krieger *et al.* 2014) and IV (Naşcu *et al.* 2012, 2014a) anaesthesia.

Advanced control strategies using either hybrid and robust multiparametric MPC or simultaneous hybrid multiparametric MPC and state estimation techniques are developed and tested. Here, we first generate a piecewise linearization of the Hill curve. The main advantage of this procedure is that the parameter space is linearized and the uncertainty in some key parameters of the Hill curve is compensated for. As a result of the linearization, the anaesthesia model is described by a piecewise affine system. This will lead to a hybrid model predictive control (hMPC) problem formulation (Bemporad and Morari 1999) and thus a mixed-integer quadratic programming (MIQP) problem formulation. However, the online implementation of hMPC involves the online solution of the MIQP problem, which introduces a high computational burden. To overcome this, the hMPC problem is solved explicitly offline via the solution of a state-of-the-art multiparametric mixed-integer quadratic programming (mp-MIQP) problem (Dua *et al.* 2002; Oberdieck and Pistikopoulos 2015).

Another important challenge in the control of DOA that is addressed in this chapter is the high inter- and intra-patient variability, which introduces a high degree of uncertainty in the system. A number of robust control strategies and a state estimation technique are developed and presented simultaneously with the multiparametric hybrid model predictive control (mp-hMPC) problem. State estimation is used for the unavailable states and to overcome issues that arise from noisy outputs. In particular, MHEs implemented in a multiparametric fashion (Darby and Nikolaou 2007; Voelker *et al.* 2013; Naşcu *et al.* 2014b) are used simultaneously with the hMPC control. The control strategies are tested on a set of 12 patients for the induction and maintenance phases of general anaesthesia.

4.3.2 Hybrid Patient Model Formulation

A feature of the PD model for DOA control is the presence of nonlinearities corresponding to the Hill curve. Due to its S-shaped profile, a piecewise linearization of the Hill curve divides BIS into three partitions, where each partition i is associated with a different linear function $BIS = C_i C_e + e_i$. The resulting piecewise affine formulation is shown in Table 4.3, where the parameters describing the PK model can be found in Table 4.2.

Table 4.3 Hybrid model for intravenous anaesthesia.

	Intravenous anaesthesia
PK model	$\dot{x}_1(t) = -[k_{10} + k_{12} + k_{13}] \cdot x_1(t) + k_{21} \cdot x_2(t) + k_{31} \cdot x_3(t) + u(t)/V_1$
	$\dot{x}_2(t) = k_{12} \cdot x_1(t) - k_{21} \cdot x_2(t)$
	$\dot{x}_3(t) = k_{13} \cdot x_1(t) - k_{31} \cdot x_3(t)$
Effect-site compartment	$\dot{C}_e(t) = k_{e0} \cdot (C_e(t) - C_p(t))$
PD model (Hill curve)	$BIS = C_i C_e + e_i$
	$0 \le C_{e1} \le \lambda_1 \delta_1$
	$\lambda_1 \delta_2 \le C_{e2} \le \lambda_2 \delta_2$
	$\lambda_2 \delta_3 \le C_{e3} \le \lambda_3 \delta_3$
	$C_e = C_{e1} + C_{e2} + C_{e3}$
	$\delta_1 + \delta_2 + \delta_3 = 1$
	$C_i = f[E_0, E_{\max}, EC_{50}, \gamma]$

$\delta \in [0,1]^3$; $C_i, e_i, i \in \{1,2,3\}$ C_i are found using the first-order Taylor expansions at the points for the linearization; $\sum_i \delta_i = 1$.

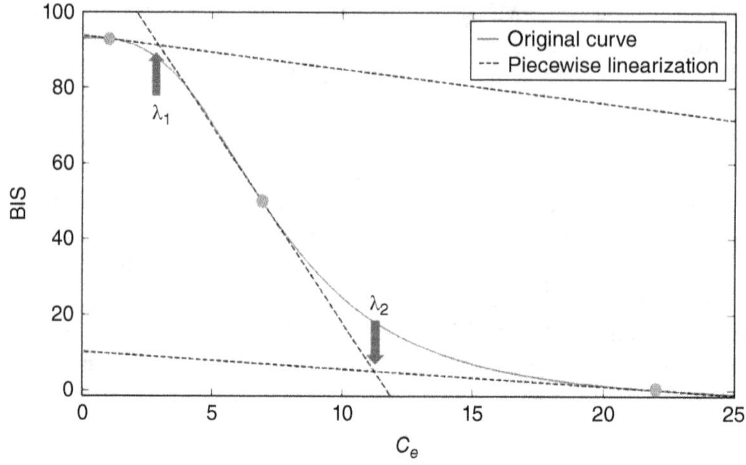

Figure 4.44 The original Hill curve and a piecewise linearized version. The red dots denote the points around which the linearization was performed, while the purple arrows show the switching points $\boldsymbol{\lambda}_1$ and $\boldsymbol{\lambda}_2$, respectively. *Source*: Naşcu *et al.* (2017). Reproduced with permission of Elsevier.

The binary variables δ thereby denote whether a certain partition is active. As a result, this system belongs to the class of hybrid systems (i.e. systems which are described by continuous as well as discrete dynamics and/or logical constraints).

Note that $\sum_i \delta_i = 1$ holds, as only one linearization is active for every drug concentration C_e, and the choice of which linearization is active is described via the switching points λ_1 and λ_2 (see Figure 4.44).

Systems which can be described via the equations presented in Table 4.3 are part of the mixed-logical dynamical (MLD) systems, which are a well-studied class of systems (Bemporad and Morari 1999; Heemels *et al.* 2001). Their basic principle is that, in addition to the commonly encountered continuous parts, discrete elements are present in the problem formulation as inputs, states, variables or outputs. Additional information can be found in Chapter 2.

4.3.3 Control Design

4.3.3.1 Hybrid Formulation of the Control Problem: Intravenous Anaesthesia

Based on the piecewise affine formulation presented in Table 4.3, the following hybrid explicit MPC can be obtained (Bemporad and Morari 1999):

$$
\begin{aligned}
&\min_{u} J=\sum_{k=1}^{N}\left(BIS_k-BIS_k^R\right)^T QR_k\left(BIS_k-BIS_k^R\right)+\sum_{k=0}^{N_u-1}\Delta u_k^T R_k \Delta u_k \\
&s.t.\ \ x_{t+1}=\begin{bmatrix} -(k_{10}+k_{12}+k_{13}) & k_{21} & k_{31} & 0 \\ k_{12} & -k_{21} & 0 & 0 \\ k_{13} & 0 & -k_{31} & 0 \\ k_{41} & 0 & 0 & -k_{e0} \end{bmatrix} x_t+\begin{bmatrix}1\\0\\0\\0\end{bmatrix} u_t \\
&\qquad BIS=C_i C_e+e_i \\
&\qquad 0\le C_{e1}\le \lambda_1\delta_1 \\
&\qquad \lambda_1\delta_2\le C_{e2}\le \lambda_2\delta_2 \\
&\qquad \lambda_2\delta_3\le C_{e3}\le \lambda_3\delta_3 \\
&\qquad C_e=C_{e1}+C_{e2}+C_{e3} \\
&\qquad \delta_1+\delta_2+\delta_3=1 \\
&\qquad C_i=f\left[E_0,\ E_{\max},\ EC_{50},\gamma\right] \\
&\qquad u_{\min}\le u\le u_{\max} \\
&\qquad x_t\in X\subseteq \Re^p,\ u_t\in U\subseteq \Re^s,\ y\in[0,1]^3
\end{aligned}
\tag{4.12}
$$

where x = states, y = outputs and u = controls, all (discrete) time-dependent vectors. The prediction horizon is denoted by N, and the control horizon by N_u. X, U are the sets of the state and input constraints that contain the origin in their interior. The weight matrix for manipulated variables $R \succ 0$ is a positive definite diagonal matrix, and QR is the weight matrix for tracked outputs. Thus, if a certain combination of integer variables is fixed, Equation (4.12) results in a convex QP. For the design of the controller, the following design parameters were used: the objective coefficients for states (x); the weight matrix for tracked outputs (y), $QR = 10^2$; and the weight matrix for manipulated variables (u), $R = 1$.

Equation (4.12) can be recast as an mp-MIQP problem, for which we have recently proposed the first exact solution reported in the literature (Oberdieck and Pistikopoulos 2015). Once the algorithm is initialized, a candidate solution is found which is fixed in the original problem, thus transforming it into an mp-QP problem. The mp-QP problem is solved using available solvers. Next, the objective values of the mp-QP problem and the upper bound in the critical region considered are compared against each other to form a new, tighter upper bound. The algorithm terminates if a termination criterion is reached. More details on the exact solution can be found in Chapter 2.

Figure 4.45 presents a typical solution of the multiparametric programming problem in the form of two-dimensional projection of the critical space. The parametric vector θ consists of: the estimated states, the current time output

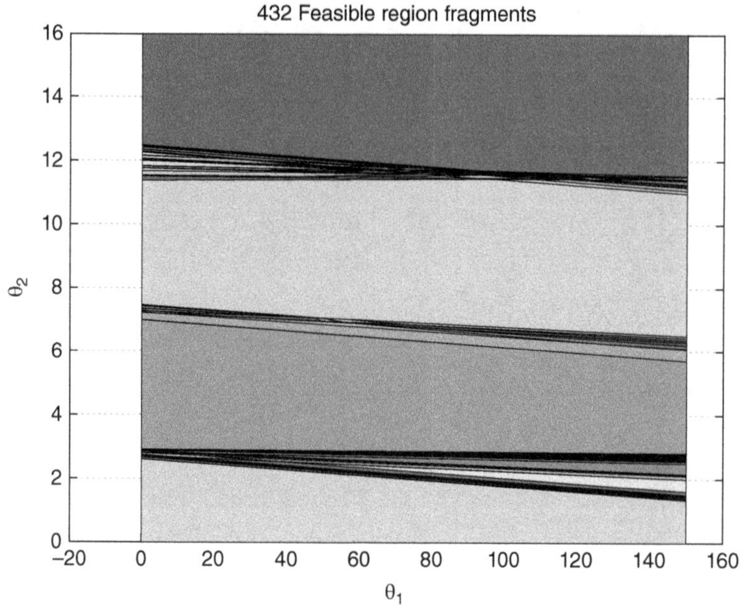

Figure 4.45 Map of critical regions – mp-hMPC. *Source*: Naşcu *et al.* (2017). Reproduced with permission of Elsevier.

and the output reference ($BIS = 50$). Here θ_1 and θ_2 represent the concentration of the effect-site compartment, C_e and the first state x_1.

4.3.3.2 Robust Hybrid mp-MPC Control Strategy: Offset Free

Another challenge for the DOA control is the high inter- and intra-patient variability, which introduces a high degree of uncertainty in the system. Thus, robust control strategies or estimation techniques are required. Robust techniques and a multiparametric MHE technique which are able to deal with these types of problems have been developed.

One key problem of inter-patient variability is the presence of an offset in the output of the process. Hence, the first (intuitive) approach is to introduce a new parameter Δy which captures this offset. In a mathematical form, it can be understood as expanding the definition of the output y_k:

$$x_{q,k+1} = x_{q,k} + \underbrace{\left(y^R - y_k\right)}_{error}, \ \forall k = 1,\ldots,N \tag{4.13}$$

and added penalties in the objective function of the problem:

$$\min_u J = x_{q,N}^{'} P_q x_{q,N} + \sum_{k=1}^{N-1} x_{q,k}^{'} Q_{q,k} x_{q,k} \tag{4.14}$$

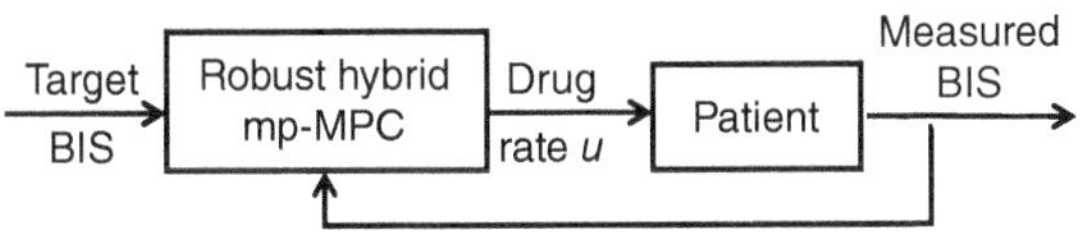

Figure 4.46 Robust hybrid mp-MPC control scheme.

Note that the offset Δy is assumed to be the same for the entire horizon. At each step, this offset is calculated and fed as a parameter to the system, thus resulting in an offset-free approach. The advantage of this approach is its simplicity (in fact, Sakizlis *et al.* [2004] proposed a similar strategy); however, it only provides a symptomatic approach, rather than tackling the underlying issue.

4.3.3.3 Control Scheme

The proposed control design scheme for the mp-hMPC and the robust control strategies is presented in Figure 4.46.

The patient is simulated using the mathematical patient model composed of the PK part (linear) (4.1) and the PD part (4.4) and (4.5). The developed robust strategies presented in Section 4.3.3.2 are implemented within the mp-MPC design. The *robust hybrid mp-MPC* block calculates the error between the *measured BIS* from the patient and the target BIS, and provides the optimal drug rate u to the patient block in order to drive it to the desired target value.

4.3.4 Results

The closed-loop control tests are performed on a set of 12 patients (Ionescu *et al.* 2011) plus an extra patient representing the nominal values of all 12 patients (PaN = patient nominal) presented in Table 4.2.

All of the designed controllers are simulated for the whole set of data presented in Table 4.2 to better understand their behaviour in different patients and analyse the variability. The performances of the controllers are evaluated in both the induction and maintenance phases of DOA. Note that the controllers are designed using the values of the nominal patient, which means that for this patient we will have the best behaviour of the controllers.

See the first paragraph of Section 4.1.6.1, "Induction phase," and of Section 4.2.4.2, "Maintenance phase," for descriptions of these phases. A standard stimulus profile is defined and presented in Figure 4.24.

4.3.4.1 No Offset Correction

In Figure 4.47 and Figure 4.48, we have the simulations of all the patients and the nominal one in the induction phase for the mp-hMPC controller without any robust techniques or state estimation. Figure 4.47 represents the BIS response of the patients, while Figure 4.48 represents the corresponding

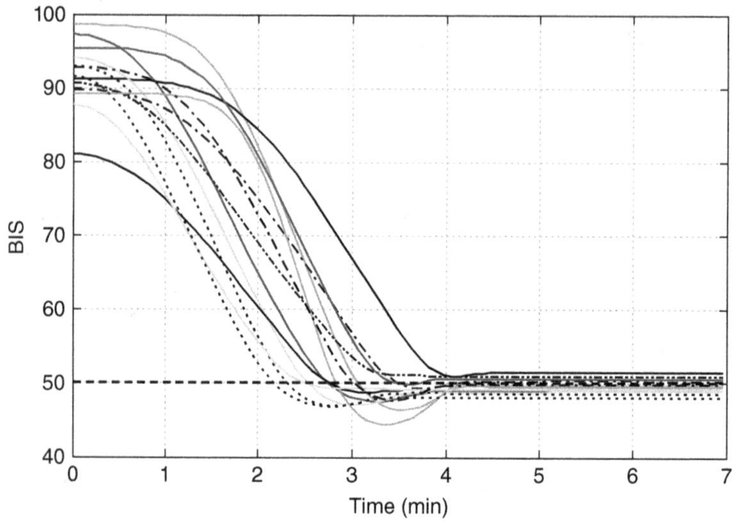

Figure 4.47 BIS output for all 13 patients without offset correction – induction phase. *Source*: Naşcu *et al.* (2017). Reproduced with permission of Elsevier.

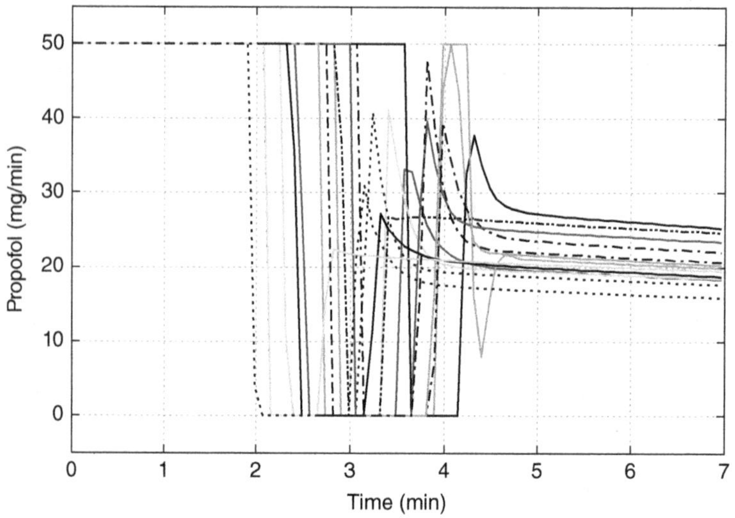

Figure 4.48 Drug infusion for all 13 patients without offset correction – induction phase. *Source*: Naşcu *et al.* (2017). Reproduced with permission of Elsevier.

control action. It can be observed that except for the nominal patient (which was used for the design of the controller), all patients present an offset from the set point. Such behaviour is explained due to the high inter- and intra-patient variability. Figure 4.49 and Figure 4.50 present the simulations of all the patients and the nominal one for the maintenance phase, and it can be observed that,

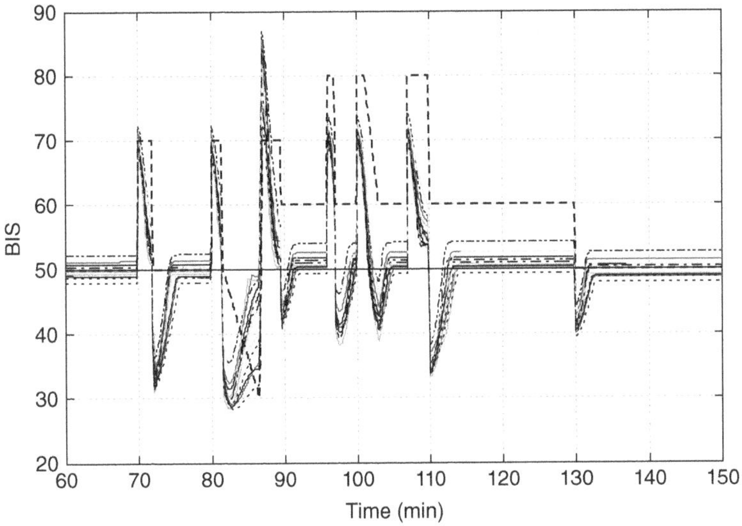

Figure 4.49 BIS output for all 13 patients without offset correction – maintenance phase. *Source*: Naşcu *et al.* (2017). Reproduced with permission of Elsevier.

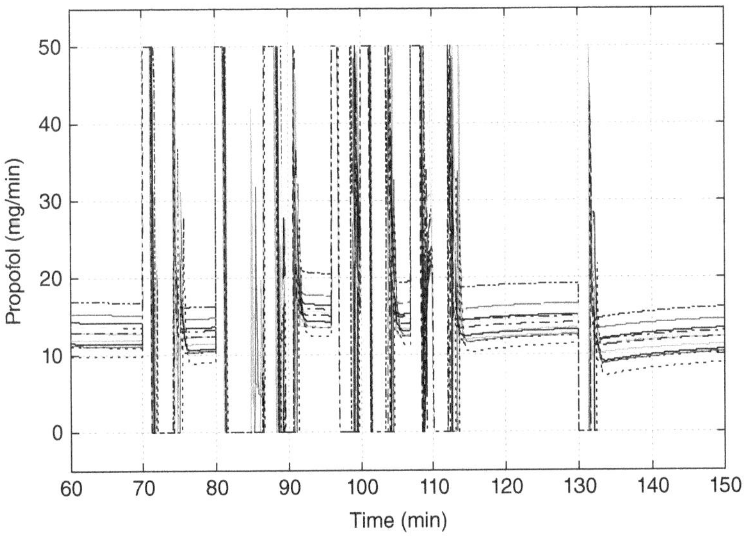

Figure 4.50 Drug infusion for all 13 patients without offset correction – maintenance phase. *Source*: Naşcu *et al.* (2017). Reproduced with permission of Elsevier.

similar to the induction phase, all patients (except the nominal one) present an offset from the set point.

The average settling time for the whole set of patients is 240 s, and the undershoot for the most sensitive patient (patient 9), representing the worst-case scenario, is 5.7%.

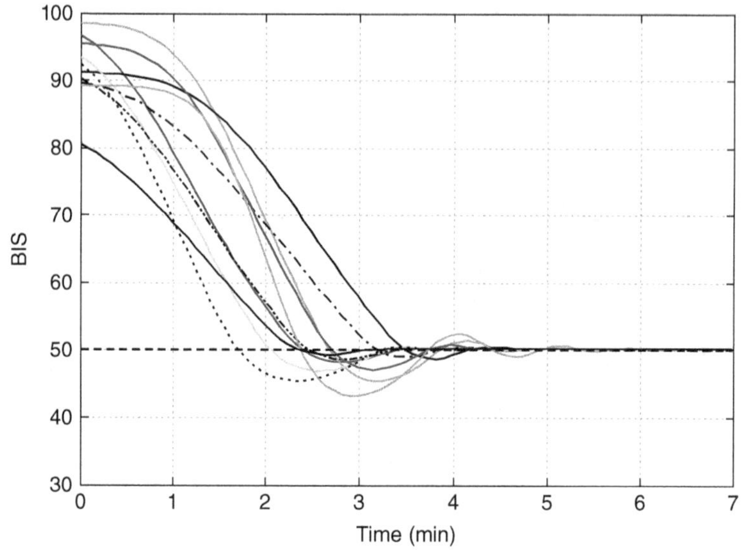

Figure 4.51 BIS output for all 13 patients – strategy 2 – induction phase. *Source*: Naşcu *et al.* (2017). Reproduced with permission of Elsevier.

4.3.4.2 Offset Free

Figure 4.51 and Figure 4.52 present the simulations of all the patients and the nominal one in the induction phase for the mp-hMPC using the offset correction. It can be observed from Figure 4.51, where we have the BIS response of the patients, that the controller is able to compensate for the offset and brings all the patients to the set point value of 50. In Figure 4.52, we have the corresponding propofol infusion rate. Simulations of some patients show very small oscillations around the steady-state values. The average settling time is 250 s.

The BIS response of all patients in the maintenance phase is depicted in Figure 4.53, while Figure 4.54 depicts the corresponding drug infusion rate. The controller compensates for disturbances, but due to pump limitations the simulations exhibit some offsets from the set point.

4.3.5 Discussion

This chapter discussed a piecewise affine formulation for a compartmental anaesthesia patient model, based on which a hybrid explicit/multiparametric MPC was proposed and developed. For the case when variability is not considered, it is shown that this requires the solution of a novel multiparametric mixed-integer quadratic problem. In the presence of variability, robust explicit MPC techniques were incorporated within the overall hybrid explicit MPC strategy. These advanced control strategies are tested on a set of 12 patients

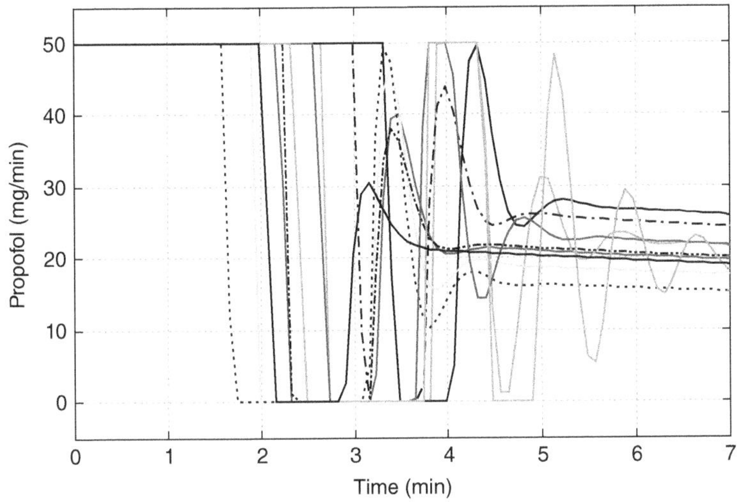

Figure 4.52 Drug infusion for all 13 patients – strategy 2 – induction phase. *Source*: Naşcu *et al.* (2017). Reproduced with permission of Elsevier.

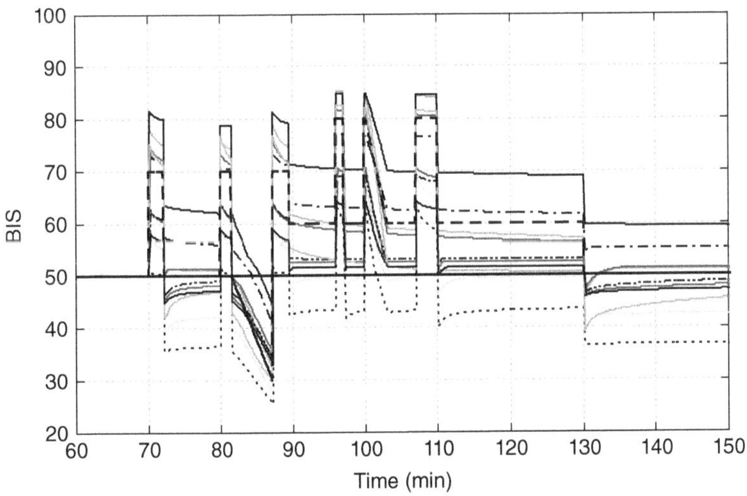

Figure 4.53 BIS output for all 13 patients – strategy 2 – maintenance phase. *Source*: Naşcu *et al.* (2017). Reproduced with permission of Elsevier.

and a nominal one for the automatic induction and control of DOA during induction and maintenance phases.

The resulting mp-hMPC controller was tested for the set of patients in the induction. For the nominal case with no offset correction, we can observe from Figures 4.47–4.50 that all patients present an offset from the desired target

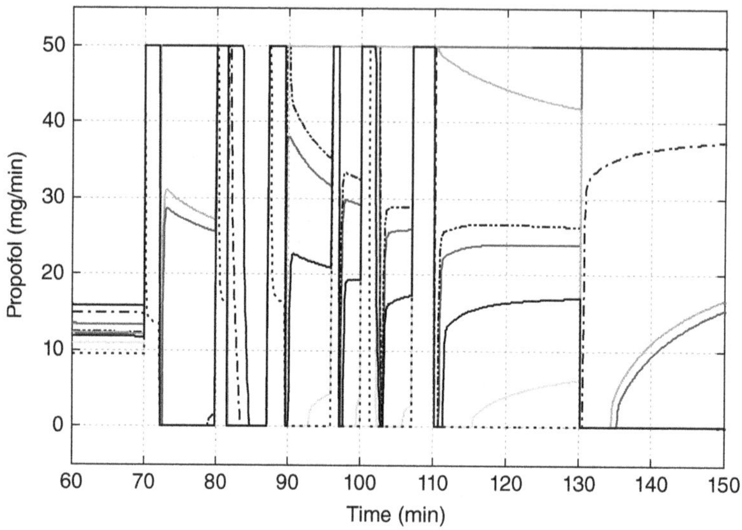

Figure 4.54 Drug infusion for all 13 patients – strategy 2 – maintenance phase. *Source:* Naşcu *et al.* (2017). Reproduced with permission of Elsevier.

value, with the exception of course of the nominal patient. This is due to the high inter- and intra-patient variability, and this can be compensated by making the control robust or using estimation techniques. Thus, robust techniques have been developed: offset correction. The strategies have been tested for the set of patients in the induction phase. It can be observed that the applied robust strategies manage to correct the offset from the nominal case, therefore improving the performances of the controller. In the induction phase, the average settling time is 250 s. The operation procedure starts after the patient reaches an adequate DOA, usually taking up to 15 min and requiring a rise time between 4 and 5 min.

Even though some patients show small oscillations around the steady-state values, the highest undershoot or overshoot is 5.8%. For DOA, undershoots or overshoots of up to 10% are acceptable provided that the set point is reached as soon as possible. This further confirms the satisfactory performance of the derived hybrid controller.

The nominal mp-hMPC as well as the mp-hMPC using offset correction are tested in the maintenance phase to see how well they can deal with disturbance rejection. In Figure 4.49 and Figure 4.53, the controller's response to a disturbance signal that mimics the events that occur in an operation theatre for all patients is shown. It can be observed that the robust controllers and the controller using mp-MHE are able to overcome the offset, especially around the value of 50, with the remaining offset due to limits imposed on the controller.

4.4 Conclusions

As the role of the anaesthetist has become more complex and indispensable to maintain the patients' vital functions before, during and after surgery, automation of drug/anaesthetic administration may reduce workload while offering additional support during critical situations. Optimization and control of the depth of anaesthesia are also important for the safety of the patient and reduction of potential side effects.

The main objective of this chapter was to develop advanced explicit/multiparametric model predictive control (mp-MPC) strategies for the IV anaesthesia process.

The first section describes the mp-MPC framework based on a mathematical model for IV anaesthesia featuring a PK and PD compartment model structure. Different strategies were applied to overcome issues related to the nonlinear part of the model, the Hill curve of the PD model. Specialized linearization techniques were employed for this purpose.

The second section describes the simultaneous mp-MPC and state estimation strategies for the IV anaesthesia. Different estimation techniques to estimate the state of each individual patient were implemented and tested. The estimators were applied simultaneously with the mp-MPC to overcome challenges related to the inter- and intra-patient variability and unmeasurable and noisy outputs.

The final section describes a piecewise linearization of the Hill curve, leading to a hybrid formulation of the patient model and thus the development of hybrid mp-MPC. Robust algorithms are implemented with the hybrid mp-MPC to deal with the inter- and intra-patient variability issue.

Some of the important aspects of this application is the high inter- and intra-patient variability and the unmeasurable data that have a high impact on the estimations and also on the performance of the controllers.

The results show a high-efficiency optimal dosage and robustness of the model predictive control algorithm to induce and maintain the desired Bispectral Index reference while rejecting typical disturbances from surgery.

References

Absalom, A. R., R. De Keyser and M. M. R. F. Struys (2011). "Closed loop anaesthesia: are we getting close to finding the Holy Grail?" *Anesthesia & Analgesia* **112**(3): 516–518.

Bailey, J. M. and W. M. Haddad (2005). "Drug dosing control in clinical pharmacology." *IEEE Control System Magazine* **25**: 35–31.

Bemporad, A. and M. Morari (1999). "Control of systems integrating logic, dynamics, and constraints." *Automatica* **35**(3): 407–427.

Clarke, D. W., C. Mohtadi and P. Tuffs (1987). "Generalized predictive control – Part I. The basic algorithm." *Automatica* **23**(2): 137–148.

Darby, M. L. and M. Nikolaou (2007). "A parametric programming approach to moving-horizon state estimation." *Automatica* **43**(5): 885–891.

De Keyser, R. (2003). "Model based predictive control." In *Encyclopaedia of Life Support Systems (EoLSS)*, UNESCO (eds.). Eolss Publishers Co Ltd, Oxford.

De Keyser, R. M. C. and A. R. Van Cauwenberghe (1985). *Extended Prediction Self-Adaptive Control.* IFAC Proceedings Series.

Dua, V., N. A. Bozinis and E. N. Pistikopoulos (2002). "A multiparametric programming approach for mixed-integer quadratic engineering problems." *Computers and Chemical Engineering* **26**(4–5): 715–733.

Eikaas, H. and J. Raeder (2009). "Total intravenous anaesthesia techniques for ambulatory surgery." *Curr Opin Anaesthesiol* **22**(6): 725–729.

Franklin, G. F., D. J. Powell and A. Emami-Naeini (2001). *Feedback Control of Dynamic Systems.* Prentice Hall PTR, Englewood Cliffs, NJ.

Haddad, W. M., T. Hayakawa and J. M. Bailey (2003). "Nonlinear adaptive control for intensive care unit sedation and operating room hypnosis." Paper presented at the American Control Conference.

Heemels, W. P. M. H., B. d. Schutter and A. Bemporad (2001). "Equivalence of hybrid dynamical models." *Automatica* **37**(7): 1085–1091.

Ionescu, C. M., R. D. Keyser, B. C. Torrico, T. D. Smet, M. M. R. F. Struys and J. E. Normey-Rico (2008). "Robust predictive control strategy applied for propofol dosing using BIS as a controlled variable during anesthesia." *IEEE Transactions on Biomedical Engineering* **55**(9): 2161–2170.

Ionescu, C., J. T. Machado, R. De Keyser, J. Decruyenaere and M. M. R. F. Struys (2015). "Nonlinear dynamics of the patient's response to drug effect during general anesthesia." *Communications in Nonlinear Science and Numerical Simulation* **20**(3):914–926.

Ionescu, C. M., I. Naşcu and R. Keyser (2011). "Robustness tests of a model based predictive control strategy for depth of anesthesia regulation in a propofol to bispectral index framework." In *International Conference on Advancements of Medicine and Health Care through Technology*, S. Vlad and R. Ciupa (eds.), vol. **36**, 234–239. Springer: Berlin Heidelberg.

Kouramas, K. I., N. P. Faísca, C. Panos and E. N. Pistikopoulos (2011). "Explicit/multi-parametric model predictive control (MPC) of linear discrete-time systems by dynamic and multi-parametric programming." *Automatica* **47**(8): 1638–1645.

Krieger, A., N. Panoskaltsis, A. Mantalaris, M. C. Georgiadis and E. N. Pistikopoulos (2014). "Modeling and analysis of individualized pharmacokinetics and pharmacodynamics for volatile anesthesia." *IEEE Transactions on Biomedical Engineering* **61**(1): 25–34.

Minto, C. F., T. W. Schnider, T. D. Egan, E. J. Youngs, H. J. Lemmens, P. L. Gambus, V. Billard, J. F. Hoke, K. H. Moore, D. J. Hermann, K. T. Muir, J. W. Mandema and S. L. Shafer (1997a). "Influence of age and gender on the pharmacokinetics and pharmacodynamics of reminfetanil. I. Model development." *Anesthesiology* **86**: 10–23.

Minto, C. F., T. W. Schnider and S. L. Shafer (1997b). "Pharmacokinetics and pharmacodynamics of remifentanil. II. Model application." *Anesthesiology* **86**(1): 24–33.

Mohtadi, C., D. Clarke and P. Tuffs (1987). "Generalized predictive control-part II. Extensions and Interpretations." *Automatica* **23**(2): 149–160.

Naşcu, I., A. Krieger, C. Ionescu and E. Pistikopoulos (2014a). "Advanced Model Based Control Studies for the Induction and Maintenance of Intravenous Anaesthesia." *IEEE Transactions on Biomedical Engineering* 01–10.

Naşcu, I., R. S. C. Lambert and E. N. Pistikopoulos (2014b). *A combined estimation and multi-parametric model predictive control approach for intravenous anaesthesia*. Conference Proceedings – IEEE International Conference on Systems, Man and Cybernetics.

Naşcu, I., I. Naşcu, C. M. Ionescu and R. De Keyser (2012). *Adaptive EPSAC predictive control of the hypnotic component in anesthesia.*

Naşcu, I., R. Oberdieck and E. N. Pistikopoulos (2017). "Explicit hybrid model predictive control strategies for intravenous anaesthesia." *Computers and Chemical Engineering.*

Naşcu, I. and E. Pistikopoulos (2017). "Modeling, Estimation and Control of the Anaesthesia Process." *Computers & Chemical Engineering.*

Nunes, C. S., T. Mendonça, J. M. Lemos and P. Amorim (2009). "Feedforward adaptive control of the Bispectral Index of the EEG using the intravenous anaesthetic drug propofol." *International Journal of Adaptive Control and Signal Processing* **23**(5): 485–503.

Oberdieck, R. and E. N. Pistikopoulos (2015). "Explicit hybrid model-predictive control: The exact solution." *Automatica* **58**: 152–159.

ParOs (2004). "Parametric optimization programming (POP)."

Pistikopoulos, E. N., N. A. Bozinis and V. Dua (1999). *POP: A Matlab Implementation of Multi-parametric Quadratic Programming Algorithm.* Centre for Process System Engineering, Imperial College, London.

PSE (2015). *gPROMS Model Developer Guide*, Version 4.1.

Sakizlis, V., V. Dua, J. D. Perkins and E. Pistikopoulos (2004). "Robust model-based tracking control using parametric programming." *Computers & Chemical Engineering* **28**(1–2): 195–207.

Schnider, T. W., C. F. Minto, P. L. Gambus, C. Andresen, D. B. Goodale, S. L. Shafer and E. J. Youngs (1998). "The influence of method of administration and covariates on the pharmacokinetics of propofol in adult volunteers." *Anesthesiology* **88**(5): 1170–1182.

Schnider, T. W., C. F. Minto, S. L. Shafer, P. L. Gambus, C. Andresen, D. B. Goodale and E. J. Youngs (1999). "The influence of age on propofol pharmacodynamics." *Anesthesiology* **90**(6): 1502–1516.

Struys, M. M. R. F., T. De Smet, S. Greenwald, A. R. Absalom, S. Bingé and E. P. Mortier (2004). "Performance evaluation of two published closed-loop control systems using Bispectral Index monitoring: a simulation study." *Anesthesiology* **100**(3): 640–647.

Struys, M. M. R. F., H. Vereecke, A. Moerman, E. W. Jensen, D. Verhaeghen, N. De Neve, F. J. E. Dumortier and E. P. Mortier (2003). "Ability of the Bispectral Index, autoregressive modelling with exogenous input-derived auditory evoked potentials, and predicted propofol concentrations to measure patient responsiveness during anesthesia with propofol and remifentanil." *Anesthesiology* **99**(4): 802–812.

Su, Q. L., M. W. Hermanto, R. D. Braatz and M. S. Chiu (2016). "Just-in-time-learning based extended prediction self-adaptive control for batch processes." *Journal of Process Control* **43**: 1–9.

Voelker, A., K. Kouramas and E. N. Pistikopoulos (2013). "Simultaneous design of explicit/multi-parametric constrained moving horizon estimation and robust model predictive control." *Computers and Chemical Engineering* **54**: 24–33.

Welch, G. and G. Bishop (2001). "An introduction to the Kalman filter: SIGGRAPH 2001 Course 8." *Proc. Comput. Graph., Annu. Conf. Comput Graph., Interact. Techn.*: 12–17.

West, N., G. A. Dumont, K. Van Heusden, C. L. Petersen, S. Khosravi, K. Soltesz, A. Umedaly, E. Reimer and J. M. Ansermino (2013). "Robust closed-loop control of induction and maintenance of propofol anesthesia in children." *Paediatric Anaesthesia* **23**(8): 712–719.

Yelneedi, S., L. Samavedham and G. P. Rangaiah (2009). "Advanced control strategies for the regulation of hypnosis with propofol." *Industrial & Engineering Chemistry Research* **48**(8): 3880–3897.

Part II

5

Part A: Type 1 Diabetes Mellitus: Modelling, Model Analysis and Optimization

Stamatina Zavitsanou[1], *Athanasios Mantalaris*[2], *Michael C. Georgiadis*[3], *and Efstratios N. Pistikopoulos*[4]

[1] Paulson School of Engineering & Applied Sciences, Harvard University, USA
[2] Department of Chemical Engineering, Imperial College London, UK
[3] Laboratory of Process Systems Engineering, School of Chemical Engineering, Aristotle University of Thesaloniki, Greece
[4] Texas A&M Energy Institute, Artie McFerrin Department of Chemical Engineering, Texas A&M University, USA

5.a Type 1 Diabetes Mellitus: Modelling, Model Analysis and Optimization

5.a.1 Introduction: Type 1 Diabetes Mellitus

Type 1 diabetes mellitus (T1DM) is a metabolic disorder that is characterized by insufficient or absent insulin circulation, elevated levels of glucose in the plasma and beta cells' inability to respond to metabolic stimulus. It results from autoimmune destruction of beta cells in the pancreas, which is responsible for secretion of insulin, the hormone that contributes to glucose distribution in the human cells.

T1DM is one of the most prevalent chronic diseases of childhood. According to the American Diabetes Association, 1 in 400–600 children and adolescents in the USA have T1DM, and the incidence is increasing worldwide (Onkamo *et al.*, 1999; Patterson *et al.*, 2009) not only in populations with high incidence such as Finland (2010: 50/100,000 a year) but also in low-incidence populations (30/100,000 a year) (see Figure 5.a.1).

T1DM can cause serious complications in the major organs of the body. Problems in the heart, kidney, eyes and nerves can develop gradually over years. The risk of the complications can be decreased only when blood glucose is efficiently regulated.

The most common treatment of T1DM is daily subcutaneous insulin injections. This method subjects the patient to several complications, such as requirement of the patient's appropriate education and adherence to a specific

Modelling Optimization and Control of Biomedical Systems, First Edition.
Edited by Efstratios N. Pistikopoulos, Ioana Naşcu, and Eirini G. Velliou.

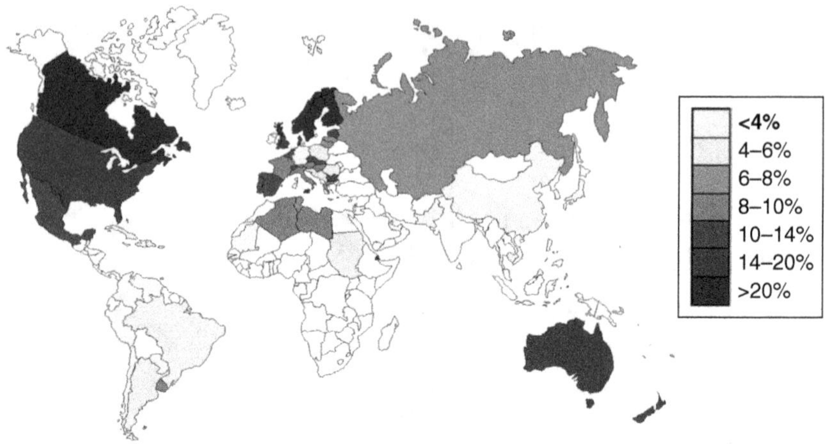

Figure 5.a.1 Incidence of type 1 diabetes mellitus (T1DM) worldwide. *Source*: Onkamo *et al.* (1999). Reproduced with permission of Springer.

lifestyle, risk of hypoglycaemia and therefore ability of the patient to manage the hypoglycaemic episodes, infection of injected sites and so on. Additionally, the patient is restricted to his treatment therapy, meaning that participation in daily activities without adhering to strict glycaemic control could provoke deviations from the normal glucose range, accompanied with medical consequences. Motivated by the challenge to improve the living conditions of a diabetic patient and actually to adapt the insulin treatment to the patient's life rather than the opposite, the idea of an automated insulin delivery system that would mimic the endocrine functionality of a healthy pancreas has been well established in the scientific society.

5.a.1.1 The Concept of the Artificial Pancreas

Currently, the most advanced insulin delivery system for patients with T1DM is an insulin pump. The insulin pump delivers a basal dose of rapid-acting insulin and several bolus doses according to the meal plan of the patient. Good glycaemic control requires 4–6 measurements of blood glucose per day. These measurements, taken either by standalone finger-stick meters or by continuous blood glucose sensors, are loaded into the pump usually by the user or by wireless connection. These measurements are an indicator of whether insulin administration needs adjustment. A wireless connection of the pump data with a personal computer offers a good programming of the pump settings.

The appropriate basal dose for a specific patient is set by the physician, and it can be modified to several profiles (e.g. weekdays and weekends). The bolus doses are set by the patient himself, depending on the meal content, and indicated by the blood glucose levels.

The automation of this therapy constitutes the concept of the artificial pancreas. Essentially, the *artificial pancreas* is a device composed of a continuous glucose monitoring system (CGMS), which reports blood glucose concentration approximately every 5 min; a controller implemented on portable and remotely programmable hardware (a microchip), which computes the appropriate insulin delivery rate according to the provided data from the sensor; and, finally, an insulin pump which infuses the previously calculated insulin amount. The insulin pump, which incorporates the controller and the CGMS, is wirelessly connected.

Many research groups worldwide have believed in this idea, and the research society has focused on the development of the key components for the realization of the artificial pancreas. Pump and CGM manufacturers, as well as the US Food and Drug Administration (FDA) and several diabetes organizations such as JDRF, are involved in projects by encouraging collaborations and solving practical issues to accelerate the design of the artificial pancreas. The state of the art on these topics related to the artificial pancreas can be found in: Kovatchev *et al.* (2010), Dassau *et al.* (2013), Thabit and Hovorka (2013), Soru *et al.* (2012), Cobelli *et al.* (2012), Breton *et al.* (2012) and Herrero *et al.* (2013).

Towards this direction, as shown in Figure 5.a.2, the development of an artificial pancreas is given in two levels (Dua *et al.*, 2006, 2009). The first level is the development of a high-fidelity mathematical model that represents in

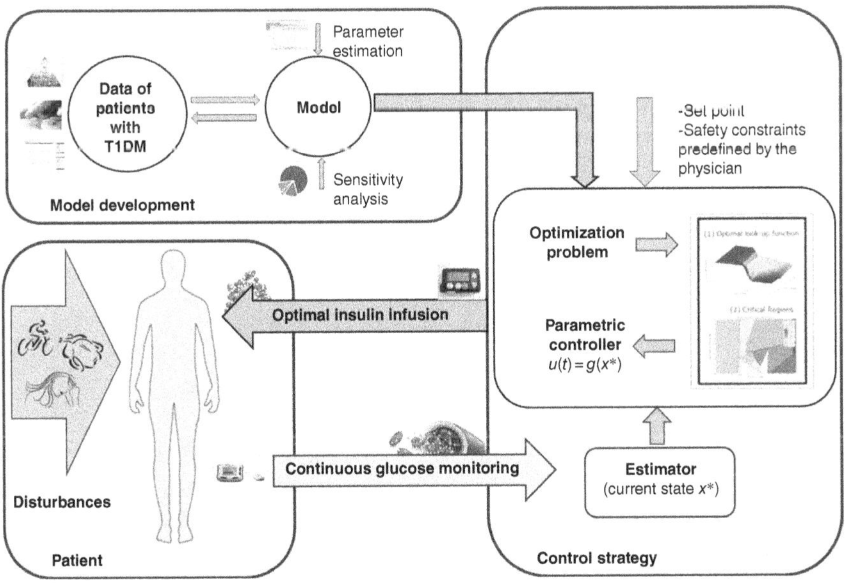

Figure 5.a.2 The framework of an automated insulin delivery system.

depth the complexity of the glucoregulatory system, presents adaptability to patient variability and demonstrates adequate capture of the dynamic response of the patient to various clinical conditions (normoglycaemia, hyperglycaemia and hypoglycaemia). This model is then used for detailed simulation and optimization studies to gain a deep understanding of the system. The second level is the design of model-based predictive controllers by incorporating techniques appropriate for the specific demands of this problem.

5.a.2 Modelling the Glucoregulatory System

In the last 25 years, a large number of models describing the glucoregulatory system have been developed. The pharmacodynamics (effect of a drug on the body) and the pharmacokinetics (effect of the body to the drug) have been approached in several ways. Firstly, compartmental models have been developed such as those of Bergman *et al.* (1981), Dalla Man (2007), Wilinska (2010) and their further extensions, which assume that the relative mechanisms and interactions of insulin and its effect on blood glucose can be represented within several compartments, which are connected through the underlying mass balances. The most common difficulty occurring in this approach is to relate the model parameters (compartment's volume, transfer rate between compartments) to physiological parameters. To overcome these difficulties, physiological models are developed. These models accurately predict the drug–body interactions by using detailed description of the body environment (tissues, organs etc.). Examples of this type of approach are Sorenshen (1978) and Parker (2000). However, this approach can lead to complicated models whose validation requires a lot of experimental effort. Alternative models such as data-driven models or hybrid models such as the one developed by Mitsis (2009) can also be used. A selection of models can be seen in Table 5.a.1. Inspired by these previous approaches and previous work in the group of Dua and colleagues (Dua & Pistikopoulos, 2005; Dua *et al.*, 2006, 2009), a physiologically based compartmental simulation model describing the glucoregulatory system has been developed.

5.a.3 Physiologically Based Compartmental Model

The proposed model describes glucose distribution in the involved body compartments, as presented in Figure 5.a.3, and the effect of insulin on glucose uptake and suppression of endogenous glucose production (EGP). At steady state, an approximation of constant physiological conditions, the blood glucose concentration equals the net balance of endogenous glucose release in the circulation and glucose uptake. When food is consumed, the contained carbohydrates break

Table 5.a.1 Mathematical models of glucose–insulin system.

Mathematical models				
Compartmental models				
Number of compartments				
Glucose kinetics	**Insulin kinetics**	**Validation**	**Comments**	**Reference**
1	2	IVGTT data	Minimal complexity Healthy subjects	Bergman *et al.* (1981)
1	2	Literature data	Minimal model for type 1 DM	Fisher (1991)
2	2	IVGTT data	Healthy subjects	Caumo (1993)
1	3	Literature data	No published data for clinical evaluation	Berger and Rodbard (1991)
1	2	Literature data	AIDA: educational tool	Lehmann and Deutsch (1992)
1	1	Literature data	Experimental data on critically ill patients	Hann *et al.* (2005)
2	2	Literature data	Average patient Circadian SI variation	Fabietti *et al.* (2006)
2	3	Literature data	Critically ill patients	Herpe *et al.* (2007)
2	3 3 effect of insulin action	Clinical study of closed-loop insulin delivery in young people with T1DM	Validated simulation environment	Wilinska *et al.* (2010)
2	2	Experiments	FDA approval	Dalla Man *et al.* (2007a, 2007b)
Physiological models				
6	6	Literature data	Average 70 kg man	Sorensen (1978)
6	6	Literature data	Average 70 kg man Includes a meal sub-model	Parker *et al.* (1999)
Models in the form of delayed differential equation				
1	2 3 delayed insulin effect	Literature data	Healthy subjects Implicit delays	Tolić *et al.* (2000)
1	2	Literature data	Healthy subjects Explicit delays	Bennett (2004)

(*Continued*)

Table 5.a.1 (Continued)

Glucose kinetics	Insulin kinetics	Validation	Comments	Reference
1	2	Literature data	Healthy subject Explicit delay	Engelborghs *et al.* (2001)
1	2	Literature data	Healthy subjects Explicit/implicit delays	Li *et al.* (2006)
1	2	Literature data	Type 1 DM Explicit delay	Chen *et al.* (2010)
Empirical models				
Volterra model		Literature data		Mitsis *et al.* (2009)
ARMA model		Literature data		Eren-Oruklu *et al.* (2009)
NARX model		Literature data		Ghosh and Maka (2009)
Compartmental-neural networks		Literature data		Mougiakakou *et al.* (2005)

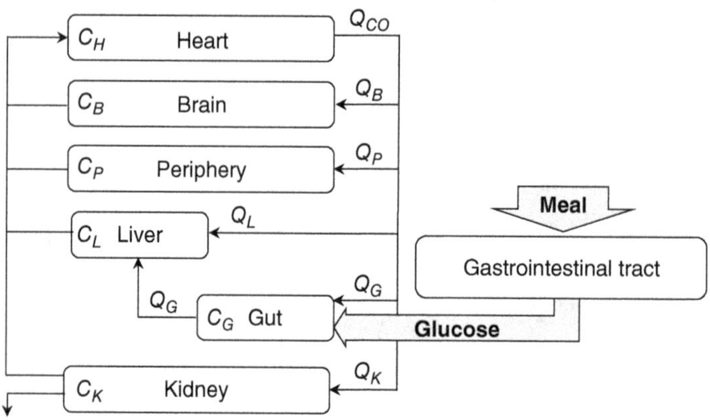

Figure 5.a.3 Structure of the physiologically based compartmental model of glucose metabolism in T1DM.

down into glucose in the gastrointestinal tract which is absorbed through the small intestine into the bloodstream. Physiologically, an increase in blood glucose triggers pancreatic insulin release, which activates glucose transporters to mediate glucose translocation into the insulin-sensitive cells (adipose tissue, and skeletal and cardiac muscles) and additionally suppresses the EGP. In T1DM, the pancreatic insulin secretion is replaced by optimal administration of exogenous insulin that mimics the pancreatic response.

For the highly perfused organs (brain, liver, gut and kidney), glucose concentration is considered to be in equilibrium with the tissue glucose concentration. The periphery compartment lumps the adipose tissue and muscle cells. Glucose transfer from the blood capillaries to the interstitial fluid and glucose uptake in the periphery are described with two compartments. Homogeneity and instant mixing are assumed for every compartment, imposing all the exiting fluxes to be in equilibrium with the compartment. For the insulin-insensitive organs, glucose uptake is assumed to be a constant ratio of the available glucose. The core of the model is described with Equations (5.a.1)–(5.a.6), and the definitions of the involved variables are presented in Table 5.a.2 and Table 5.a.3.

The driving force for glucose transport into the compartments is the blood–tissue concentration difference. The concentration in every organ is given by mass balances in every compartment.

Brain (B):

$$V_{g,B}\frac{dC_B}{dt}=Q_B(C_H-C_B)-u_B \tag{5.a.1}$$

Table 5.a.2 Variables of glucose metabolism model.

Symbol	Definition	Units
Q_i	Blood flow	dL/min
Q_{CO}	Cardiac output	mL/min
C_i	Glucose concentration	mg/dL
$V_{g,i}$	Accessible glucose volume of compartment *i*	dL
u_i	Glucose uptake	mg/min
$r_{u,i}$	Ratio of glucose uptake	–
$r_{CO,i}$	Ratio of cardiac output	–
excretion	Excretion rate	mg/min
EGP	Endogenous glucose production	mg/min
Ra	Rate of glucose appearance	mg/min
p	Rate constant defined as the rate of loss of solute from blood to tissue	dL/min
I_p	Plasma insulin	pmol/L
I_d	Delayed insulin signal	pmol/L
M_L	Liver glucose mass	mg/kg

Table 5.a.3 Variable subscript denotation.

Subscript	Denotation	Subscript	Denotation
i	Organ compartment	H	Heart
B	Brain	P	Periphery
K	Kidney	P_t	Periphery tissue
L	Liver	P,ISF	Interstitial periphery
G	Gut		

Kidney (K):

$$V_{g,K}\frac{dC_k}{dt}=Q_K(C_H-C_K)-u_K-excretion \tag{5.a.2}$$

Liver (L):

$$V_{g,L}\frac{dC_L}{dt}=Q_L\cdot C_H+Q_G\cdot C_G-Q_L\cdot C_L-u_L+BW\cdot EGP \tag{5.a.3}$$

Gut (G):

$$V_{g,G}\frac{dC_G}{dt}=Q_G(C_H-C_G)-u_G+BW\cdot R_a \tag{5.a.4}$$

Heart (H):

$$V_{g,H}\frac{dC_H}{dt}=Q_B\cdot C_B+Q_L\cdot C_L+Q_P\cdot C_P+Q_K\cdot C_K-Q_{CO}\cdot C_H-u_H \tag{5.a.5}$$

Periphery (P):

$$V_{g,Pc}\frac{dC_P}{dt}=Q_P(C_H-C_P)-p(C_P-C_{Pt}) \tag{5.a.6.1}$$

$$V_{g,P,ISF}\frac{dC_{Pt}}{dt}=p(C_P-C_{Pt})-u_P \tag{5.a.6.2}$$

$$u_P=(\lambda_o)\cdot C_{Pt} \tag{5.a.6.3}$$

where the C_i is the glucose concentration (mg/dL) in i compartment, $V_{g,i}$ the accessible glucose volume (dL) of i compartment, Qi the blood flow (dL/min) in i compartment, u_i the glucose uptake (mg/min), EGP the endogenous glucose production (mg/kg/min), Ra the rate of glucose appearance in the blood (mg/kg/dL) and λ_o the rate of glucose uptake (dL/min).

Table 5.a.4 Ratio of cardiac output at rest.

Tissue	($r_{co,i}$)	Reference
Brain	0.11	Ferrannini and DeFronzo (2004)
Liver	0.20	Ferrannini and DeFronzo (2004)
Kidneys	0.13	Ferrannini and DeFronzo (2004)
Gut	0.15	Ferrannini and DeFronzo (2004)
Periphery	0.40	Ferrannini and DeFronzo (2004)

Table 5.a.5 Ratio of glucose uptake.

Tissue	($r_{u,b,i}$)	Reference
Brain	0.45	Ferrannini and DeFronzo (2004)
Liver	0.13	Ferrannini and DeFronzo (2004)
Kidneys	0.02	Ferrannini and DeFronzo (2004)
Gut	0.07	Ferrannini and DeFronzo (2004)
Periphery	0.30	Ferrannini and DeFronzo (2004)

For Equations (5.a.1)–(5.a.6), the blood flow in every organ i is described with Equation (5.a.7). The ratio of cardiac output perfusing every organ is presented in Table 5.a.4.

$$Q_i = r_{CO,i} \cdot Q_{CO} \tag{5.a.7}$$

Similarly, the glucose uptake in every organ is described with Equation (5.a.8), and the ratio of glucose uptake $r_{u,i}$ is presented in Table 5.a.5.

$$u_i = r_{u,i} \cdot Total\_uptake \tag{5.a.8}$$

In the remainder of this section, the sub-models of glucose metabolism functions are described in more detail.

5.a.3.1 Endogenous Glucose Production (EGP)

Approximately 80% of glucose is produced endogenously in the liver through gluconeogenesis and glucogenolysis, and 20% in the cortex of the kidney mainly through gluconeogenesis (Cano, 2002; Gerich, 2010). In this study, due to limited data availability, it is assumed that glucose is produced entirely by the liver. In T1DM, the rate of EGP depends on adequate control of the disease

(Roden & Bernroider, 2003). When referring to intensive insulin therapy, it can be assumed that EGP is approximately the same as in normal humans (Davis *et al.*, 2000). The model describing the EGP in T1DM and used in Equations (5.a.2) and (5.a.3) is adapted from Dalla Man *et al.* (2007). M_L (mg/kg) denotes the liver glucose mass, and I_d (pmol/l) denotes the delayed insulin signal described by a chain of two compartments (I_1, I_d). The model parameters are estimated using available literature data (Boden *et al.*, 2003).

$$EGP = \left(k_{p1} - k_{p2} \cdot M_L - k_{p3} \cdot I_d\right) \tag{5.a.9}$$

$$\frac{dI_1}{dt} = k_i \cdot \left(I_1 - I_p\right) \tag{5.a.10}$$

$$\frac{dI_d}{dt} = k_i \cdot \left(I_d - I_1\right) \tag{5.a.11}$$

5.a.3.2 Rate of Glucose Appearance (Ra)

The model describing the rate of glucose appearing in the circulation when food is consumed is adopted from Dalla Man *et al.* (2006).

5.a.3.3 Glucose Renal Excretion (Excretion)

In diabetes, the threshold of renal glucose reabsorption is exceeded when glucose concentrations increase above 180 mg/dl and glucose gets excreted by the kidney. It is assumed that renal glucose excretion (mg/min) increases proportionally to increasing blood glucose concentration (Rave *et al.*, 2006; Wilinska *et al.*, 2010).

$$E(t) = \begin{cases} CL_{renal}\left(G_K - 180\right) & \text{If} \quad G_K > 180 mg/dL \quad (5.a.12) \\ 0 & \text{If} \quad G_K \leq 180 mg/dL \quad (5.a.13) \end{cases}$$

where CL_{renal} (dl/min) is renal glucose clearance.

5.a.3.4 Glucose Diffusion in the Periphery

Glucose distribution and uptake in the periphery compartment are modelled according to the structure presented in Figure 5.a.4.

It is assumed that glucose is extracted from the arterial flux with a rate factor given in the current literature (Crone, 1965; Regittnig *et al.*, 2003).

$$p = Q_P \cdot \left(1 - \exp\left(-PS/Q_P\right)\right) \tag{5.a.14}$$

where *PS* is the permeability across the capillary wall, a product of permeability of exchange surface to glucose *P* and exchange surface area *S*. This rate

Figure 5.a.4 Detailed glucose uptake in the periphery.

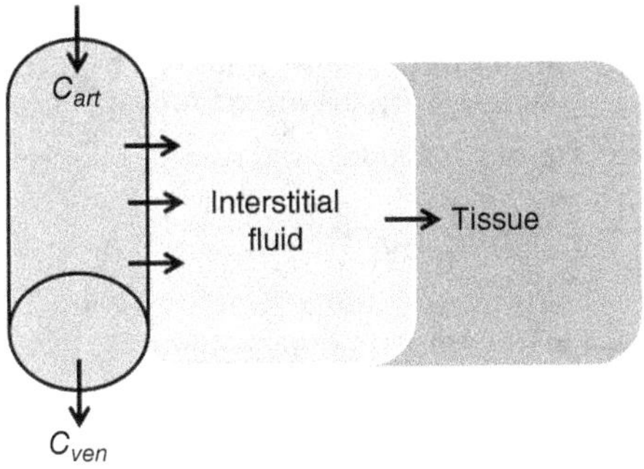

factor can increase in case of increased blood flow to the periphery or increased perfusion due to increased capillary exchange area (e.g. during exercise). According to Gudbjörnsdóttir *et al.* (2003), PS was increased significantly during a one-step hyperinsulinemic clamp. Equation (5.a.15) describes the influence of insulin on glucose permeability across the capillary wall:

$$\frac{dPS}{dt} = -k_{2,PS}PS + k_{1,PS} \cdot I_p \tag{5.a.15}$$

When glucose enters the interstitial fluid, it is absorbed by the tissues to provide them with energy (5.a.6). The rate of uptake, λ_o (dL/min), is dependent on insulin concentration in the blood.

$$\frac{d\lambda_o}{dt} = -k_2\lambda_o + k_1 \cdot (I_p - I_{basal}) \qquad with\ \lambda_o(0) = \lambda_{basal} = \frac{k_1}{k_2} I_{basal} \tag{5.a.16}$$

$$S_I = k_1 / k_2 \tag{5.a.17}$$

where *SI* represents the patient's sensitivity to insulin.

5.a.3.5 Adaptation to the Individual Patient

5.a.3.5.1 Total Blood Volume

The total blood volume (dL) is adapted to the patient's height, weight and gender to account for the differences between obese and underweight patients and for males and females. The formula used for men is (Wennesland *et al.*, 1959):

$$TBV_M = 0.285h + 0.316BW - 2.820 \tag{5.a.18}$$

And for women (Brown *et al.* 1962):

$$TBV_F = 0.1652h + 0.3846BW - 1.369 \tag{5.a.19}$$

The height (h) is in centimetres and weight (BW) in kilograms.

5.a.3.5.2 Cardiac Output

The cardiac output (mL/min) can be efficiently approximated as a proportional relationship to the patient's weight *BW* (kg) according to Equation (5.a.20) (Ederle *et al.*, 2000):

$$Q_{CO} = 224BW^{3/4} \tag{5.a.20}$$

5.a.3.5.3 Compartmental Volume

Plasma proteins comprise approximately 8% of the plasma volume, and the erythrocytes about 38% of the total packed red blood cells volume or haematocrit (*Hemat*) (Ferrannini & DeFronzo, 2004). This percentage of the total blood volume is inaccessible to glucose. Consequently, the accessible glucose volume in every compartment is determined as:

$$V_{g,i} = \left(1-\left(0.08\cdot(1-Hemat)+0.38Hemat\right)\right)\cdot(V_{V,i}+V_{C,i}) \tag{5.a.21}$$

The blood volume of every compartment i is defined as the sum of venous and capillary volume. The glucose venous volume equals 60% of total blood volume, and the capillary volume 10% of total blood volume (Gerich *et al.*, 2001; Ederle, 2011). The compartmental venous and capillary volumes are defined as:

$$V_{V,i} = r_{f,i}\cdot 0.6TBV \tag{5.a.22}$$

$$V_{C,i} = r_{c,i}\cdot 0.1TBV \tag{5.a.23}$$

where $r_{f,i}$ refers to the ratio of total venous volume in compartment i and is calculated with Equation (5.a.24):

$$r_{f,i} = Q_i / \Sigma Q_i \tag{5.a.24}$$

and $r_{c,i}$ refers to the ratio of total capillary volume, respectively (Sorensen, 1978). and is presented in Table 5.a.6.

Table 5.a.6 Ratio of capillary volume.

Tissue	$(r_{c,i})$
Brain	0.071
Liver	0.18
Kidneys	0.08
Gut	0.13
Periphery	0.53

5.a.3.5.4 Peripheral Interstitial Volume

The total regional volume for the adipose tissue is defined as:

$$V_P = V_{Capillary,P} + V_{Interstitial,P} + V_{Intracellular,P} \quad (5.a.25)$$

According to Oh and Uribarri (2006), the interstitial volume represents 28% of the total body water, while the intracellular volume is 60%. Hence, $V_{Intracellular} = 0.47 V_{Interstitial}$.

According to Deurenberg *et al.* (1991):

$$m_{AT} = (1.2BMI - 10.8sex + 0.23age - 5.4)0.01m \quad (5.a.26)$$

With

$$d_i = m_i / V_i \quad (5.a.27)$$

The interstitial volume of the muscles and the adipose tissue is considered to be 10% of the total tissue volume according to Johnson (2003) and Eckel (2003), respectively. Muscle mass is considered to be approximately 40% of the total body weight (5.a.28), according to Ackland *et al.* (2009).

$$m_{muscles} = 0.4BW \quad (5.a.28)$$

The peripheral volume of the interstitial fluid is calculated with Equations (5.a.25)–(5.a.29), using Table 5.a.7:

$$V_{g,P,ISF} = V_{Interstitial,AT} + V_{Interstitial,musc} \quad (5.a.29)$$

5.a.3.6 Insulin Kinetics

Insulin kinetics comprises the mechanisms involved from the moment insulin is administered in the subcutaneous tissue until it is fully eliminated from the body. Several models have been proposed in the literature (Kraegen & Chisholm, 1984; Nucci & Cobelli, 2000; Tarín *et al.*, 2005; Kuang & Li, 2008), with compartmental modelling being the most common approach. In this study, the structure to describe insulin kinetics is investigated when an insulin pump is used. Four alternative compartmental models are presented here

Table 5.a.7 Density of muscles and adipose tissue.

Tissue	Density (kg/L)	Reference
Adipose tissue (d_{AT})	0.92	Gallagher *et al.* (1998)
Muscles ($d_{muscles}$)	1.04	Gallagher *et al.* (1998)

(see Table 5.a.8) that describe experimental data of insulin kinetics and compare in terms of identifiability and parameter accuracy, as discussed in Section 5.a.4.

The variable and parameter definitions for both models are shown in Table 5.a.9.

5.a.4 Model Analysis

In this section, the most suitable model for insulin kinetics is selected by performing a series of analysis tests. Experimental data obtained in the literature are used to estimate the model parameters. Additionally, the suggested structure of the EGP sub-model is evaluated in terms of reliability, using again experimental data from the literature to estimate the model parameters and confirm the model's accuracy. Consecutively, the previously presented entire mathematical model of glucose metabolism is analysed in order to identify the most influential parameters that contribute to the model's uncertainty. This uncertainty originates from the high intra- and inter-patient variability that dominates the system. Global sensitivity analysis, parameter estimation and accuracy tests are performed to evaluate the model's ability to represent the physiology.

5.a.4.1 Insulin Kinetics Model Selection

The values of the parameters of the four models of insulin kinetics are identified via parameter estimation, performed in gPROMS (PSE, 2011b), using experimental data obtained from the literature (Boden *et al.*, 2003) The solution method used in gPROMS to obtain the optimal parameter estimates is to minimize the maximum log-likelihood objective function by solving a nonlinear optimization problem.

Figure 5.a.6 shows the plasma insulin concentration profiles produced by the suggested models versus the experimental data. Generally, we can conclude that all models describe relatively well the experimental data. However, a more in-depth analysis reveals the strengths and the weaknesses of each model.

A Pearson's chi-squared test (x^2) (PSE, 2011b) is performed (Table 5.a.10) to confirm the results indicated by Figure 5.a.5. For $\mathrm{k = N - p}$ degrees of freedom, where N is the number of experimental data and p the number of parameters, the x^2 value is obtained for a 95% confidence level. The calculated x^2 smaller than the reference x^2 value indicates that the fit of the considered model is good.

The Akaike criterion (AIC; Akaike, 1974) is applied in order to select the most appropriate model that represents the experimental data. The test is presented in Equation (5.a.39):

$$\mathrm{AIC = N\ln(WRSS) + 2K} \tag{5.a.39}$$

Table 5.a.8 Model equations of three proposed insulin kinetics models and a reference model; schematic representation of the models.

Scheme	Equations		Model
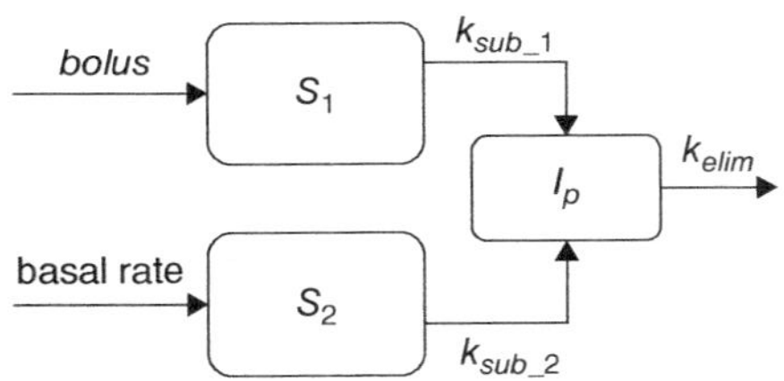	$\frac{dS_1}{dt} = bolus - k_{sub\_1} \cdot S_1$ $\frac{dS_2}{dt} = basal - k_{sub\_2} \cdot S_2$ $\frac{dI_p}{dt} = \frac{k_{sub\_1} \cdot S_1 + k_{sub\_2} \cdot S_2}{V_{dist}} - k_{e\lim} \cdot I_p$	(5.a.30) (5.a.31) (5.a.32)	Model 1
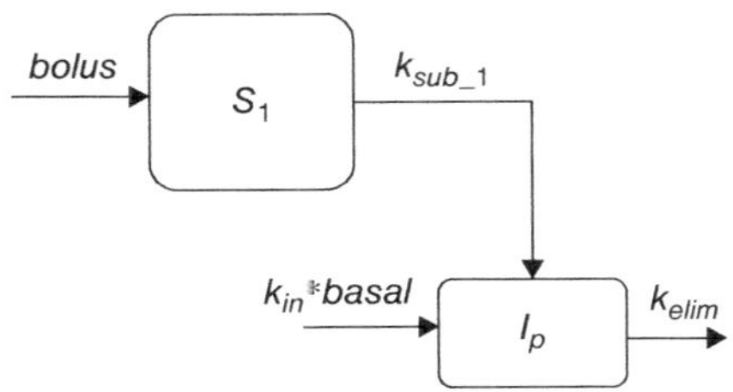	$\frac{dS_1}{dt} = bolus - k_{sub\_1} \cdot S_1$ $\frac{dI_p}{dt} = \frac{k_{sub\_1} \cdot S_1}{V_{dist}} + k_{in} basal - k_{e\lim} \cdot I_p$	(5.a.33) (5.a.34)	Model 2

(Continued)

Table 5.a.8 (Continued)

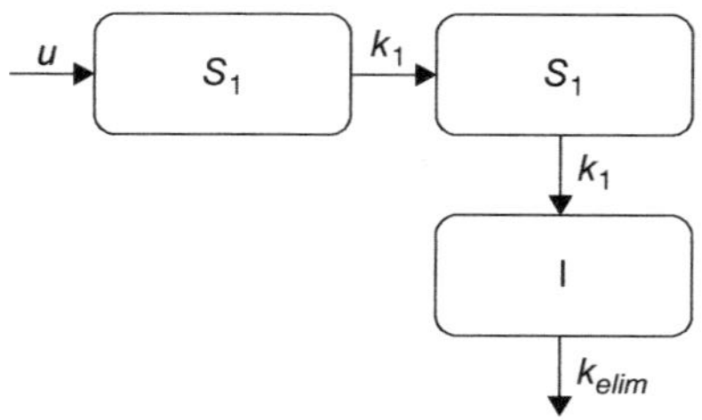	$$\frac{dS_2}{dt}=k_{sub\_1}S_1-k_{sub\_1}S_2 \quad (5.a.35)$$ $$\frac{dS_1}{dt}=u-k_{sub\_1}S_1 \quad (5.a.36)$$ $$\frac{dI}{dt}=k_{sub\_1}S_2-k_{e\lim}I \quad (5.a.37)$$ $$I_p=\frac{I}{V_{dist}} \quad (5.a.38)$$	Model 3
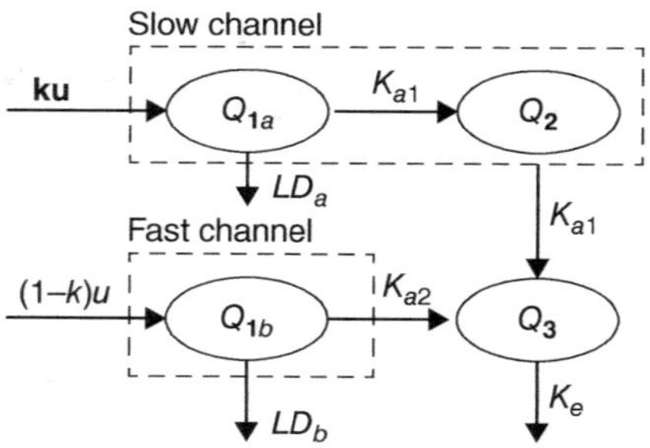	(Wilinska, 2005)	Reference model

Table 5.a.9 Variable and parameter definition of Models 1, 2 and 3.

Symbol	Definitions
S_1, S_2	Insulin mass (mU) in the subcutaneous compartments
I	Insulin mass (mU) in the plasma compartment
k_{sub_1}, k_{sub_2}	Intercompartmental transfer rate constant (min^{-1})
k_{elim}	Elimination rate constant (min^{-1})
V_{dist}	Insulin distribution volume (L/kg)
u, basal, bolus	Continuous insulin infusion (U/min)

Table 5.a.10 Goodness of fit of proposed models and model selection.

	Model 1	Model 2	Reference model	Model 3
Pearson's chi-squared test (x^2)	23.404	9.1041	10.729	9.0727
x^2 Value (95%, k)	27.587	27.587	23.685	28.869
Akaike criterion	11.28	13.19	9.95	5.13

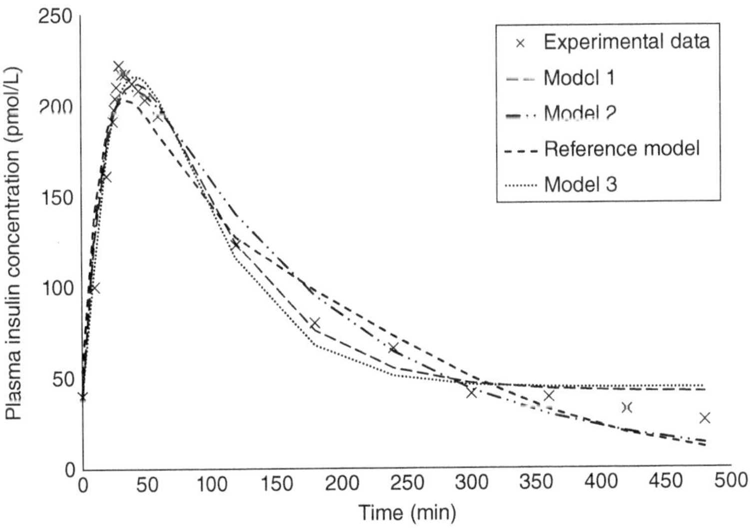

Figure 5.a.5 Comparison of Models 1, 2 and 3 and a reference model with experimental data.

Table 5.a.11 Optimal mean parameter estimates and standard deviations reported in parentheses.

	Model 1	Model 2	Reference model	Model 3
k_{sub1}	0.36(±0.28)	0.59(±0.3)	–	0.025(±0.0017)
k_{sub2}	0.016(±0.048)	0.042(±0.012)	–	–
k_{elim}	0.017(±0.0065)	0.011(±0.004)	0.024(±0.014)	0.418 (±0.338)
V_{dist}	3.46(±1.57)	2.76(±0.5)	0.859(±0.63)	0.087(±0.072)
k	–	–	0.130(±0.1)	
k_{a1}	–	–	0.0661(±0.019)	–
k_{a2}	–	–	0.0035(±0.0006)	–
k_m	–	–	69.25(±28.87)	–
V_{maxLD}	–	–	0.116(±0.068)	–

where N denotes the number of data points, K the number of parameters and $WRSS$ the weighted residuals sum of squares.

The Akaike values, as shown in Table 5.a.10, indicate that the most suitable model to describe the available experimental data is Model 3, when compared to the other three models. Model 3 is a trilinear compartment, which involves two compartments to describe insulin absorption through the subcutaneous tissue and a single compartment for insulin in the plasma. This model of insulin kinetics has been widely used in the literature (Wilinska *et al.*, 2005, 2010). Table 5.a.11 presents the optimal estimated values of all the model parameters.

The values of the estimated parameters for the reference model and Model 3 are in good accordance with the literature (Wilinska *et al.*, 2005).

5.a.4.2 Endogenous Glucose Production: Parameter Estimation

The experimental data used for parameter estimation are obtained from Boden *et al.* (2003). The purpose of this experiment was to study the mechanisms of endogenous glucose production during insulin excess and insulin deficiency, while maintaining blood glucose concentration constant. Therefore, the parameter related to the effect of glucose on the suppression of EGP, k_{p2}, was kept constant and equal to the mean value obtained from Dalla Man *et al.* (2007).

Figure 5.a.6 shows that the model fits well with the experimental data, and the values of the estimated model parameters can be seen in Table 5.a.12. A t-test (PSE, 2011b) is performed that indicates accurate estimates of the parameters since the t-value is larger than the reference t-value for the 95% confidence level. Additionally, the confidence interval shows the precision of

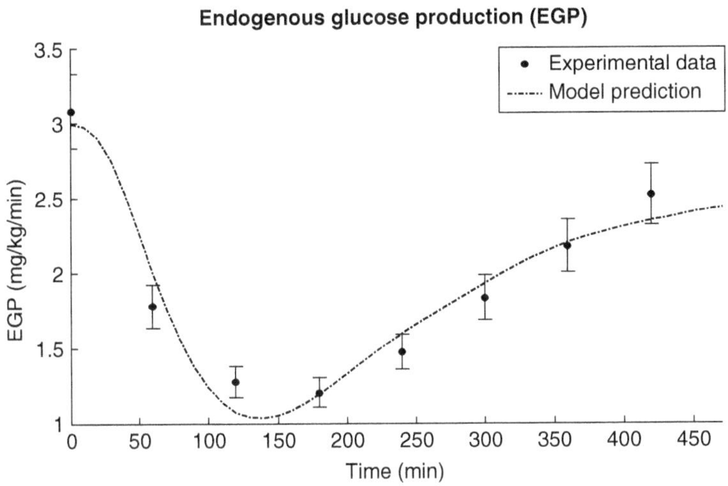

Figure 5.a.6 Effect of subcutaneous insulin injection on endogenous glucose production.

Table 5.a.12 Parameter estimation results.

Symbol	Optimal estimate (mean ± SD)	Confidence interval* (95%)	95% *t*-value
$\mathbf{k_i}$	0.024 ± 0.0034	0.0085	2.82
$\mathbf{kp_1}$	3.058 ± 0.17	0.42	7.33
$\mathbf{kp_3}$	0.014 ± 0.0022	0.0053	2.7

Reference *t*-value (95%): 1.94.

the estimated values for the corresponding parameters and is calculated with Equation (5.a.40), considering the confidence level $a = 95\%$.

$$\text{Confidence Interval} = \pm \frac{t_a}{2}(n-1)\frac{SD}{\sqrt{n}} \tag{5.a.40}$$

5.a.4.3 Global Sensitivity Analysis

The model's reliability is evaluated with the performance of global sensitivity analysis (GSA). The uncertain factors that have a relative influence on the model's measurable output are determined and provide information on the proposed model's structure, in an effort to reduce the model's uncertainty by examining the most influential parameters. GSA has been performed with graphical user interface/high-dimensional model representation (GUI-HDMR) software (Ziehn & Tomlin, 2009) which uses an expansion of the random sampling HDMR (RS-HDMR) method. The sampling was performed by simulating the model in gPROMS via the gO:MATLAB interface, developed by Krieger *et al.* (2014). The sensitivity index (SI) is scaled between 0 and 1, indicating that a SI equal to 0 refers to a non-influential parameter. The parameters values vary between their upper and lower bounds, and for every GSA, a set of 20,000 Sobol distributed points within the range were used to calculate the SI for specified time points. Sobol's sampling set is preferred because it provides evenly uniform distributed points of the input space. The sum of all the SI converges to 1. In this study, the effect of the parameters on blood glucose concentration was evaluated in two cases. In the first case, the SIs were calculated for all the parameters to investigate their influence in a system with respect to intra- and inter-patient variability. In the second case, only the parameters related to intra-patient variability were included, assuming that the weight, the organ volumes, the insulin distribution and the meal absorption can be considered constants for an individual patient and were fixed at their default values. The results are presented in Table 5.a.13.

5.a.4.3.1 Individual Model Parameters

The model parameters are shown in Table 5.a.13. The range of the parameters Q_{co} and $V_{g,i}$ is calculated from Equations (5.a.18)–(5.a.23) when considering the body weight of 50–115 kg, height of 150–190 cm and age of 18–80 years. The default values are set for a male patient of 170 cm height, 52 years old and 94 kg. The range of the parameters related to the Ra and EGP is adapted from the Uva/Padova Simulator. The default values of the parameters for these subsystems were set at the mean value. The ratio of cardiac output and the ratio of glucose uptake were considered to vary ±5%, a value chosen when performing a series of stochastic simulation studies, while the default values were obtained from Table 5.a.4 and Table 5.a.5. The range and the default value of the parameters for insulin kinetics were obtained from Wilinska *et al.* (2005). A big variation of the default value in the parameters k_1, k_2 was assumed to evaluate the prediction ability of the model. Finally, a ±20% variation was assumed for $k_{1,PS}$ and $k_{2,PS}$. The initial guess of the values of the parameters k_1, k_2, and $k_{1,PS}$, $k_{2,PS}$ was selected when performing a set of stochastic simulation studies in comparison with the simulation results provided by the Simulator.

A meal containing 50 g of carbohydrates and a 10 U bolus were given at 420 min. The time points in Table 5.a.13 refer to 1 h and 5 h after meal consumption, and they were chosen to investigate the influence of the parameters when the sub-models of meal absorption and bolus insulin kinetics are active, all the external disturbances are absorbed and the system is relatively balanced. For the first case, the most influential parameters are the k_1, k_2, k_{p3}, k_{abs} and $r_{u,L}$ at 480 min and k_1, k_2, $r_{u,L}$ and $r_{u,H}$ at 720 min. Hence, the parameters related to

Table 5.a.13 Model parameters' default values and range, and SIs for all parameters and for those related to intra-patient variability calculated with the GUI-HDMR toolbox.

			Sensitivity Index				
			All parameters		Intra-patient parameters		
Symbol	Default	Range	480 min	720 min	480 min	720 min	Units
k_{a1}	1.66×10^{-02}	$(1.0-2.66)\times10^{-2}$	0	0	–	–	min^{-1}
V_{dist}	5.38×10^{-02}	$(1.16-25.08)\times10^{-2}$	1.12E-06	5.07E-07	–	–	L/Kg
k_{elim}	3.02×10^{-01}	$(6.79-134.55)\times10^{-2}$	0	0	–	–	min^{-1}
k_1	3.00×10^{-04}	$(0.40-1.00)\times10^{-03}$	**0.263565**	**0.340726**	**0.791256**	**0.445745**	dL^2 per pmol· min^2
k_2	2.00×10^{-01}	$(0.50-5.00)\times10^{-01}$	**0.096337**	**0.418659**	**0.154249**	**0.565001**	min^{-1}
k_{p1}	$5.38\times10^{+00}$	$(3.56-7.20)\times10^{+00}$	0	0	7.93E-06	3.21E-05	mg/kg/min
k_{p2}	5.23×10^{-03}	$(2.44-8.02)\times10^{-03}$	0	0.000721	0	0	min^{-1}
k_{p3}	1.43×10^{-02}	$(0.46-2.39)\times10^{-02}$	**0.301874**	0.005473	**0.11209**	0.039743	mg/kg/min per pmol/L
k_i	0.78×10^{-02}	$(0.29-1.62)\times10^{-02}$	3.51E-06	4.19E-05	0	0.000163	$min - ^{1}$
k_{2_PS}	4.00×10^{-03}	$(3.20-4.80)\times10^{-03}$	0.015557	0.004761	0	3.37E-05	min^{-1}
k_{1_PS}	5.00×10^{-04}	$(4.00-6.00)\times10^{-04}$	0.000932	0.000138	3.64E-05	2.27E-05	dL^2 per pmol· min^2
k_{max}	3.01×10^{-01}	$(0.21-5.82)\times10^{-01}$	0	0	–	–	min^{-1}
k_{min}	4.00×10^{-02}	$(2.19-5.82)\times10^{-02}$	0	0.000127	–	–	min^{-1}
k_{abs}	8.84×10^{-03}	$(0.28-1.49)\times10^{-02}$	**0.160871**	1.67E-05	–	–	min^{-1}
k_{gri}	4.00×10^{-02}	$(2.19-5.82)\times10^{-02}$	0	8.23E-05	–	–	min^{-1}
b	7.95×10^{-01}	$(6.27-9.62)\times10^{-01}$	3.63E-05	0.001582	–	–	–
d	2.15×10^{-01}	$(0.92-3.37)\times10^{-01}$	0	0.001022	–	–	–

(*Continued*)

Table 5.a.13 (Continued)

Symbol	Default	Range	Sensitivity Index				Units
			All parameters		Intra-patient parameters		
			480 min	720 min	480 min	720 min	
CL_{renal}	5.00×10^{-05}	$(4.00-6.00)\times10^{-01}$	1.33E-04	0	6.32 E-05	0	dL/min
Q_{co}	$6.04\times10^{+03}$	$(3.76-7.02)\times10^{+03}$	0.003759	0.003217	6.69E-05	2.64E-05	mL/min
V_K	$3.90\times10^{+00}$	$(2.24-4.86)\times10^{+00}$	0.000289	0.000225	–	–	dL
V_G	$4.44\times10^{+00}$	$(2.55-5.54)\times10^{+00}$	0.012974	0.008315	–	–	dL
V_P	$1.09\times10^{+01}$	$(0.63-1.37)\times10^{+01}$	0	0.000374	–	–	dL
V_B	$3.06\times10^{+00}$	$(1.76-3.82)\times10^{+00}$	6.78E-05	0.010876	–	–	dL
V_L	$5.62\times10^{+00}$	$(3.23-7.02)\times10^{+00}$	0	0.00272	–	–	dL
V_H	$1.34\times10^{+01}$	$(1.27-1.35)\times10^{+01}$	3.38E-05	0.000164	–	–	dL
$r_{co,K}$	1.78×10^{-01}	$(1.69-1.87)\times10^{-01}$	1.33E-05	0.000539	1.16E-05	0.006560	–
$r_{co,G}$	1.95×10^{-01}	$(1.85-2.05)\times10^{-01}$	0.067301	0.004797	0	3.68E-05	–
$r_{co,P}$	4.39×10^{-01}	$(4.17-4.61)\times10^{-01}$	4.28E-05	0.00048	6.26E-05	0.003743	–
$r_{co,B}$	1.38×10^{-01}	$(1.31-1.45)\times10^{-01}$	0.000107	0.003262	1.29E-05	2.34E-06	–
$r_{co,L}$	2.44×10^{-01}	$(2.32-2.56)\times10^{-01}$	2.75E-05	0.018778	1.04E-05	0.000398	–
$r_{u,K}$	2.00×10^{-02}	$(1.90-2.10)\times10^{-02}$	0	0.00064	0.000774	0.003097	–
$r_{u,G}$	7.00×10^{-02}	$(6.65-7.35)\times10^{-02}$	0.02257	0.001398	0.008611	0.003169	–
$r_{u,L}$	1.30×10^{-01}	$(1.24-1.37)\times10^{-01}$	**0.052423**	**0.137398**	**0.027909**	0.001827	–
$r_{u,H}$	1.80×10^{-02}	$(1.71-1.89)\times10^{-02}$	0.000969	**0.033467**	0.000112	**0.025256**	–

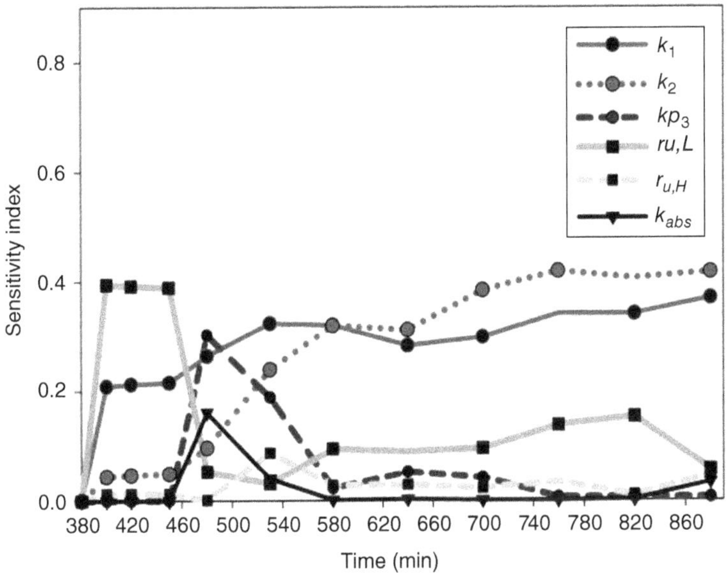

Figure 5.a.7 Time-varying SIs when all parameters are considered.

glucose absorption from the periphery k_1, k_2 as a function of insulin concentration (5.a.16) are the most critical since they are related to the patient's sensitivity to insulin and therefore their ability to absorb glucose. For the second case, the parameters k_1, k_2, $r_{u,L}$ and $r_{u,H}$ are the most influential.

The time-varying parameters for the two cases defined in Table 5.a.13 are shown in Figure 5.a.7 and Figure 5.a.8. Only the parameters with the highest sensitivities are included in the graphs. For both cases, the sensitivities of parameters k_1 and k_2 remain high throughout the performance analysis, and both are increased after meal and bolus administration. The sensitivity of k_{p3}, as expected, increases during bolus administration and decreases at the postprandial state when insulin concentration decreases after the bolus peak. Additionally, for k_{abs}, a parameter that indicates how fast the blood glucose is absorbed from the small intestine, the sensitivity increases with meal consumption and decreases when glucose has been absorbed. For the ratio of glucose absorption from the liver, the sensitivity is high at the fasting state and decreases relatively at the postprandial state, while the ratio of glucose absorption from the heart increases after meal consumption, indicating that both of these parameters influence glucose regulation in accordance to Equations (5.a.3) and (5.a.5).

As a conclusion, it can be stated that the parameters with the most influential role are those related to insulin effect on glucose. The parameters related to

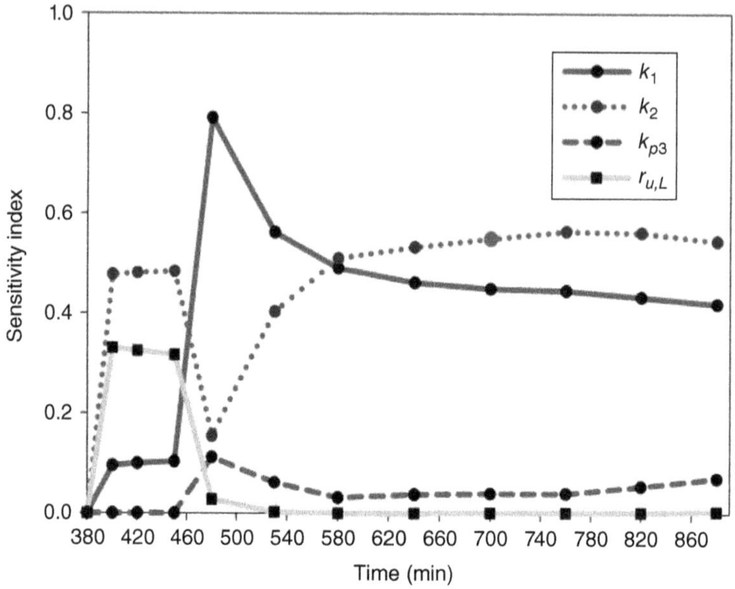

Figure 5.a.8 Time-varying SIs when intra-patient variability-related parameters are considered.

insulin distribution, absorption and elimination through the subcutaneous tissue, as well as the parameters related to glucose distribution in the various compartments, can be considered as non-influential compared to the insulin effect–related parameters.

5.a.4.4 Parameter Estimation

The performance of the proposed model is evaluated with detailed simulation studies performed in gPROMS, and its prediction ability is verified when compared with data of 10 adult patients provided by the UVa/Padova T1DM Simulator. To demonstrate the prediction ability of the proposed model, a specific diet plan of 45 g of carbohydrates for breakfast, 70 g for lunch and 70 g for dinner and the appropriate insulin regimen for each patient is set, and the simulation results are shown for the 10 patients. The same conditions are applied in the Simulator, and the blood glucose and plasma insulin concentration profiles are used as experimental data to estimate the most influential model parameters (presented in Table 5.a.14). The parameters of the Ra and EGP sub-models are also estimated for each patient to obtain patient-specific glucose–insulin dynamics. The default values are used for the remaining nonsignificant parameters.

Table 5.a.14 Optimal parameter estimates, presented as mean (lower-upper) value for the 10 patients.

Symbol	Value	Units	Symbol	Value	Units
k_1	1.58×10^{-04} $(2.11-38.4)\times10^{-05}$	dL^2 per pmol·min	k_i	8.15×10^{-03} $(0.294-1.34)\times10^{-02}$	min^{-1}
k_2	2.35×10^{-02} $(1.17-4.53)\times10^{-02}$	min^{-1}	k_{p1}	5.65 $(3.97-7.2)$	mg/kg/min
$r_{u,L}$	1.7×10^{-01} $(0.6-1.9)\times10^{-01}$	–	k_{p2}	4.73×10^{-03} $(2.44-7.72)\times10^{-03}$	min^{-1}
k_{max}	3.53×10^{-02} $(2.19-5.82)\times10^{-02}$	min^{-1}	k_{p3}	1.49×10^{-02} $(0.0551-2.39)\times10^{-02}$	mg/kg/min per pmol/L
k_{min}	7.62×10^{-03} $(0.373-1.16)\times10^{-02}$	min^{-1}	k_{elim}	1.36 $(0.2996-2.1433)$	min^{-1}
k_{abs}	1.14×10^{-01} $(0.214-5.82)\times10^{-01}$	min^{-1}	k_{sub}	1.86×10^{-02} $(1.21-2.46)\times10^{-02}$	min^{-1}
b	8.27×10^{-01} $(7.36-9.29)\times10^{-01}$	–	V_{dist}	1.54×10^{-02} $(1.00-5.16)\times10^{-02}$	L/kg
d	1.91×10^{-02} $(0.98-3.32)\times10^{-01}$	–			

5.a.5 Simulation Results

The performance of the proposed model is evaluated with detailed simulation studies performed in gPROMS (PSE, 2011a), and its predictability is verified when compared with data provided by the UVa/Padova T1DMS Simulator (Kovatchev *et al.*, 2011). To demonstrate the predictability of the proposed model, a specific diet plan of 45 g of carbohydrates for breakfast, 70 g for lunch and 70 g for dinner is set, and the simulation results are shown for three patients. The insulin regimen is predefined for each patient, as shown in Table 5.a.14. The same conditions are applied in the Simulator, and the blood glucose and plasma insulin concentration profiles are used as experimental data to estimate the most influential model parameters. Hence, the individual parameters of model 2 for insulin kinetics and *k*1, *k*2, *ru*, *L* and *ke* of glucose metabolism are estimated as shown in Table 5.a.14. The reported confidence interval for each value is a measure of the estimated precision, indicating that the smaller the interval, the more reliable the estimated value is.

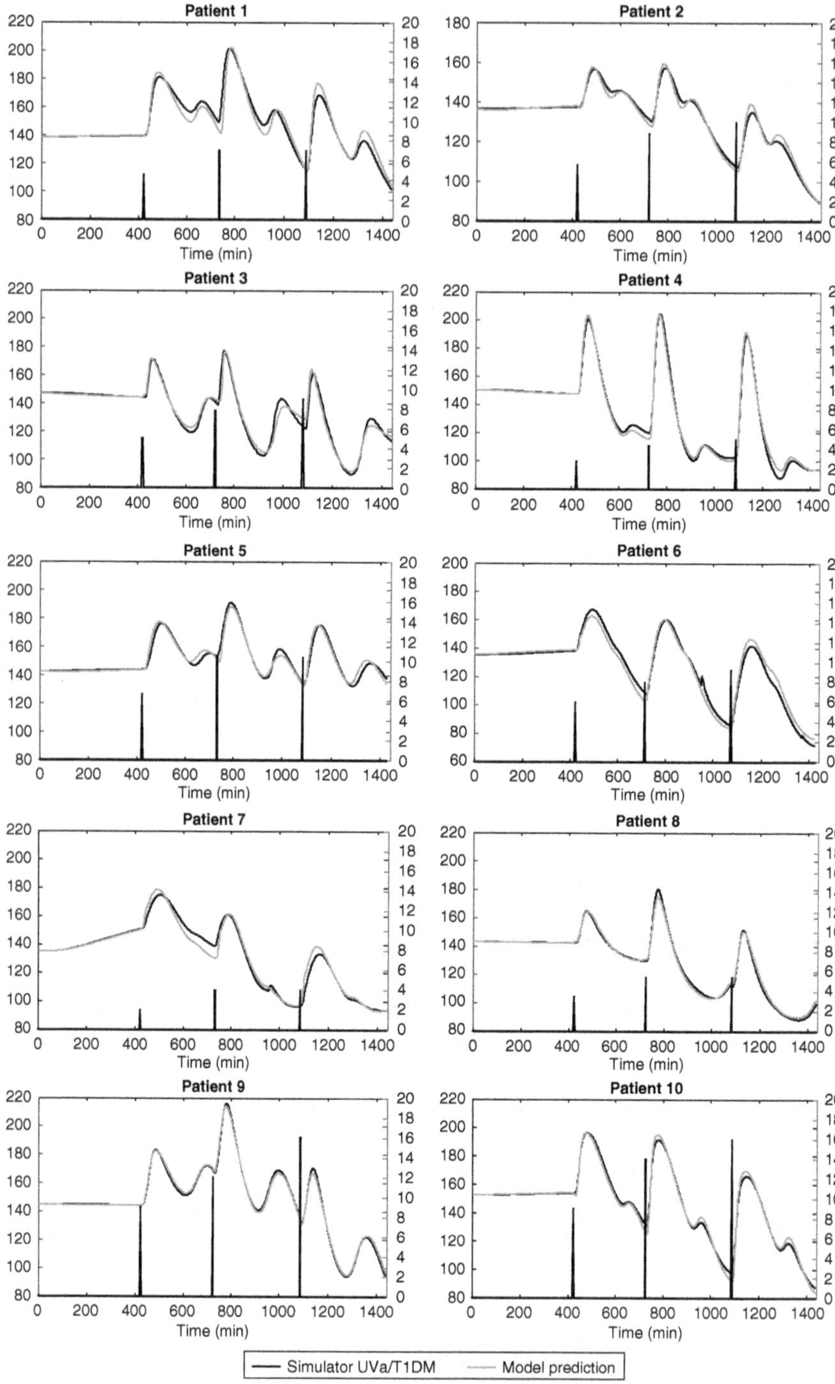

Figure 5.a.9 Comparison of blood glucose concentration (mg/dL) as predicted from the proposed model with the Simulator, for the 10 adults when a meal plan of 45 g, 70 g and 70 g of carbs are considered at 420 min, 720 min and 1080 min, respectively. The insulin infusion (U) is shown at the right axis for every patient.

The glucose profiles of the proposed model compared to the Simulator are shown in Figure 5.a.9 for the 10 patients.

The simulation results indicate that the proposed model can predict accurately the blood glucose concentration profile in the fasting, prandial and postprandial states. The good fit of the model to the UVa/Padova Simulator shows that the estimates of the most influential parameters, as identified from the model analysis, are well adjusted for all cases.

5.a.6 Dynamic Optimization

One of the great challenges of an automated system is the delayed insulin absorption and action. That means that there is a time lag between the time insulin is given and the time to cause the maximum effect. This time lag is related to the type of insulin used, the route of administration, the detection of a glucose fluctuation and the patient's sensitivity to insulin. The difference in the glycaemic response produced by the same dose of insulin in different individuals indicates that there is a high intra-patient variability involved in glucose–insulin interactions. When this variability is low, then a more predictable glycaemic response can be determined, which is important for a closed-loop system. In order to reduce the factors that cause variability and deteriorate the prediction of the glycaemic response, open-loop simulation analysis and optimization studies are performed to gain deep knowledge of the particular system and use the conclusions as a guideline for closed-loop studies.

In this study, the UVa/Padova T1DM Simulator (Kovatchev *et al.*, 2011; and see Appendix 5A) is used as the process model, which has been approved by the FDA to substitute animal trials in the pre-clinical testing of certain control strategies in T1DM. Simulation studies are performed to quantify the delayed insulin effect on 10 adult patients. This analysis has motivated the performance of patient-specific optimization studies, to find the optimal timing of insulin dosing to maintain the patient's glycaemic target. An alternative to bolus dosing regimens is investigated in order to be incorporated in the closed-loop insulin delivery strategy, and the results are presented.

5.a.6.1 Time Delays in the System

Time delay in a system is the time that intervenes from the instant the input, the control or a force is applied until the instant the effect is observed. In this particular system, the input is the insulin dose, and the effect is the decrease in the blood glucose concentration. Figure 5.a.10 reveals the complexity of blood glucose regulation when subcutaneous rapid-acting insulin is used. Rapid-acting insulin is a human insulin analogue that, due to its chemical structure, reduces aggregation of insulin molecules and therefore accelerates the absorption

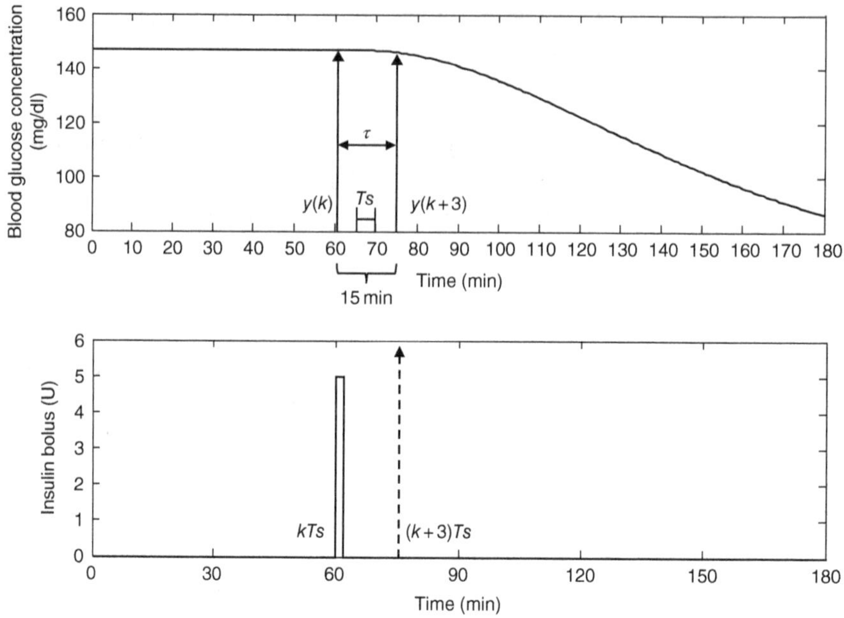

Figure 5.a.10 Delayed insulin effect. *Source*: Zavitsanou *et al.* (2014). Reproduced with permission of IEEE.

process. Assuming that the sampling time *Ts* is 5 min (available measurements of glucose concentration in the blood from the sensor), it can be noticed that insulin requires up to 15 min to initiate the decrease of blood glucose concentration, practically to observe a 1 mg/dl change of the concentration. This time involves the absorption of rapid-acting insulin through the subcutaneous tissue and insulin action that can take up to 1–3 h for its maximum effect.

In Figure 5.a.11, 1 U bolus of rapid-acting insulin is given at 60 min in four patients. It can be noticed that the time to observe a 10 mg/dl decrease of blood glucose concentration is not equal for the four patients. This can be explained by the fact that every patient responds differently to insulin and has a different ability to increase the body's glucose uptake from the various tissues. This can be quantified with the insulin sensitivity index. The more sensitive to insulin the patient is, the less amount of insulin is required. Patients 2 and 4 with a high insulin SI require less time for their blood glucose to be decreased than patients 1 and 3.

In Figure 5.a.12 for two patients, who high and low insulin sensitive, three bolus doses are given at 400 min without considering meal consumption. It can be noticed that the time required for glucose to be decreased by 10 mg/dl is dependent on the amount of bolus given. The delayed insulin effect decreases, while the amount of insulin bolus increases. This implies that the time delay property cannot be considered constant for an individual patient.

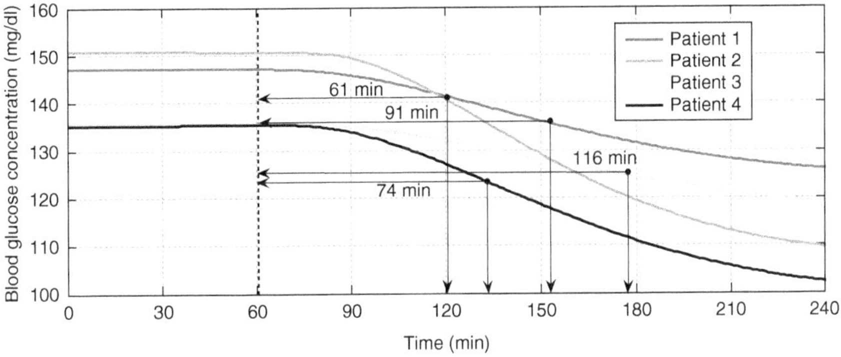

Figure 5.a.11 Patient-dependent time delay. *Source*: Zavitsanou *et al.* (2014). Reproduced with permission of IEEE.

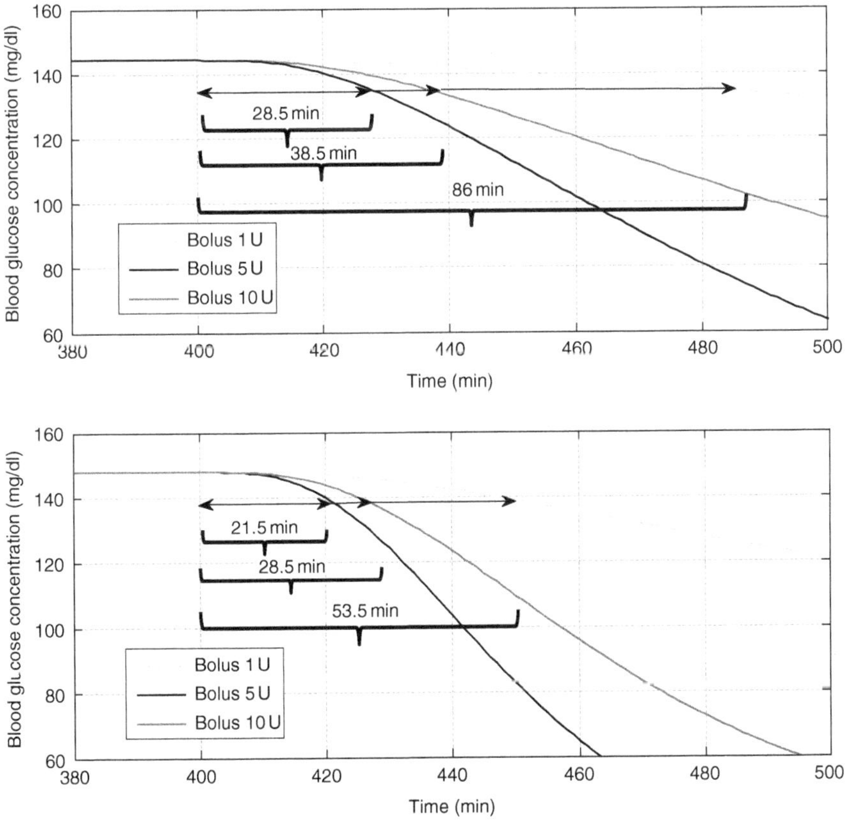

Figure 5.a.12 Time delay dependent on patient and bolus. *Source*: Zavitsanou *et al.* (2014). Reproduced with permission of IEEE.

In conclusion, the dynamic system involves inherent time delays which are the delayed insulin absorption and action and also the approximately 10 min delayed glucose appearance in the blood after food consumption due to interstitial glucose kinetics, meaning the route from the mouth to the small intestine and then to the blood. Apart from these delays, there are additional technical delays which involve the delayed detection of blood glucose concentration change because the continuous glucose monitoring devices calculate blood glucose concentration by measuring interstitial fluid glucose concentration (Keenan *et al.*, 2009). Hence, the time lag of the displayed glucose value and the real blood glucose value consists of the time lag between the Insulin Sensitivity Factor (ISF) and blood glucose accounting for the processing requirements as well.

5.a.6.2 Dynamic Optimization of Insulin Delivery

From the previous analysis, it has been evident that in order for patients to maintain their blood glucose close to their glycaemic target, the timing of the bolus insulin administration must be optimally decided to achieve safe glycaemic regulation. It has also been evident that each patient presents a unique response to insulin and therefore must be treated differently. Hence, patient-specific optimization studies are performed to obtain the optimal insulin profile that minimizes the time glucose is outside of the normal range. The mathematical formulation of the optimization problem has the following general form:

$$\min_{d_i} \int_0^{t_f} (w_1 + w_2)\,dt \tag{5.a.41}$$

s.t.

$$G = f\left(\dot{x}(t), x(t), y(t), u(t), d_i\right) \tag{5.a.42}$$

$$\sum_{i=1}^{N_{\text{int}}} d_i = 1, \qquad d_i \in \{0,1\} \tag{5.a.43}$$

$$w_1 \geq \varepsilon, \qquad w_1 \geq G - G_{\max} \tag{5.a.44.a}$$

$$w_2 \geq 0, \qquad w_2 \geq G_{\min} - G \tag{5.a.44.b}$$

where tf is the time horizon, G is the blood glucose concentration described by the nonlinear process model specific for every patient (Kovatchev *et al.*, 2011; and see Appendix 5A), and $Gmax$ (140 mg/dL) and $Gmin$ (70 mg/dL) are the upper and lower glucose concentration bounds. Equation (5.a.44.a) is a soft constraint, as opposed to (5.a.44.b) which is a hard constraint to prevent from any severe health complications related to hypoglycaemia. At $t_0 = 400$ min, a breakfast meal of 50 g of carbohydrates was given to the 10 patients. The optimal amount of insulin, appropriate to compensate for the forthcoming glucose increase due to the meal intake, was provided by the Simulator when closed-loop studies were performed and was chosen for every patient. The optimization studies were performed in gPROMS (PSE, 2011c). A window of 4 h before the meal was considered to include any extreme low insulin sensitive patient, and this time span was discretized every 2 min, which is the time the pump requires to deliver an insulin bolus (hence, Nint = 120). A time-invariant, binary variable *di* was considered to be 0 if no bolus was given or 1 at time *i* if a bolus was given. The mixed-integer nonlinear programming problem was solved using the approach described in Bansal *et al.* (2003) as implemented in gPROMS. An augmented penalty strategy is employed to increase the possibility to obtain a global solution (PSE, 2011c).

The optimization results are presented in Figure 5.a.13 for six patients. The grey line shows the optimized glucose profile, while the black line shows the simulated profile when the bolus is given simultaneously with a meal. The optimal timing of insulin administration for every patient is summarized in Table 5.a.15. When the bolus is given at the optimal time, the glucose profile is improved in terms of maintenance of the concentration within the normal range for all the patients. In Table 5.a.15, the area between the upper glucose bound and the glucose profile is calculated. The difference of the values between the simulated and optimized curves indicates that a superior regulation of glucose is achieved when insulin infusion scheduling is considered. Additionally, hypoglycaemic events are not observed for any of the patients, despite the considerable difference in timing between them. This is related to the sensitivity of the patient to insulin, as mentioned, and for the specific optimal dose the patient would not reach the lower glucose bound.

5.a.6.3 Alternative Insulin Infusion

An alternative to bolus dosing is considered as a piecewise constant infusion rate that holds a specific value for 5 min time intervals. The profile is calculated with an optimizing criterion, the minimum range of glucose outside the normal bounds. Figure 5.a.14, for patient 1, includes the optimized glucose profile when the bolus is given at the time calculated with the previous optimization problem (a), and the glucose profile when a piecewise approach is considered (d) with a time frame of 32 min (Table 5.a.15); both are compared with the glucose profile when a bolus is given simultaneously with a meal (b). The two approaches produce the same effect on glucose, indicating that a stepwise infusion could be considered as a possible mechanism since it provides flexibility and can be

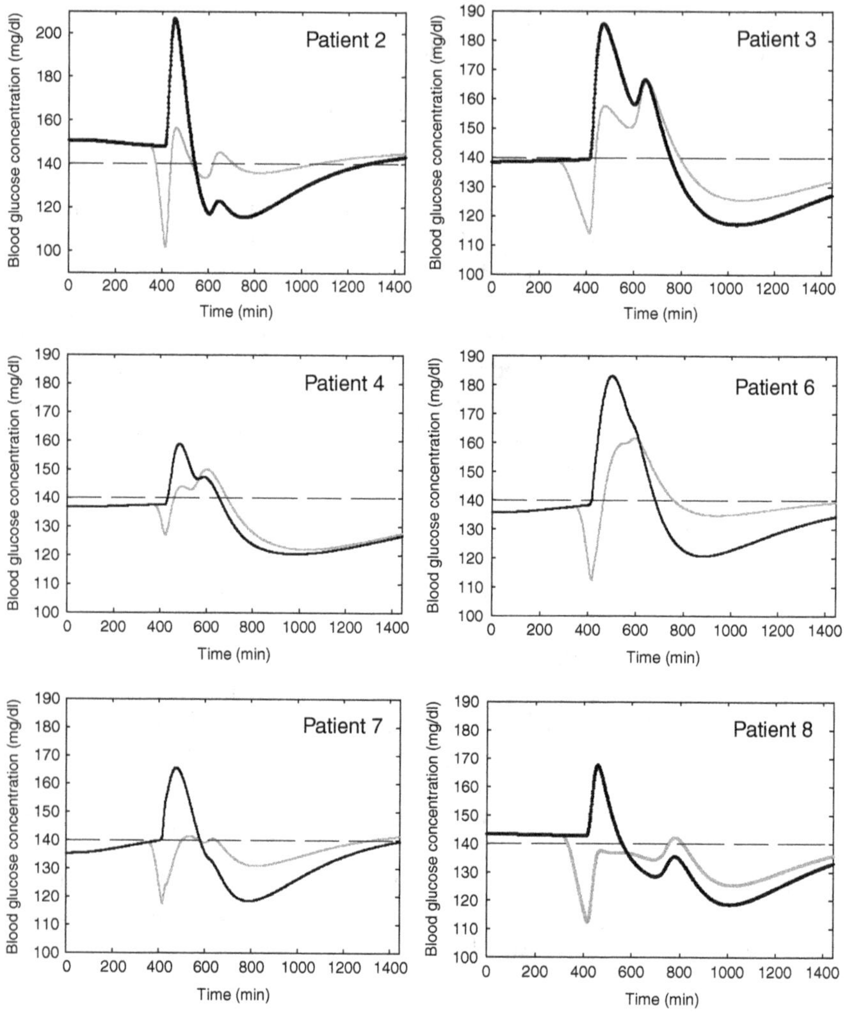

Figure 5.a.13 Optimization (grey line) and simulation (black line) glucose profiles. *Source*: Zavitsanou *et al.* (2014). Reproduced with permission of IEEE.

better adjusted in an automated delivery system. In Figure 5.a.14, for patient 5, in order to avoid a long time frame (62 min) which can be restricting from a control point of view, a time frame of 30 min is considered. The glucose profiles are compared, and additionally the profile when bolus is given 30 min in advance (c) is included. The stepwise approach (d) and the 30 min bolus in advance (c) produce comparatively the same results. This approach, although it is not optimal, can still be regarded as a considerable alternative for control design.

Table 5.a.15 Area under the curve (outside the normal range).

	Simulated glucose curve	Optimized glucose curve	Optimal time of bolus before meal
Patient 1	5.1747e+03	4.4825e+03	32 min
Patient 2	6.7083e+04	5.8923e+03	66 min
Patient 3	8.3306e+03	5.1267e+03	140 min
Patient 4	2.1919e+03	1.3213e+03	36 min
Patient 5	4.0646e+03	1.7180e+03	62 min
Patient 6	2.0961e+05	3.9445e+03	62 min
Patient 7	2.0833e+05	2.7859e+02	52 min
Patient 8	5.9726e+04	1.2093e+03	100 min

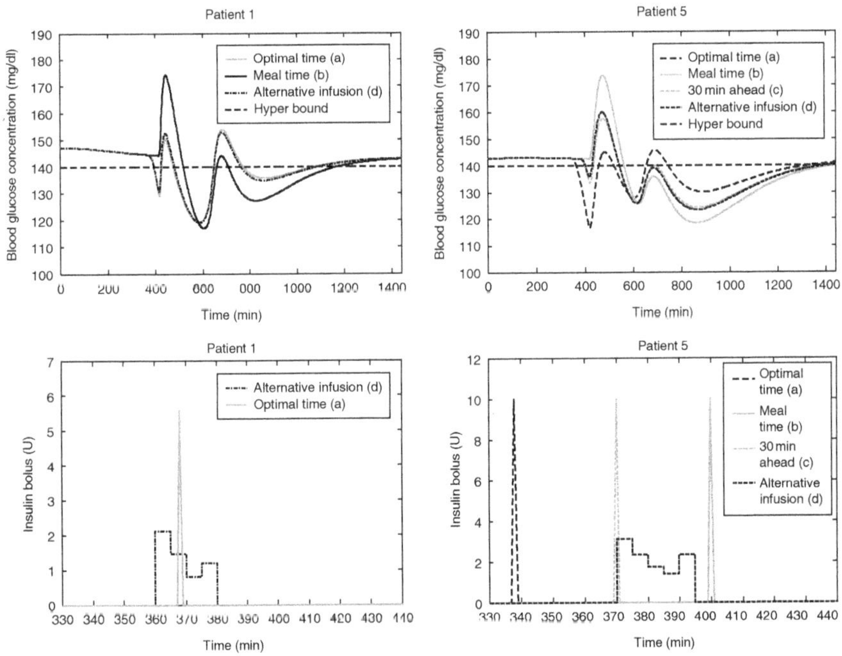

Figure 5.a.14 Optimal glucose profiles when insulin is given as a bolus and as a piecewise constant infusion. *Source*: Zavitsanou *et al.* (2014). Reproduced with permission of IEEE.

5.a.6.4 Concluding Remarks

Exogenous insulin administration causes delayed effect on glucose regulation. The involved time lags have been quantified for 10 patients, and it has been shown that for the same insulin dose, the delayed effect on glucose is patient dependent. Therefore, patient-specific, in terms of appropriate insulin dosing for each patient, optimization studies were performed to find the optimal timing to give the bolus dose. An alternative, stepwise insulin regimen has been considered, and the optimization results indicate that it could provide a considerable alternative for closed-loop applications.

Part B: Type 1 Diabetes Mellitus: Glucose Regulation

Stamatina Zavitsanou, Athanasios Mantalaris, Michael C. Georgiadis, and Efstratios N. Pistikopoulos

5.b Type 1 Diabetes Mellitus: Glucose Regulation

5.b.1 Glucose–Insulin System: Typical Control Problem

T1DM is a lifelong disease, and therefore its treatment with exogenous insulin should have the minimal impact on the patient's lifestyle. It is necessary to develop novel drug delivery techniques that suggest a structure of drug administration which ensures the therapeutic efficacy and safety of the patient, and take into consideration the patient's comfort and convenience. Motivated by the challenge to improve the living standard of a diabetic patient, the idea of an artificial pancreas that mimics the endocrine functionality of a healthy pancreas has been well established in the scientific society. See Section 5.a.1.1, "The concept of the artificial pancreas," for discussion.

The blood glucose–insulin system can be formulated as a typical control system. The *plant* is the glucoregulatory system itself, the *manipulated variable* is insulin and the *controlled variable* is blood glucose concentration, as presented in Figure 5.b.1. The system undergoes external disturbances such as meal consumption, exercise, illness, stress and so on. The two fundamental components of a control system are the model and the control strategy.

Several control methodologies have been suggested in the literature (Doyle *et al.*, 2014; Thabit & Hovorka, 2014), such as PID, model predictive control (MPC) and fuzzy logic. MPC theory has been widely established as a possible choice for this particular application. Table 5.b.1 highlights the studies on MPC where its performance has been clinically evaluated in patients with T1DM.

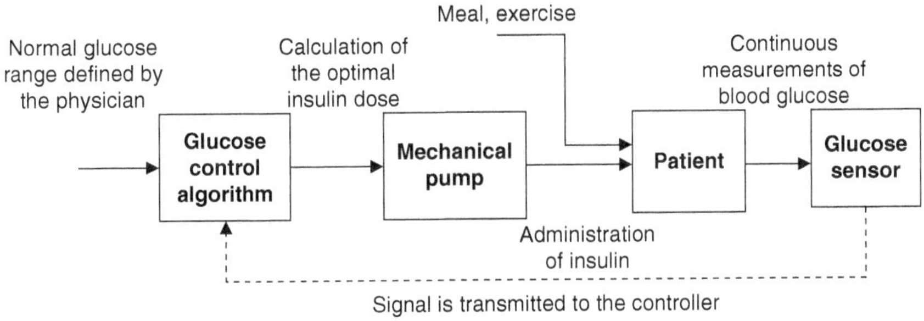

Figure 5.b.1 Model-based control structure.

Table 5.b.1 Selected clinical studies that evaluate MPC as a control strategy to regulate BG concentration in T1DM.

Clinical studies: references	Summary
Hovorka *et al.* (2010, 2014) Elleri *et al.* (2012, 2013, 2011)	The MPC design of this study is based on Bequette (2005), using an internal model (Hovorka *et al.*, 2002).
Kovatchev *et al.* (2010, 2013)	The linear MPC design is described in Magni *et al.* (2007). The model used for validation is found in Dalla Man *et al.* (2007) but modified adequately for T1DM. This model is linearized at average population basal conditions. The MPC specifications are tailored to each patient. An interface and safety module are included in Patek *et al.* (2012).
Russell *et al.* (2012)	Bihormonal closed-loop system (El-Khatib *et al.*, 2010) insulin administration with MPC control and glucagon with PD. The internal model is ARMAX with identified model parameters.
Dassau *et al.* (2013)	The linear mp-MPC design is described in Percival *et al.* (2010), the model used is a transfer function with patient-specific parameters. More details on the explicit MPC can be found in Dua *et al.* (2006).
Breton *et al.* (2012)	Range correction module and safety supervision module (Kovatchev *et al.*, 2009)

ARMAX, autoregressive–moving-average (model); BG, blood glucose; MPC, model predictive control; mp-MPC, multiparametric model predictive control; PD, pharmacodynamics; T1DM, type 1 diabetes mellitus.

Although the applied MPC theory for glucose regulation has reduced the occurrence of hypoglycaemic episodes in most clinical studies (Doyle *et al.*, 2014), the challenge remains when the patient is examined in free living conditions (Bequette, 2012), subjected to unannounced disturbances such as a meal. This involves the risk of direct prandial hyperglycaemia that leads to aggressive insulin action and possible postprandial hypoglycaemia. Another important issue is the high intra- and inter-patient variability that dominates the system. To address this problem, patient-specific approximations of the original system (Magni *et al.*, 2009; van Heusden *et al.*, 2012) and control specifications are considered. Although this approach has minimized the effect of intra-patient variability on predictability of the internal model and therefore reliability of the regulation, inter-patient variability remains an important source of uncertainty that requires advanced control techniques such as robust control (Sakizlis *et al.*, 2004; Pistikopoulos *et al.*, 2009) or complementary components (Breton *et al.*, 2012) to incorporate the effect and restrict its impact on the system. In this part, the involved steps of closed-loop insulin delivery are presented, while emphasis is given on the importance of developing a reliable, patient-specific approximate model for MPC.

5.b.2 Model Predictive Control Framework

The general framework according to which the controller to regulate the BG concentration is designed is presented in Figure 5.b.2, as adapted from Pistikopoulos (2009). It involves the development of a high-fidelity model that accurately predicts the glucose–insulin dynamics in T1DM, the simplification of the original model with system identification or model order reduction techniques to derive a reliable approximation of the system dynamics, and finally the design of the appropriate control strategy. In the MPC formulation, one of the key components is the approximate model; it needs to be relatively simple to facilitate the computational complexity, but also very informative to enclose the system dynamics. The involved steps are described analytically in the remainder of this section.

5.b.2.1 "High-Fidelity" Model

The mathematical model used in this study as a virtual patient for closed-loop control validation studies and to derive simplified or approximate models necessary for model-based control is the model developed by the Cobelli group

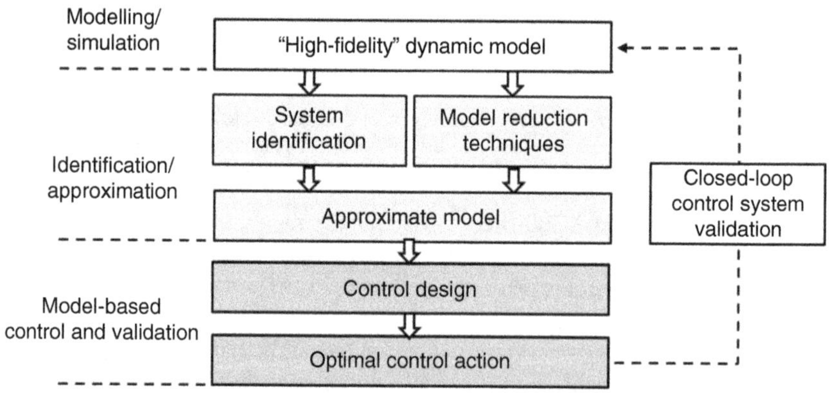

Figure 5.b.2 Framework for MPC controller design.

(Dalla Man *et al.*, 2007a, 2007b) and approved for pre-clinical closed-loop studies from the FDA as the Uva/Padova Simulator. The model is simulated and fully validated in gPROMS (Appendix 5A) using individual patient parameters obtained from the UVa/Padova Simulator for 10 adults.

5.b.2.2 The Approximate Model

Sources of nonlinearity in the model of glucose–insulin interactions can be found not only in nonlinear expressions of specific variables (e.g. gastric emptying) but most importantly in nonlinear dependencies among variables (e.g. insulin-dependent peripheral glucose absorption). Sources of nonlinearity originated from insulin action on glucose uptake from the periphery and overall effect on blood glucose decrease. Another challenging inherent source of nonlinearity in this system is the involved time delays. The time that intervenes from the instant the input is applied until the instant the effect on glucose is observed is not proportional to insulin dosage.

The internal model used to predict the future output $G(t)$ depending only on past inputs $u(k-1)$, $u(k-2)$, …, is usually considered to be linear because this constitutes the calculation of the optimal insulin infusion relatively simplified in a MPC framework.

5.b.2.2.1 Linearization

The model of the UVa/Padova Simulator is linearized (Appendix 5C). The linear model involves 12 states:

$$x=\begin{bmatrix}G_p & G_t & X_{disp} & Q_{sto1} & Q_{sto2} & Q_{gut} & I_{del1} & I_{del2} & I_{sc1} & I_{sc2} & I_l & I_p\end{bmatrix}'$$

When the model is linearized at the steady state, an approximation of constant physiological conditions, the glucose concentration does not coincide with the profile of the original model in the presence of meal disturbances and insulin boluses, resulting in large offset. To overcome the difficulty to find stable equilibrium points during meal consumption and insulin absorption, which trigger the system away from the steady state, and to capture the dynamics of the system during fasting, and prandial and postprandial conditions of different meal sizes and insulin boluses, a series of parameter estimation studies are performed to estimate the values of specific parameters of the linear model related to meal and insulin absorption that are described with nonlinear equations (Appendix 5C). The parameter estimation studies are performed in gPROMS (PSE, 2011b) and involve the design of patient-specific *in silico* experiments of different meal plans and insulin regimens that take into consideration:

1) Effect of one meal on BG concentration – no bolus is considered (Experiment A)
2) Effect of one bolus on BG concentration – no meal is considered (Experiment B)

Table 5.b.2 Estimated parameters of linearized model for 10 adults.

	Experiments	Patient 1	Patient 2	Patient 3	Patient 4	Patient 5
A(2,2)	C, D, E	−0.10475	−0.12086	−0.30005	−0.1493	−0.21861
A(2,3)	C, D, E	−0.01763	−0.01307	−0.01682	−0.065476	−0.00679
A(5,5)	A	−0.00516	−0.01893	−0.00897	−0.018062	−0.00361
A(6,4)	A	0.01314	0.00157	0.00314	0.009870	0.01552
$B_d(2,1)$	C, D, E	1.627	0.17399	2.2616	3.9858	0.50847
	Reduced model					
A(9,9)	B, E	−0.043407	−0.018901	−0.020299	−0.018992	−0.025647
A(9,10)	B, E	2.891E-5	0.0	−1.073E-6	−6.346E-6	2.208E-5
A(10,9)	B, E	0.003991	0.001407	0.001059	0.002412	0.004241
A(10,10)	B, E	−0.026149	−0.021764	−0.01929	−0.021041	−0.04864
		Patient 6	**Patient 7**	**Patient 8**	**Patient 9**	**Patient 10**
A(2,2)	C, D, E	−0.07788	−0.13319	−0.08963	−0.17237	−0.21344
A(2,3)	C, D, E	−0.024081	−0.049294	−0.01747	−0.00375	−0.01069
A(5,5)	A	−0.015856	−0.014347	−0.00743	−0.02216	−0.01217
A(6,4)	A	−0.000119	6.386E-4	0.00442	−3.458E-7	−1.407E-5
$B_d(2,1)$	C, D, E	3.516	1.5165	1.4351	0.72648	2.1189
	Reduced model					
A(9,9)	B, E	−0.012480	−0.020122	−0.025168	−0.018841	−0.017661
A(9,10)	B, E	−4.983E-8	−1.460E-8	−2.548E-8	−1.053E-7	−3.308E-8
A(10,9)	B, E	8.275E-4	0.001284	0.001802	0.001677	0.001356
A(10,10)	B, E	−0.018837	−0.013686	−0.014726	−0.030188	−0.022085

3) Effect of one meal and bolus given simultaneously (Experiment C)
4) Steady state – no bolus and meal are considered (Experiment D)
5) Day simulation with different meal sizes and bolus doses (Experiment E).

The values of the estimated parameters are presented in Table 5.b.2 for the 10 adults.

5.b.2.2.2 Physiologically Based Model Reduction

In order to reduce the computational complexity in a control application caused by the relatively large size of the previously presented 12-state linear physiological model, physiologically based model order reduction is used to mathematically transform the model equations to provide the same dynamical behaviour but in a smaller system. The involved time delays of the system, both in glucose absorption from food and in insulin absorption through the subcutaneous tissue, do not allow the lumping of many compartments and further simplification of the model. The equations to be reduced are the states $[I_{sc1}\ I_{sc2}\ I_l\ I_p]'$. The compartments I_{sc1} and I_p are forced to be left unmodified since they are used in other equations in the model (Appendix 5B).

The linear equations are described with the general formulation:

$$\frac{dy}{dt} = Ky \tag{5.b.1}$$

where $y = [I_{sc1}\ I_{sc2}\ I_l\ I_p]^T$ and $K = \begin{bmatrix} k_d + k_{a1} & 0 & 0 & 0 \\ k_d & -k_{a2} & 0 & 0 \\ 0 & 0 & -(m_1 + m_3) & m_2 \\ k_{a1} & k_{a2} & m_1 & -(m_2 + m_4) \end{bmatrix}$

The new system is described by Equation (5.b.1), where $\hat{K}$ is the new set of parameters with rank $n = 2$. The new parameters are found by solving a maximum likelihood parameter estimation problem in gPROMS (PSE, 2011b) that determines the values of the new set of parameters that maximize the probability that the new mathematical equations will predict the dynamics of the original model that is used to specify suitable experiments obtained from the experiments.

$$\frac{d\hat{y}}{dt} = \hat{K}y \tag{5.b.2}$$

The set of the reduced equations is defined as:

$$\frac{d}{dt}\begin{bmatrix} I_{sc1} \\ I_p \end{bmatrix} = \begin{bmatrix} A(9,9) & A(9,10) \\ A(10,9) & A(10,10) \end{bmatrix}\begin{bmatrix} I_{sc1} \\ I_p \end{bmatrix}$$

The values of the parameters are presented in Table 5.b.2 for the 10 patients.

Hence, the states of the reduced model are:

$$x_{red} = \begin{bmatrix} G_p & G_t & X_{disp} & Q_{sto1} & Q_{sto2} & Q_{gut} & I_{del1} & I_{del2} & I_{sc1} & I_p \end{bmatrix}'$$

Further reduction of the model states leads to loss of the system dynamics. The model is discretized with $t_s = 5$ min. Figure 5.b.3 compares the dynamic model with the state-space reduced-order model, and Figure 5.b.4 shows the accuracy of the linearized model.

The model accuracy is calculated using Equation (5.b.3), and for patient 2 it is 81%.

$$Fit = \left(1 - \frac{\sum_{i=1}^{N} |y_i - \widehat{y_i}|}{\sum_{i=1}^{N} |y_i - \overline{y}|}\right) \times 100 \tag{5.b.3}$$

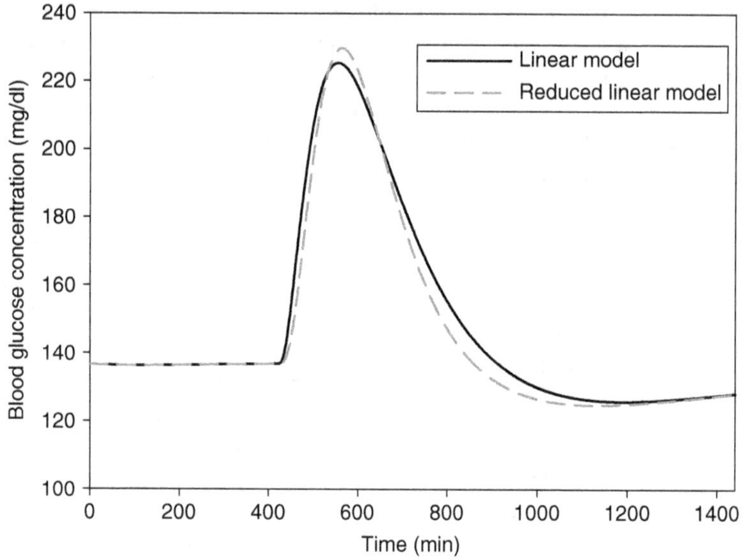

Figure 5.b.3 Comparison of full-state and reduced linearized model for patient 2.

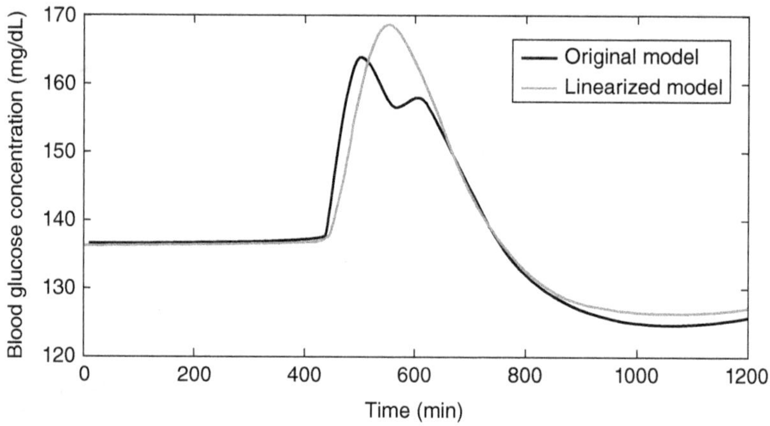

Figure 5.b.4 Comparison of original model and linearized model when 50 g of carbs are consumed and a 5 U bolus is given to patient 2.

The advantage of the linearized model over the previously presented model is that the states represent the states of the original model, and therefore the deep knowledge of the system's behaviour facilitates the design of control studies. Yet again, this model is large, which makes the design of explicit MPC relatively difficult.

5.b.3 Control Design

5.b.3.1 Model Predictive Control

The suggested control strategy appropriate for glucose regulation with manipulating insulin infusion refers to MPC theory (Dua *et al.*, 2006, 2009). The reference point of all the MPC methodologies is the use of a mathematical representation of the controlled system (model) to predict the system's output/states, for a finite time horizon (prediction horizon). The model is used for the formulation of an optimization problem that minimizes an appropriately chosen objective function. Decision variables of this problem are the future values of the manipulated variable (insulin) in a generally smaller future time horizon, the control horizon. When the optimal sequence of the future control laws is determined, only the first value is applied on the system and the optimization problem is then reformulated and solved at the next time instant, when new system information is available. The basic concept of MPC is illustrated in Figure 5.b.5.

The appropriate current control action is obtained by solving on-line, at each sampling instant, the finite-horizon open-loop optimal control problem:

$$\begin{aligned} \min_{x,y,u} J = & \sum_{k=1}^{N-1} x_k' Q_k x_k + \sum_{k=1}^{N-1} \left(y_k - y_k^R\right)' QR_k \left(y_k - y_k^R\right) \\ & + \sum_{k=1}^{M-1} u_k' R_k u_k + \sum_{k=1}^{M-1} \Delta u_k' R_{1_k} \Delta u_k \\ \text{s.t.} \quad & x_{k+1} = Ax_k + Bu_k + Ed_k + Gw_k \\ & y_k = Cx_k + \upsilon_k \\ & u_{min} \le u_k \le u_{max} \\ & y_{mn} \le y_k \le y_{max} \\ & \Delta u_{min} \le u_{k+1} - u_k \le \Delta u_{max} \end{aligned} \tag{5.b.4}$$

where $x \in \mathbb{R}^n$ is the state, $n = 10$; $u \in \mathbb{R}^m$ is the input, $m = 1$, the insulin infusion in this case; $d \in \mathbb{R}^m$ is the meal disturbance; $y \in \mathbb{R}^p$ is the output; $p = 1$ is the blood glucose concentration; and $k \in \mathbb{I}_{\ge 0}$ T is a nonnegative integer representing the sample number. $A \in \mathbb{R}^{n\times n}$ is the state transition matrix, $B \in \mathbb{R}^{n\times m}$ is the input matrix, $E \in \mathbb{R}^{n\times m}$ is the disturbance matrix and $C \in \mathbb{R}^{p\times n}$ is the output matrix. N is the prediction; M is the control horizon; and Q, QR, R and R_1 are the controller's tuning parameters, which are the weight matrix for the states, the output (glucose), the control input (insulin) and the change of control input, respectively. The variables $w_k \in \mathbb{R}^g$ and υ are introduced to account for process and observation uncertainty, respectively. Matrix $G \in \mathbb{R}^{n\times g}$ is used to fine-tune the effect of the process uncertainty on the states. In order to account for this type of uncertainty, an optimal state estimation is determined for

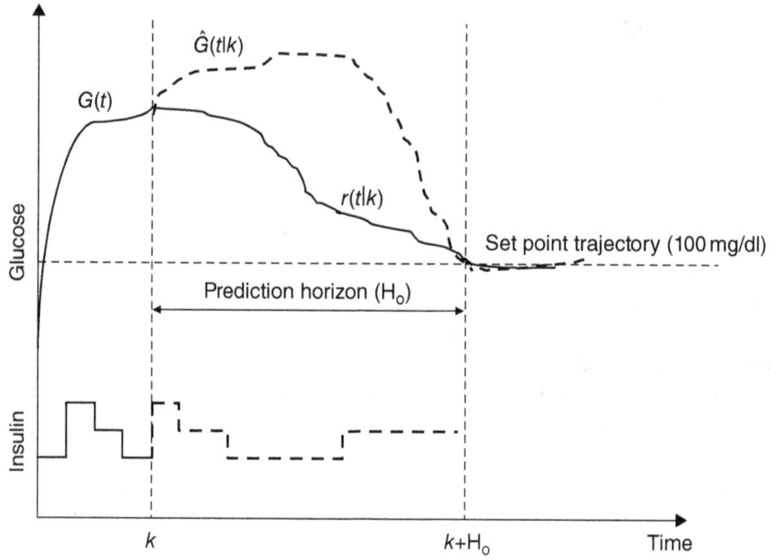

Figure 5.b.5 Basic scheme of discrete MPC.

reliable glucose regulation. A linear Kalman filter is a suitable state estimator for the examined system, assuming a zero-mean normally distributed process disturbance w_k and measurement noise υ_k with covariance matrices Q and R.

$$w(0) \sim N(0,Q) \text{ and } \upsilon(0) \sim \mathrm{N}(0,R).$$

Details on the Kalman filter can be found in Appendix 5B.

5.b.3.2 Proposed Control Design

The general proposed control design is presented in Figure 5.b.6. Depending on the nature of the meal disturbances d_i, as defined in Table 5.b.3, different components are activated.

Hence, the control designs presented in Table 5.b.4 are evaluated for different types of meal disturbances d_i.

5.b.3.3 Prediction Horizon

Because the system involves high input and disturbance delays, in order to predict the after-effect of a given input at time k (see Figure 5.a.10), the prediction horizon should be at least equal to the time lag. However, every patient has different glucose–insulin dynamics, and the time delay factor should be considered patient specific. Although the time lag is dependent on the insulin dose

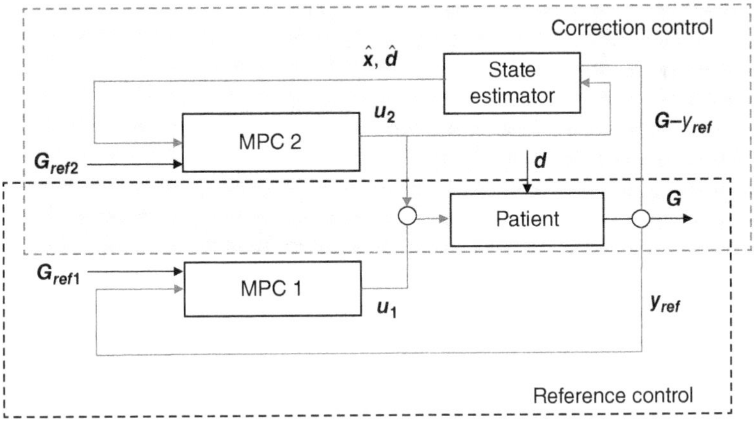

Figure 5.b.6 Proposed control strategy to compensate for unknown meal disturbances consisting of two controllers: the reference control that regulates glucose for a reference meal plan, and the correction control that regulates the difference of the glucose between a real and reference meal plan.

Table 5.b.3 Meal disturbance types.

Symbol	Disturbance type
d_p	Predefined
d_a	Announced
d_u	Unknown

Table 5.b.4 Control designs.

CD_1	Online MPC with predefined disturbance
CD_2	Online MPC with announced disturbance
CD_3	Reference and correction MPC for unmeasured disturbance
CD_4	Online MPC with unmeasured disturbance

(see Section 5.a.6), for the insulin infusion rates considered in the closed-loop system, an assumption of constant patient-specific time delays can be reasonable. Table 5.b.5 shows the prediction horizon of the 10 patients that was calculated as the average time to observe a 1 mg/dL change in blood glucose concentration when simulation studies of a step change of 0.5 U/h insulin dose from 0 to 5 U/h were performed.

Table 5.b.5 Prediction horizon for the 10 patients.

Patient	OH	Patient	OH
Patient 1	13 ≙ 65 min	Patient 6	10 ≙ 50 min
Patient 2	11 ≙ 55 min	Patient 7	8 ≙ 40 min
Patient 3	7 ≙ 35 min	Patient 8	11 ≙ 55 min
Patient 4	10 ≙ 50 min	Patient 9	14 ≙ 70 min
Patient 5	13 ≙ 65 min	Patient 10	13 ≙ 65 min

5.b.3.4 Control Design 1: Predefined Meal Disturbance

When a patient follows an exact meal plan, meaning that the exact amount and the time of the meal are known in advance, this information is introduced in the general regulation design as a predefined meal plan and the MPC 1 in Figure 5.b.6 is activated (the MPC specifications are presented in Table 5.b.6).

5.b.3.5 Control Design 2: Announced Meal Disturbance

When information concerning the amount of meal is provided at the time, the disturbance is considered as announced, and the control strategy involves again the MPC 1 of Figure 5.b.6 (see Table 5.b.6 for specifications).

5.b.3.6 Control Design 3: Unknown Meal Disturbance

In the case of unmeasured meal disturbances, there is no information concerning the amount and time of the meal. In this case, a nominal controller reacts aggressively to regulate the glucose deviation from the reference point, which means increased insulin infusion as long as glucose violates the constraints. But this control action involves the risk of postprandial hypoglycaemia due to insulin after-effects and also immediate prandial hyperglycaemia. Therefore, a different control design is proposed to compensate for unknown disturbances, as illustrated in Figure 5.b.6. It consists of the patient model; an MPC controller, acting as a reference regulator; a second MPC controller, acting as the correction control; and a state estimator. The proposed control design regulates the glucose concentration when a reference meal plan is considered and additionally responds appropriately to compensate for the deviation from the reference meal when a different-sized meal is consumed.

MPC 1: Reference Control

The desired glucose value G_{ref1} is set by the endocrinologist for every patient. A predefined reference meal plan is considered to trigger the control action.

Feedback about the current state is obtained by the model output y_{ref}, as calculated when the reference tracking problem with an announced disturbance is solved and the optimal insulin infusion is applied.

MPC 2: Correction Control

MPC 2 aims to find the optimal insulin infusion rate to regulate the difference of glucose as a real measurement coming from the patient, *G*, and glucose as calculated when solving the reference control problem, y_{ref}. This difference can be regarded as an unmeasured disturbance of the system that leads to an offset in the set point, $G_{ref,2}=0$. So the correction control is described as a disturbance rejection problem. In order to remove the offset and the nonzero disturbances, the original system is augmented with a disturbance model. In order to reduce the computational effort, the states describing the meal absorption $[Q_{sto1}\ \ Q_{sto2}\ \ Q_{gut}]'$ are removed from the state-space model. The output feedback of the patient is obtained as the difference between the actual measurement and the reference control output ($G - y_{ref}$), and the state feedback is obtained by a state estimator that provides information about the current state of the patient and the additional disturbance.

The matrix $B_d \in \mathbb{R}^{n \times n_d}$ is chosen to be the B_d matrix of Section 5.b.3, and matrix $C_d = I \in \mathbb{R}^{n_d \times n_d}$. The new derived augmented linear model $(n=7, n_d=1)$ is detectable (see Equation [5.b.5]), which means that the states will converge to the real states when a Kalman filter is used; hence, this strategy can be employed:

$$rank\begin{bmatrix} I-A & -B_d \\ C & C_d \end{bmatrix} = n + n_d = 8 \tag{5.b.5}$$

The estimated states are:

$$\hat{x} = \begin{bmatrix} G_p & G_t & X_{disp} & I_{del1} & I_{del2} & I_{sc1} & I_p & d \end{bmatrix}'$$

and $y_2 = G - y_{ref}$. The control specifications for MPC 1 and MPC 2 are presented in Tables 5.b.6 and 5.b.7. The desired glucose value y_{ref1} is set at 100 mg/dL for all patients.

Table 5.b.6 Inequality constraints.

Variable	Value	Variable	Value	Variable	Value
y_{min}	80 mg/dL	u_{min}	0	Δu_{min}	0
y_{max}	140 mg/dL	u_{max}	TDD * U/min	Δu_{max}	0.005 U/min

TDD, Total daily dose of each patient.

Table 5.b.7 Specifications of MPC 2 and the Kalman filter.

Variable	Value	Variable	Value	Variable	Value
$y_{2,min}$	−10 mg/dL	u_{min}	0 U/min	$\hat{Q}$	100
$Y2_{,max}$	10 mg/dL	u_{max}	0.02 U/min	$\hat{R}$	5

5.b.3.7 Control Design 4: Unknown Meal Disturbance

The performance of a single MPC controller when unknown meal disturbances are imposed is evaluated with CD_4 (the MPC specifications are presented in Table 5.b.6).

5.b.4 Simulation Results

In this section, the control designs are evaluated for predefined, announced and unknown disturbances for 10 adults with T1DM, provided by the Simulator. The model is developed in gPROMS, while the control designs are in MATLAB and gO:MATLAB to exchange data between the two environments.

5.b.4.1 Predefined and Announced Disturbances

The results are illustrated in Figure 5.b.7. Meals of 45, 70 and 60 g of carbs are consumed at 420, 720 and 1080 min, respectively.

5.b.4.2 Unknown Disturbance Rejection

In this section, the CD_3 control design as explained before is evaluated. The ability of the controller to maintain the blood glucose concentration in the normal range is tested for large meal sizes of 75, 100 and 90 g of carbohydrates given for breakfast at 7:00 am, lunch at 13:00 pm and dinner at 18:00 pm, respectively. The reference meal plan is 20, 30 and 25 g, respectively. The results are compared to the CD_4 for the same meal sizes and presented in Table 5.b.8 and Table 5.b.9.

Table 5.b.8 shows that with CD_3, on average 54% of the time is spent within the normal glucose values, while with CD_4 the percentage of time spent in the normal range is 45%. With CD_3 there is no event of hypoglycaemia and the minimum observed glucose value is 71 mg/dL for adult 9, in opposition to CD_4 for which an average 3.1% of the time is spent in hypoglycaemia with a minimum observed glucose value of 43 mg/dL. Additionally, the time spent in

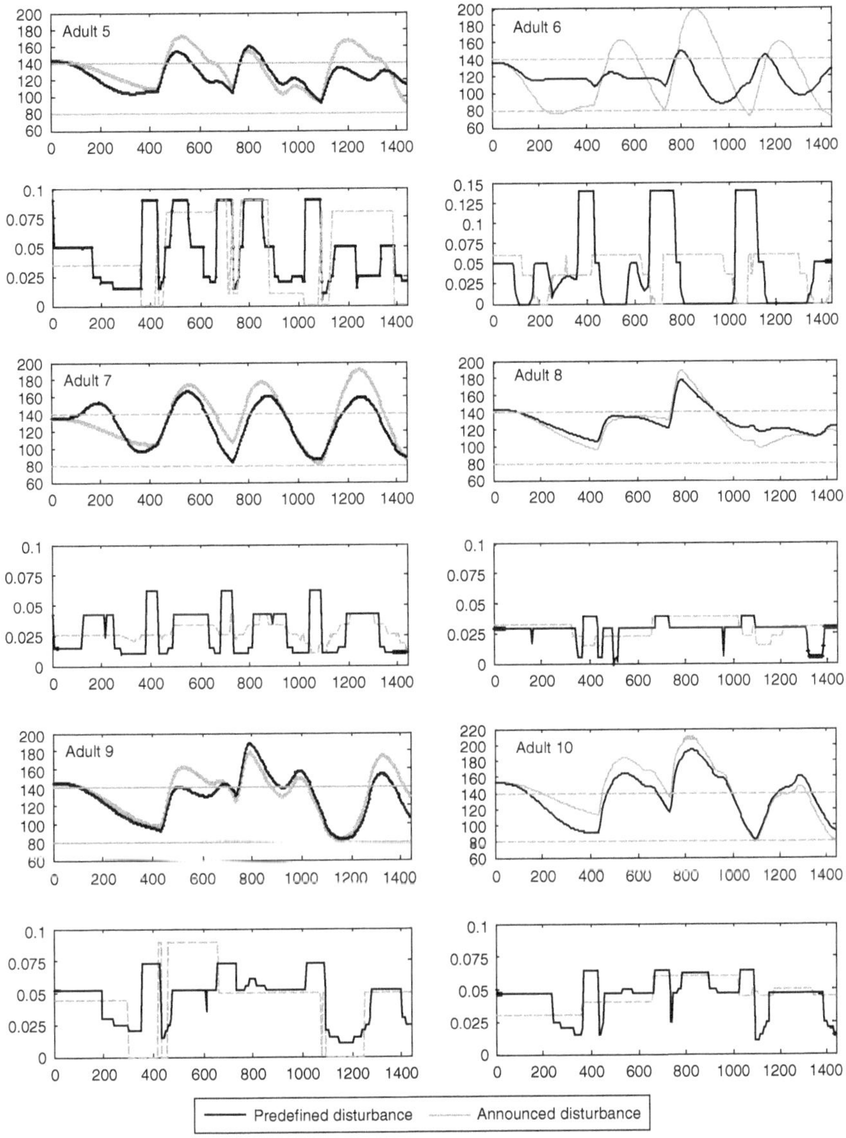

Figure 5.b.7 MPC control for 10 adults of UVa/Padova Simulator for predefined (**CD₁**) and announced meal disturbances (**CD₂**). Upper graphs: blood glucose concentration (mg/dL) profiles; lower graphs: control action, insulin (U/min).

Table 5.b.8 CD_3 (predefined meal plan).

	% Time <70	% Time <80	% Time 80<G<140	% Time 140<G<180	% Time 180<G<250	% Time >250	G_{min} (mg/dL)	G_{max} (mg/dL)
Adult1	0	1.7	44.4	38.9	14.9	0	72	247
Adult2	0	0	68.4	28.4	3.1	0	83	187
Adult3	0	5.9	36.4	49.3	8.3	0	76	233
Adult4	0	0	52.8	44.8	2.4	0	87	181
Adult5	0	1.7	59.7	18.4	20.1	0	76	228
Adult6	0	5.5	56.6	14.9	22.9	0	76	226
Adult7	0	0	67.0	20.1	12.8	0	82	205
Adult8	0	6.2	57.6	28.8	7.3	0	75	226
Adult9	0	2	57.6	27	12.5	0	71	250
Adult10	0	2.1	45.5	24.0	21.9	6.5	76	276
Mean	0	2.5	54.6	29.46	12.62	0.65	77.4	225.9
SD	0	2.4	10.1	11.4	7.4	2.0	5.0	28.9

Table 5.b.9 CD_4 (unmeasured).

	% Time <70	% Time <80	% Time 80<G<140	% Time 140<G<180	% Time 180<G<250	% Time >250	G_{min} (mg/dL)	G_{max} (mg/dL)
Adult1	0.3	2.4	54.5	15.3	24.3	3.5	69	269
Adult2	0	2.1	54.5	25.7	17.7	0	74	205
Adult3	4.5	5.9	44.4	38.2	10.1	1.4	43	256
Adult4	7.6	10.7	20.5	46.2	22.6	0	59	249
Adult5	0	0	49.6	23.6	26.7	0	83	237
Adult6	4.8	6.2	48.2	11.4	31.5	2.4	53	252
Adult7	6.6	12.8	52.8	9.1	16.1	8.6	58	288
Adult8	7.6	9.7	50.3	32.3	6.9	0	59	208
Adult9	0	0	44.7	35	14.2	5.9	85	283
Adult10	0	2.8	31.0	30.0	30.2	6.2	73	295
Mean	3.1	5.3	45.0	26.7	20.0	2.8	65.6	254.2
SD	3.4	4.6	11.0	12.0	8.4	3.1	13.4	31.2

hyperglycaemia (>180 mg/dL) is much higher for CD_4, with 22.8% of the time spent in the hyperglycaemic range, while for CD_3 the respective percentage is 13.3%. The glucose profile and the control action with both CD_3 and CD_4 are presented in Figure 5.b.8 for adult 6 for illustrative purposes.

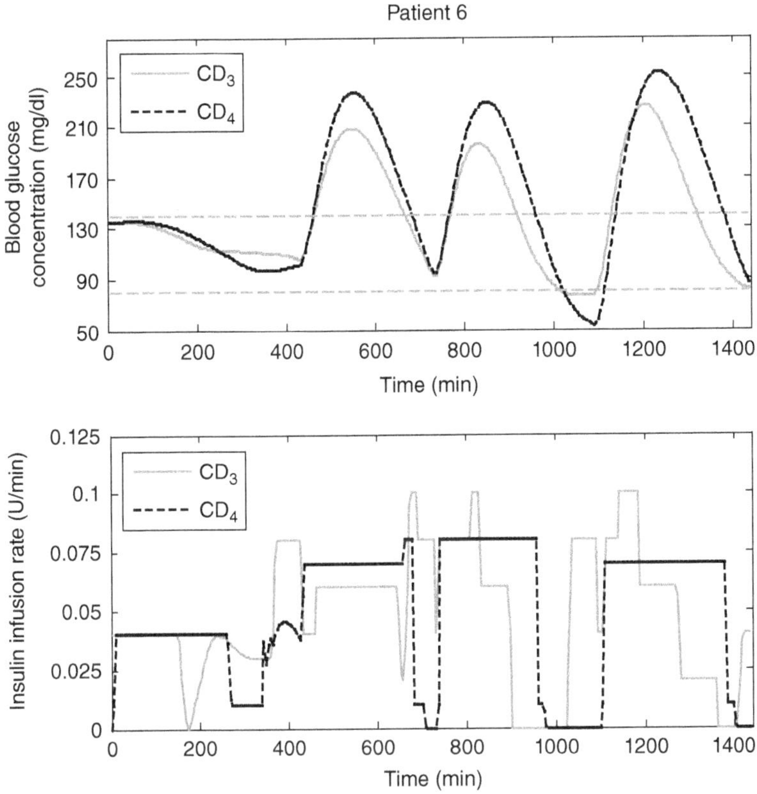

Figure 5.b.8 Comparison of glucose regulation with control designs 3 and 4 for adult 6. The meals are given at 120, 720 and 1080 min and contain 75, 100 and 90 g of carbohydrates, respectively.

5.b.4.3 Variable Meal Time

Figure 5.b.9 shows the glucose profile for adult 6 when a meal of 50 g is given 30 min before, 30 min after and simultaneously with the predefined 30 g reference meal. It can be noticed that good glycaemic control is achieved in all cases with no occurring event of hypoglycaemia. When the meal is consumed 30 min before the predetermined meal time, prandial hyperglycaemia is occurring since insulin is not acting yet.

5.b.4.4 Concluding Remarks

The closed-loop control validation studies show that the proposed control design CD_3 can efficiently regulate the blood glucose concentration when tested for large meal sizes. There is no reported event of hypoglycaemia, while

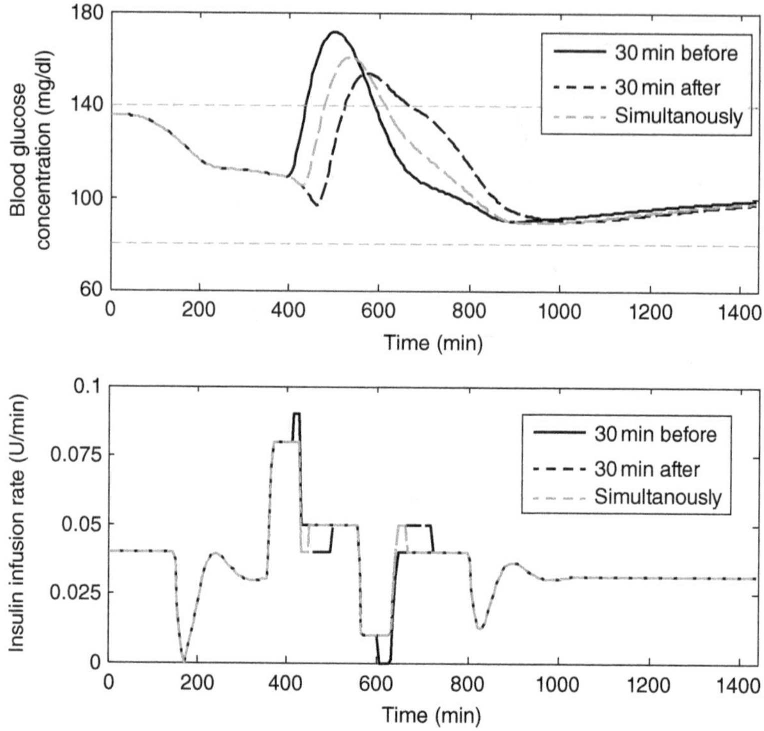

Figure 5.b.9 Evaluation of $\mathbf{CD_3}$ when a meal of 50 g is given 30 min in advance, 30 min after and simultaneous with the reference meal of 30 g.

the mean maximum glucose value is 226 mg/dL. When this control design is compared with CD_4, it becomes obvious that superior control can be achieved when the feedforward action of the MPC controller is enhanced in the presence of unknown meal disturbances. Further closed-loop validation studies are required to verify the reliability of the proposed control performance. Hence, the proposed control strategy can be regarded as a potential strategy to compensate for the unknown meal disturbances since the validation studies performed for the UVa/Padova Simulator model indicate promising closed-loop glucose regulation.

5.b.5 Explicit MPC

The previously presented control strategy involves the on-line solution, at each sampling instant, of the finite horizon open-loop optimal control problem. In order to overcome the significant online computations involved in the

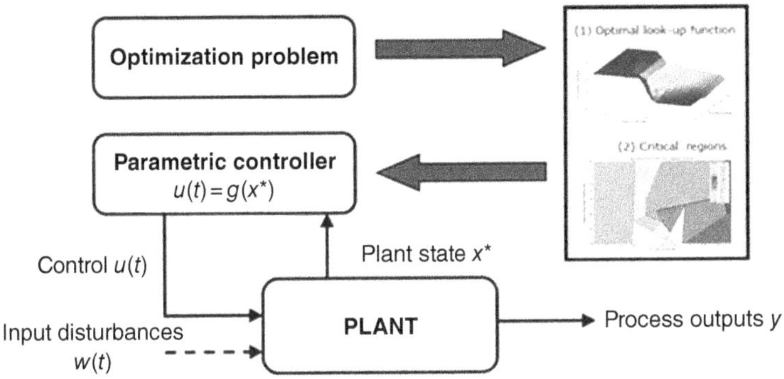

Figure 5.b.10 Multiparametric MPC.

closed-loop control implementation, an alternative solution of the on-line optimization problem has been proposed (Pistikopoulos *et al.*, 2007) which lies on a parametric optimization-based approach. In essence, the online problem is replaced by the off-line derivation of the explicit mapping of the optimal decisions in the space of the plant uncertainty, as presented in Figure 5.b.10. This approach is presented using a simplified version of the original model for illustrative reasons.

The design of the controller involves the following steps:

1) The derivation of a reduced-order linear state-space model
2) The design of a multiparametric model predictive controller.

5.b.5.1 Model Identification

A reduced-order, discrete-time, state-space model is designed with the System Identification Toolbox of MATLAB (R2010b) and by using the simulation data of the glucose–insulin model. A linearized, discrete-time, state-space model is mathematically represented by the following form:

$$x(k+1) = Ax(k) + Bu(k) \tag{5.b.6}$$

$$y(k) = Cx(k) + Du(k) \tag{5.b.7}$$

For the particular system, the *matrices* A, B, C and D are:

$$A = \begin{bmatrix} 0.99827 & 0.0052979 \\ 0.0066081 & 0.97978 \end{bmatrix} \qquad B = \begin{bmatrix} -0.0001448 & 0.0014586 \\ 0.00055274 & -0.0056066 \end{bmatrix}$$

$$C = \begin{bmatrix} 1866.4 & 15.02 \end{bmatrix} \qquad D = [0]$$

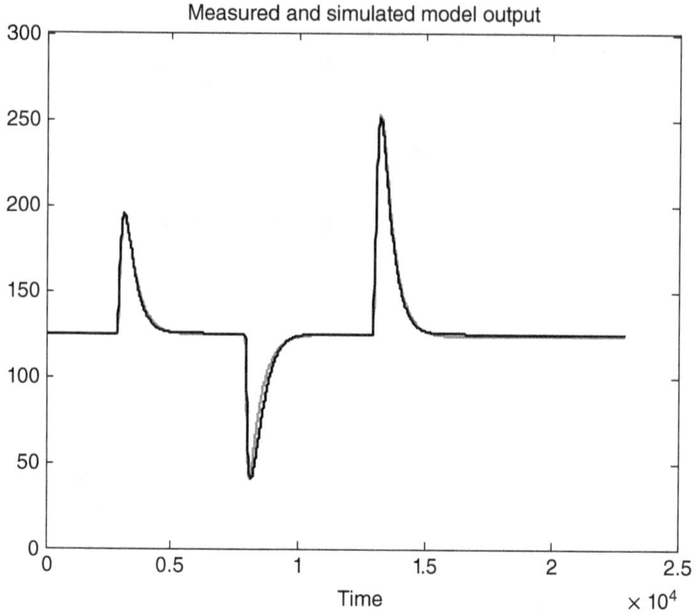

Figure 5.b.11 Comparison of the original and state-space model.

Figure 5.b.11 compares the dynamic model with the state-space, reduced-order model. The state-space model approximates well the dynamic behaviour, with an error of 11%.

A nominal mp-MPC is designed (Pistikopoulos *et al.*, 2002) using the linear state-space model presented previously, with constant matrices A, B, C and D. The following MPC formulation is considered for the glucose–insulin system:

$$\begin{aligned}
&\min_{u_{t+1},\ldots,u_{t+N_u}} J=\sum_{i=1}^{N_y} Q\left(y_i-y_{ref,i}\right)^2+\sum_{j=0}^{N_u} R_1\left(\Delta u_j\right)^2\\
&\text{s.t.}\quad x(t+1)=Ax(t)+Bu(t)\\
&\qquad\quad y(t)=Cx(t)+Du(t)\\
&1\le u_1\le 60, 0\le u_2\le 6\,60\le y_i\le 140\\
&\qquad\quad y_i=G(t+i), i=1,\ldots N_y\\
&\qquad\quad u_i=I(t+j), j=0,\ldots N_{u-1}
\end{aligned} \tag{5.b.8}$$

where u_i is the manipulated variables (insulin, meal), y_i is the controlled variable (glucose), y_{ref} is the optimal glucose profile, N_y is the total horizon and N_u is the control horizon (8,1). Food is considered as an input, with a specific modelled profile to test the response of the system.

For the case of constant system matrices, the optimization problem is a multiparametric quadratic programming (mp-QP) problem and can be solved with

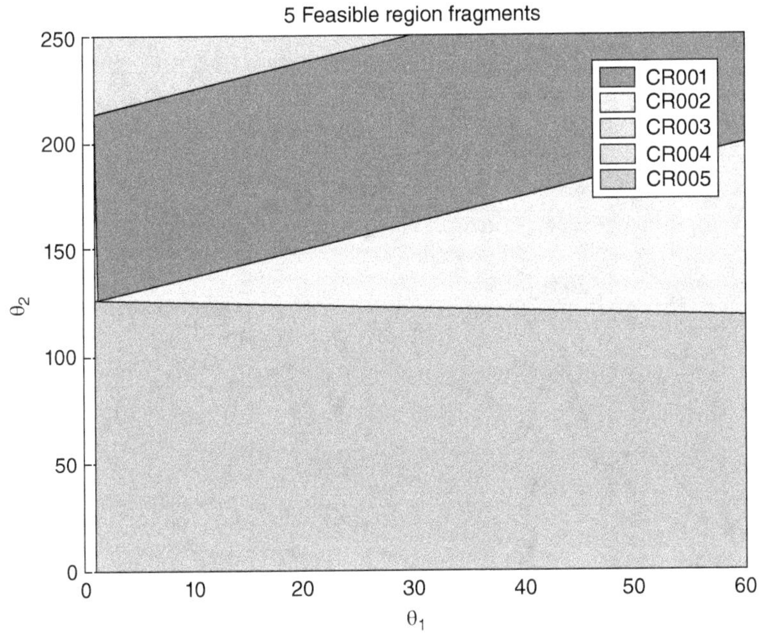

Figure 5.b.12 Critical regions for mp-MPC.

standard multiparametric programming techniques (Pistikopoulos *et al.*, 2007). In this study, Parametric Optimization Software was used (ParOS, 2003) to obtain the explicit controller description, which is the optimal map of the control variables as a function of the parameters of the system. This optimal map consists of 61 critical regions and the corresponding control laws. Each of the critical regions is described by a number of linear inequalities $A_i x \leq b_i$ and its corresponding control is piecewise linear $U_f = K_i x + c_i$, where i is the index of solutions. A two-dimensional representation of the critical regions is shown in Figure 5.b.12.

At 600 min and 1200 min, two meals are introduced in the system of 50 mg and 90 mg of carbohydrates. The second graph shows insulin concentration in the blood after subcutaneous administration.

The performance of the controller is illustrated in Figure 5.b.13.

5.b.5.2 Concluding Remarks

The advantages of using the mp-MPC control method in drug delivery systems are:

- ✓ Suitable for portable applications
- ✓ Testing off-line of different scenarios to ensure the patient's safety
- ✓ Advantages of MPC over other control designs.

Further *in silico* validation is required to improve the mp-MPC performance in the context of the proposed framework.

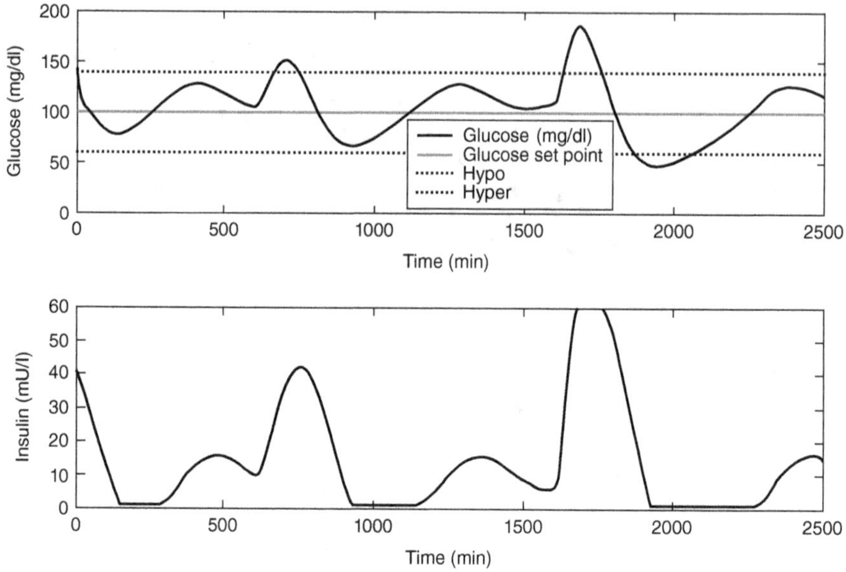

Figure 5.b.13 Closed-loop control performance.

Appendix 5.1

A5.1 Model of UVa/Padova Simulator

Glucose Metabolism

A two-compartment model is used to model the glucose subsystem, with G_p and G_t representing the plasma and tissue glucose mass (mg/kg), respectively.

$$\frac{dG_p}{dt} = EGP + R_a - U_{ii} - E - k_1 G_p + k_2 G_t \qquad G_p(0) = G_{pb} \quad (5.A.1)$$

$$\frac{dG_t}{dt} = k_1 G_p - k_2 G_t - U_{id} \qquad G_t(0) = G_{tb} \quad (5.A.2)$$

$$G = \frac{G_p}{V_g} \qquad G(0) = G_b \quad (5.A.3)$$

where *EGP* (mg/kg/min) is the endogenous glucose production, *Ra* (mg/kg/min) is the rate of glucose appearance in the bloodstream after meal consumption, U_{ii} and U_{id} (mg/kg/min) are insulin-independent and insulin-dependent glucose utilization, *E* (mg/kg/min) is the renal glucose excretion, $k_1, k_2(\text{min}^{-1})$ are rate parameters of glucose kinetics and *Vg* (dL) is the glucose distribution volume.

Rate of Glucose Appearance (Ra) from Meal

$$\frac{dQ_{sto1}}{dt} = -k_{kgri}Q_{sto1} + D\delta(t) \qquad Q_{sto1}(0) = 0 \qquad (5.A.4)$$

$$\frac{dQ_{sto2}}{dt} = -k_{empt}Q_{sto2} + k_{gri}Q_{sto1} \qquad Q_{sto2}(0) = 0 \qquad (5.A.5)$$

$$\frac{dQ_{gut}}{dt} = -k_{abs}Q_{gut} + k_{empt}Q_{sto2} \qquad Q_{gut}(0) = 0 \qquad (5.A.6)$$

$$Q_{sto} = Q_{sto1} + Q_{sto2} \qquad Q_{sto}(0) = 0 \qquad (5.A.7)$$

$$R_a = \frac{fk_{abs}Q_{gut}(t)}{BW} \qquad \mathrm{Ra}(0) = 0 \qquad (5.A.8)$$

$$k_{empt} = k_{min} + \frac{k_{max} - k_{min}}{2}\left\{\tanh\left(a_1\left(Q_{sto} - b \cdot D(t)\right)\right) - \tanh\left(b_1\left(Q_{sto} - d \cdot D(t)\right)\right) + 2\right\} \qquad (5.A.9)$$

$$\mathrm{a}_1 = \frac{5}{2 \cdot D \cdot (1 - b)} \qquad (5.A.10)$$

$$\mathrm{b}_1 = \frac{5}{2 \cdot D \cdot d} \qquad (5.A.11)$$

where Q_{sto1}, Q_{sto2} (mg) = the glucose mass in solid and liquid phase, respectively; Q_{sto} (mg) = the overall glucose mass in the stomach; Q_{gut} (mg) is the glucose mass in the small intestine; k_{empt}(min^{-1}) is the rate of gastric emptying; a_1 and b_1 are model parameters; k_{max}, k_{min} (min^{-1}) are the max and min gastric emptying; k_{abs} (min^{-1}) is the rate constant of intestinal absorption; k_{gri} is the rate constant of grinding; *f* (dimensionless) is the fraction of intestinal absorption; *b* and *d* are percentages of the dose; and *D* (mg) is the amount of ingested meal.

Endogenous Glucose Production (EGP)

$$EGP = k_{p1} - k_{p2}G_p - k_{p3}I_{del2} \qquad \mathrm{EGP}(0) = \mathrm{EGP_b} \qquad (5.A.12)$$

$$\frac{dI_{del1}}{dt} = -k_i\left(I_{del1} - I_p\right) \qquad \mathrm{I_1}(0) = \mathrm{I_{pb}} \qquad (5.A.13)$$

$$\frac{dI_{del2}}{dt} = -k_i\left(I_{del2} - I_{del1}\right) \qquad \mathrm{I_d}(0) = \mathrm{I_{pb}} \qquad (5.A.14)$$

where I_{del2} (pmol/l) = the delayed insulin signal (chain of two compartments), k_{p1} = (mg/kg/min) the extrapolated EGP at zero glucose and insulin, k_{p2} (min^{-1}) = the liver glucose effectiveness, k_{p3} (mg/kg/min per pmol/l) = the insulin action on the liver and k_i (min^{-1}) = the rate parameter for the delay between insulin signal and action.

Insulin-Dependent Glucose Utilization

$$U_{id} = \frac{V_m G_t}{k_{m0} + G_t} \qquad U_{id}(0) = U_{idb} \qquad (5.A.15)$$

$$V_m = V_{m0} + V_{mx} X_{disp} \qquad V_m(0) = V_{m0} \qquad (5.A.16)$$

$$\frac{dX_{disp}}{dt} = -p_{2u} X_{disp} + p_{2u}(I_p - I_{BASAL}) \qquad X_{disp}(0) = 0 \qquad (5.A.17)$$

$$U_{tot} = U_{ii} + U_{id} \qquad (5.A.18)$$

where X_{disp} (pmol/L) = insulin in the interstitial fluid, V_{m0} (mg/kg/min) and k_{m0} (mg/kg) are the Michaelis–Menten related parameters, V_{mx} (mg/kg/min per pmol/liter) is the disposal of insulin sensitivity and p_{2u} (min^{-1}) is the rate constant of insulin action on peripheral glucose utilization.

Glucose Renal Excretion

$$E = \begin{cases} k_{e1}(G_p - k_{e2}) & \text{if } G_p > k_{e2} \\ 0 & \text{if } G_p \le k_{e2} \end{cases} \qquad E(0) = 0 \qquad (5.A.19)$$

where k_{e1} (min^{-1}) is the glomerular filtration rate, and k_{e2} (mg/kg) is the glucose renal threshold.

Insulin Kinetics

$$\frac{dI_{sc1}}{dt} = -(k_d + k_{a1}) \cdot I_{sc1} + u(t) \qquad I_{sc1}(0) = I_{sc1ss} \qquad (5.A.20)$$

$$\frac{dI_{sc2}}{dt} = k_d I_{sc1} - k_{a2} I_{sc2} \qquad I_{sc2}(0) = I_{sc2ss} \qquad (5.A.21)$$

$$\frac{dI_l}{dt} = -(m_1 + m_3) I_l + m_2 I_p \qquad I_l(0) = I_{lb} \qquad (5.A.22)$$

$$\frac{dI_p}{dt} = -(m_2 + m_4) I_p + m_1 I_l + k_{a1} I_{sc1} + k_{a2} I_{sc2} \qquad I_p(0) = I_{pb} \qquad (5.A.23)$$

$$I = \frac{I_p}{V_i} \qquad I(0) = I_b \qquad (5.A.24)$$

$$m_3 = \frac{HE_b m_1}{1 - HE_b} \qquad (5.A.25)$$

where I_l (pmol/kg) is insulin mass in the liver; I_p (pmol/kg) is insulin mass in the plasma; I (pmol/l) is the plasma insulin concentration; I_{sc1} (pmol/kg) is the amount of non-monomeric insulin in the subcutaneous space; I_{sc2} (pmol/kg) is the amount of monomeric insulin in the subcutaneous space; $u(t)$ (pmol/kg/min) is the exogenous insulin infusion rate; m_1, m_2, m_3, m_4 (min^{-1}) are the rate parameters of insulin kinetics; V_I (L/kg) is the insulin distribution volume; k_d (min^{-1}) is the rate constant of insulin dissociation; k_{a1} (min^{-1}) is the rate constant of non- monomeric insulin absorption; and k_{a2} (min^{-1}) is the rate constant of monomeric insulin absorption.

Appendix 5.2

The optimal state estimation is calculated using the algorithm given in Equations (5.B.1)–(5.B.5), which includes a two-step approach (Chui & Chen, 2008). The first step predicts the state and covariance estimates using the model equations and previous estimates, while the second updates the prediction using information from the observations.

Time Update/"Prediction"

State prediction:

$$\hat{x}_k^- = A\hat{x}_{k-1} + Bu_{k-1} \tag{5.B.1}$$

Projection of the error covariance:

$$P_k^- = AP_{k-1}A^T + Q \tag{5.B.2}$$

Measurement update/"correction"

Computation of the Kalman gain:

$$K_k = P_k^- H^T \left(HP_k^- H^T + R\right)^{-1} \tag{5.B.3}$$

State estimate update:

$$\hat{x}_k = \hat{x}_k^- + K_k\left(z_k - H\hat{x}_k^-\right) \tag{5.B.4}$$

Error covariance update:

$$P_k = \left(I - K_k H\right)P_k^- \tag{5.B.5}$$

Appendix 5.3

Linearized Model

Matrices A, B, C, D and B_d. B_d accounts for the basal level of the states.

$$A=\begin{bmatrix} -(k_{p2}+k_1) & k_2 & 0 & 0 & 0 & k_{abs}f/BW & -k_{p3} & 0 & 0 & 0 & 0 & 0 \\ k_1 & -k_2-f(V_m,k_{km0},G_t) & -f(V_{mx},k_{km0},G_t) & 0 & 0 & 0 & 0 & 0 & 0 & 0 & 0 & 0 \\ 0 & 0 & -p_{2u} & 0 & 0 & 0 & 0 & 0 & 0 & 0 & 0 & p_{2u}/V_i \\ 0 & 0 & 0 & -k_{gri} & 0 & 0 & 0 & 0 & 0 & 0 & 0 & 0 \\ 0 & 0 & 0 & k_{gri} & -k_{empt} & 0 & 0 & 0 & 0 & 0 & 0 & 0 \\ 0 & 0 & 0 & k_{est} & k_{empt} & -k_{abs} & 0 & 0 & 0 & 0 & 0 & 0 \\ 0 & 0 & 0 & 0 & 0 & 0 & -k_i & k_i & 0 & 0 & 0 & 0 \\ 0 & 0 & 0 & 0 & 0 & 0 & 0 & -k_i & 0 & 0 & 0 & k_i/V_i \\ 0 & 0 & 0 & 0 & 0 & 0 & 0 & 0 & k_d+k_{a1} & 0 & 0 & 0 \\ 0 & 0 & 0 & 0 & 0 & 0 & 0 & 0 & k_d & -k_{a2} & 0 & 0 \\ 0 & 0 & 0 & 0 & 0 & 0 & 0 & 0 & 0 & 0 & m_1+m_3 & m_2 \\ 0 & 0 & 0 & 0 & 0 & 0 & 0 & 0 & k_{a1} & k_{a2} & m_1 & m_2+m_4 \end{bmatrix}$$

$$B=\begin{bmatrix} 0 & 0 \\ 0 & 0 \\ 0 & 0 \\ 0 & \frac{1}{D} \\ 0 & 0 \\ 0 & 0 \\ \frac{6000}{BW} & 0 \\ 0 & 00 \\ 0 & 0 \\ 0 & 0 \end{bmatrix},\ B_d=\begin{bmatrix} k_{p1}-U_{ii} \\ k_{est1} \\ -p_{2u}{}^{*}I_{BASAL} \\ 0 \\ 0 \\ 0 \\ 0 \\ 0 \\ 0 \\ 0 \\ 0 \\ 0 \end{bmatrix},C=\begin{bmatrix} \frac{1}{V_g} & 0 & 0 & 0 & 0 & 0 & 0 & 0 & 0 & 0 & 0 & 0 \end{bmatrix},D=\begin{bmatrix} 0 & 0 \end{bmatrix}$$

References

Ackland, T.R., Elliott, B. and Bloomfield, J. (2009). *Applied Anatomy and Biomechanics in Sport*, 2nd ed. Champaign, IL: Human Kinetics Press.

Akaike, H. (1974). A new look at the statistical model identification. *IEEE Transactions on Automatic Control*, AC-16(6), 716–723.

Bansal, V., Sakizlis, V., Ross, R., Perkins, J.D. and Pistikopoulos, E.N. (2003). New algorithms for mixed-integer dynamic optimization. *Computers & Chemical Engineering*, 27(5), 647–668.

Bennett, D. (2004). Asymptotic properties of a delay differential equation model for the interaction of glucose with plasma and interstitial insulin. *Applied Mathematics and Computation*, 151(1), 189–207.

Bequette, B.W. (2005). A critical assessment of algorithms and challenges in the development of a closed-loop artificial pancreas. *Diabetes Technology & Therapeutics*, 7(1), 28–47.

Bequette, B.W. (2012). Challenges and recent progress in the development of a closed-loop artificial pancreas. *Annual Reviews in Control*, 36(2), 255–266.

Berger, M.P. and Rodbard, D. (1991). A pharmacodynamic approach to optimizing insulin therapy. *Computer Methods and Programs in Biomedicine*, 34(4), 241–253.

Bergman, R.N., Phillips, L.S. and Cobelli, C. (1981). Physiologic evaluation of factors controlling glucose tolerance in man. *Journal of Clinic Investigation*, 68, 1456–1467.

Boden, G., Cheung, P. and Homko, C. (2003). Effects of acute insulin excess and deficiency on gluconeogenesis and glycogenolysis in type 1 diabetes. *Diabetes*, 52, 133–137.

Breton, M., Farret, A., Bruttomesso, D., Anderson, S., Magni, L., Patek, S., Dalla Man, C., Place, J., Demartini, S., Del Favero, S., Toffanin, C., Hughes-Karvetski, C., Dassau, E., Zisser, H., Doyle, F.J., De Nicolao, G., Avogaro, A., Cobelli, C., Renard, E. and Kovatchev, B. (2012). Fully integrated artificial pancreas in type 1 diabetes: modular closed-loop glucose control maintains near normoglycemia. *Diabetes*, 61(9), 2230–2237.

Brown, B.E., Hopper, J., Hodges, J.L., Bradley, B., Wennesland, R. and Yamauchi, H. (1962). Red cell, plasma, and blood volume in healthy women measured by radiochromium cell labeling and hematocrit. *Journal of Clinical Investigation*, 41(12), 2182–2190.

Cano, N. (2002). Bench-to-bedside review: glucose production from the kidney. *Critical Care*, 6, 317–321.

Caumo, A. (1993). Hepatic glucose production estimation by deconvolution during the labeled IVGTT: with a new minimal model. *Modeling in Physiology*, 264(27), E829–E841.

Chen, C.-L., Tsai, H.-W. and Wong, S.-S. (2010). Modeling the physiological glucose-insulin dynamic system on diabetics. *Journal of Theoretical Biology*, 265(3), 314–322.

Chui, C. K. and Chen, G. (2008). *Kalman Filtering*. Rijeka, Croatia: InTech.

Cobelli, C., Renard, E., Kovatchev, B.P., Keith-Hynes, P., Ben Brahim, N., Place, J., Del Favero, S., Breton, M., Farret, A., Bruttomesso, D., Dassau, E., Zisser, H., Doyle III, F.J., Patek, S.D. and Avogaro, A. (2012). Pilot studies of wearable outpatient artificial pancreas in type 1 diabetes. *Diabetes Care*, 35(9), 65–67.

Crone, C. (1965). Facilitated transfer of glucose from blood into brain tissue. *Journal of Physiology*, 181, 103–113.

Dalla Man, C., Camilleri, M. and Cobelli, C. (2006). A system model of oral glucose absorption: validation on gold standard data. *IEEE Transactions on Biomedical Engineering*, 53(12 Pt 1), 2472–2478.

Dalla Man, C., Raimondo, D.M., Rizza, R. and Cobelli, C. (2007). GIM, simulation software of meal glucose-insulin model. *Journal of Diabetes Science and Technology*, 1(3), 323–330.

Dalla Man, C., Rizza, R. and Cobelli, C., 2007. Meal simulation model of the glucose-insulin system. *IEEE Transactions on Biomedical Engineering*, 54(10), 1740–1749.

Dassau, E., Zisser, H., Harvey, R., Percival, M.W., Grosman, B., Bevier, W., Atlas, E., Miller, S., Nimri, R., Jovanovič, L. and Doyle III, F.J. (2013). Clinical evaluation of a personalized artificial pancreas. *Diabetes Care*, 36(4), 801–809.

Davis, S.N., Fowler, S. and Costa, F. (2000). Hypoglycemic counterregulatory responses differ between men and women with type 1 diabetes. *Diabetes*, 49, 65–72.

Deurenberg, P., Weststrate, J.A. and Seidell, J.C. (1991). Body mass index as a measure of body fatness: age- and sex-specific prediction formulas. *British Journal of Nutrition*, 65, 105–114.

Doyle, F.J. III, Huyett, L.M., Bok, J.L., Zisser, H. and Dassau, E. (2014). Closed-loop artificial pancreas systems: engineering the algorithms. *Diabetes Care*, 37(5), 191–197.

Dua P. and Pistikopoulos E.N. (2005). Modelling and control of drug delivery systems. *Computers & Chemical Engineering*, 29(11–12), 2290–2296.

Dua, P., Doyle, F.J. and Pistikopoulos, E.N. (2006). Model-based blood glucose control for Type 1 diabetes via parametric programming. *IEEE Transactions on Biomedical Engineering*, 53(8), 1478–1491.

Dua, P., Doyle, F.J. and Pistikopoulos, E.N. (2009). Multi-objective blood glucose control for type 1 diabetes. *Medical & Biological Engineering & Computing*, 47(3), 343–352.

Eckel, R.H. (2003) *Obesity: Mechanisms and Clinical Management: Mayo Clinic Proceedings*. Philadelphia: Lippincott Williams & Wilkins.

Ederle, J., Blanchard, S. and Bronzino, J. (2000). Cardiovascular mechanics. In *Introduction to Biomedical Engineering*, ed. J. Ederle, S. Blanchard and J. Bronzino, 467–536. San Diego, CA: Academic Press.

Ederle, J. (2011). Compartmental modeling. In *Introduction to Biomedical Engineering*, ed. J. Ederle and J. Bronzino, 3rd ed., 360–447. Oxford: Elsevier.

El-Khatib, Firas H, Steven J Russell, David M Nathan, Robert G Sutherlin, and Edward R Damiano. 2010. A bihormonal closed-loop artificial pancreas for type 1 diabetes. *Science Translational Medicine*, 2(27): 27ra27. doi:10.1126/scitranslmed.3000619.

Elleri, D., Allen, J.M., Biagioni, M., Kumareswaran, K., Leelarathna, L., Caldwell, K., Nodale, M., Wilinska, M.E., Acerini, C.L., Dunger, D.B. and Hovorka, R. (2012) Evaluation of a portable ambulatory prototype for automated overnight closed-loop insulin delivery in young people with type 1 diabetes. *Pediatric Diabetes*, 13(6), 449–453.

Elleri, D., Allen, J.M., Kumareswaran, K., Leelarathna, L., Nodale, M., Caldwell, K., Cheng, P., Kollman, C., Haidar, A., Murphy, H.R., Wilinska, M.E., Acerini, C.L., Dunger, D.B. and Hovorka, R. (2013). Closed-loop basal insulin delivery over 36 hours in adolescents with type 1 diabetes: randomized clinical trial. *Diabetes Care*, 36, 838–844.

Elleri, D., Allen, J.M., Nodale, M., Wilinska, M.E., Mangat, J.S., Larsen, A.M.F., Acerini, C.L., Dunger, D.B. and Hovorka, R. (2011). Automated overnight closed-loop glucose control in young children with type 1 diabetes. *Diabetes Technology & Therapeutics*, 13(4), 419–424.

Engelborghs, K., Lemaire, V., Bélair, J. and Roose, D. (2001). Numerical bifurcation analysis of delay differential equations arising from physiological modeling. *Journal of Mathematical Biology*, 42(4), 361–385.

Eren-Oruklu, M., Cinar, A., Quinn, L. and Smith, D. (2009). Adaptive control strategy for regulation of blood glucose levels in patients with type 1 diabetes. *Journal of Process Control*, 19(8), 1333–1346.

Fabietti, P.G., Canonico, V., Federici, M.O., Benedetti, M.M. and Sarti, E. (2006). Control oriented model of insulin and glucose dynamics in type 1 diabetics. *Medical & Biological Engineering & Computing*, 44(1–2), 69–78.

Ferrannini, E. and DeFronzo, R.A. (2004). Insulin action in vivo: glucose metabolism. In *International Textbook of Diabetes Mellitus*, ed. R.A. DeFronzo, E. Ferrannini, H. Keen and P. Zimmet, 3rd ed., 277–301. Chichester: Wiley.

Fisher, M.E. (1991). A semiclosed-loop algorithm for the control of blood glucose levels in diabetics. *IEEE Transactions on Biomedical Engineering*, 38(1), 57–61.

Gallagher, D., Belmonte, D., Deurenberg, P., Wang, Z., Krasnow, N., Pi-Sunyer, F.X. and Heymsfield, S.B. (1998). Organ-tissue mass measurement allows modeling of REE and metabolically active tissue mass. *American Journal of Physiology, Endocrinology and Metabolism*, 275, E249–E258.

Gerich, J.E. (2010). Role of the kidney in normal glucose homeostasis and in the hyperglycaemia of diabetes mellitus: therapeutic implications. *Diabetic Medicine : A Journal of the British Diabetic Association*, 27(2), 136–142.

Gerich, J.E., Meyer, C., Woerle, H.J. and Stumvoll, M. (2001). Renal gluconeogenesis: its importance in human glucose homeostasis. *Diabetes Care*, 24(2), 382–391.

Ghosh, S. and Maka, S. (2009). A NARX modeling-based approach for evaluation of insulin sensitivity. *Biomedical Signal Processing and Control*, 4(1), 49–56.

Gudbjörnsdóttir, S., Sjostrand, M., Strindberg, L., Wahren, J. and Lonnroth, P. (2003). Direct measurements of the permeability surface area for insulin and glucose in human skeletal muscle. *Journal of Clinical Endocrinology & Metabolism*, 88(10), 4559–4564.

Hann, C.E., Chase, J.G., Lin, J., Lotz, T., Doran, C.V. and Shaw, G.M. (2005). Integral-based parameter identification for long term dynamic verification of a glucose-insulin system model. *Computer Methods and Programs in Biomedicine*, 77(3), 259–270.

Herpe, T. Van, Espinoza, M., Haverbeke, N., Moor, B. De and den Berghe, G. Van (2007). Glycemia prediction in critically ill patients using an adaptive modeling approach. *Journal of Diabetes Science and Technology*, 1(3), 348–356.

Herrero, P., Georgiou, P., Oliver, N., Reddy, M., Johnston, D. and Toumazou, C. (2013). A composite model of glucagon-glucose dynamics for in silico testing of bihormonal glucose controllers. *Journal of Diabetes Science and Technology*, 7(4), 941–951.

Hovorka, R., Allen, J.M., Elleri, D., Chassin, L.J., Harris, J., Xing, D., Kollman, C., Hovorka, T., Larsen, A.M.F. and Nodale, M. (2010). Manual closed-loop insulin delivery in children and adolescents with type 1 diabetes: a phase 2 randomised crossover trial. *The Lancet*, 375(9716), 743–751.

Hovorka, R., Elleri, D., Thabit, H., Allen, J.M., Leelarathna, L., El-Khairi, R., Kumareswaran, K., Caldwell, K., Calhoun, P., Kollman, C., Murphy, H.R., Acerini, C.L., Wilinska, M.E., Nodale, M. and Dunger, D.B. (2014). Overnight closed-loop insulin delivery in young people with type 1 diabetes: a free-living, randomized clinical trial. *Diabetes Care*, 37(5): 1204–1211.

Hovorka, R., Shojaee-Moradie, F., Carroll, P.V., Chassin, L.J., Gowrie, I.J., Jackson, N.C., Tudor, R.S., Umpleby, M. and Jones, R.H. (2002). Partitioning glucose distribution/transport, disposal, and endogenous production during IVGTT. *American Journal of Physiology, Endocrinology and Metabolism*, 282(5), E992–E1007.

Johnson, L.R. (2003). *Essential Medical Physiology*, 3rd ed. San Diego, CA: Academic Press.

Keenan, D.B., Mastrototaro, J.J., Voskanyan, G. and Steil, G.M. (2009). Delays in minimally invasive continuous glucose monitoring devices:a review of current technology. *Journal of Diabetes Science and Technology*, 3(5), 1207–1214.

Kovatchev, B.P., Breton, M.D., Dalla Man, C. and Cobelli, C. (2011). In silico preclinical trials: a proof of concept in closed-loop control of type 1 diabetes. *Journal of Diabetes Science and Technology*, 3(1), 44–55.

Kovatchev, B., Cobelli, C., Renard, E., Anderson, S., Breton, M., Patek, S., Clarke, W., Bruttomesso, D., Maran, A., Costa, S., Avogaro, A., Man, C.D., Facchinetti, A., Magni, L., Nicolao, G. De, Place, J. and Farret, A. (2010). Multinational study of subcutaneous model-predictive closed-loop control in type 1 diabetes mellitus: summary of the results. *Journal of Diabetes Science and Technology*, 4(6), 1374–1381.

Kovatchev, B., Patek, S., Dassau, E., Doyle, F.J., Magni, L., Nicolao, G. De and Cobelli, C. (2009). Control to range for diabetes: functionality and modular architecture. *Journal of Diabetes Science and Technology*, 3(5), 1058–1065.

Kovatchev, B.P., Renard, E., Cobelli, C., Zisser, H. C., Keith-Hynes, P., Anderson, S.M., Brown, S.A., Chernavvsky, D.R., Breton, M., Farret, A., Pelletier, M.-J., Place, J., Bruttomesso, D., Del Favero, S., Visentin, R., Filippi, A., Scotton, R., Avogaro, A. and Doyle III, F. J. (2013). Feasibility of outpatient fully integrated closed-loop control. *Diabetes Care*, 36, 1851–1858.

Kraegen, E.W. and Chisholm, D.J. (1984). Insulin responses to varying profiles of subcutaneous insulin infusion: kinetic modelling studies. *Diabetologia*, 26, 208–213.

Krieger, A., Panoskaltsis, N., Mantalaris, A., Georgiadis, M.C. and Pistikopoulos, E.N. (2014), Modeling and analysis of individualized pharmacokinetics and pharmacodynamics for volatile anesthesia. *IEEE Transactions on Biomedical Engineering*, 61(1), 25–34.

Kuang, Y. and Li, J. (2008) Systemically modeling the dynamics of plasma insulin in subcutaneous injection of insulin analogues for type 1 diabetes. *Mathematical Biosciences and Engineering*, 6(1), 41–58.

Lehmann, E.D. and Deutsch, T. (1992). A physiological model of glucose-insulin interaction in type 1 diabetes mellitus. *Journal of Biomedical Engineering*, 14(3), 235–242.

Li, J. and Kuang, Y. (2009). Systemically modeling the dynamics of plasma insulin in subcutaneous injection of insulin analogues for type 1 diabetes. *Mathematical Biosciences and Engineering*, 6(1), 41–58.

Li, J., Kuang, Y. and Mason, C.C. (2006). Modeling the glucose-insulin regulatory system and ultradian insulin secretory oscillations with two explicit time delays. *Journal of Theoretical Biology*, 242(3), 722–735.

Magni, L., Raimondo, D.M., Bossi, L., Dalla Man, C., De Nicolao, G., Kovatchev, B. and Cobelli, C. (2007). Model predictive control of type 1 diabetes: an in silico trial. *Journal of Diabetes Science and Technology*, 1(6), 804–812.

Magni, L., Raimondo, D.M., Dalla Man, C., De Nicolao, G., Kovatchev, B. and Cobelli, C. (2009). Model predictive control of glucose concentration in type I diabetic patients: an in silico trial. *Biomedical Signal Processing and Control*, 4(4), 338–346.

Mitsis, G.D., Markakis, M.G. and Marmarelis, V.Z. (2009). Nonlinear modeling of the dynamic effects of infused insulin on glucose: comparison of compartmental with Volterra models. *IEEE Transactions on Biomedical Engineering*, 56(10), 2347–2358.

Mougiakakou, S., Prountzou, K. and Nikita, K. (2005). A real time simulation model of glucose-insulin metabolism for type 1 diabetes patients. In *27th*

Annual International Conference of the IEEE Engineering in Medicine and Biology Society, 298–301.

Nucci, G. and Cobelli, C. (2000). Models of subcutaneous insulin kinetics: a critical review. *Computer Methods and Programs in Biomedicine*, 62(3), 249–257.

Oh, S.M. and Uribarri, J. (2006). Electrolytes, water and acid-base balance. In *Modern Nutrition in Health and Disease*, ed. M.S. Shils, M. Shike, C.A. Ross, B. Caballero and R.J. Cousins, 10th ed., 149–194. New York: Lippicott Williams & Wilkins.

Onkamo, P., Väänänen, S., Karvonen, M. and Tuomilehto, J. (1999). Worldwide increase in incidence of type I diabetes: the analysis of the data on published incidence trends. *Diabetologia*, 42(12), 1395–1403.

Parametric Optimization Solutions (ParOS) Ltd (2003). *Parametric Optimization Programming (POP) Toolbox.*

Parker, R.S., Doyle, F.J. and Peppas, N. (1999). A model-based algorithm for blood glucose control in type I diabetic patients. *IEEE Transactions on Biomedical Engineering*, 46(2), 148–157.

Patek, S.D., Magni, L., Dassau, E., Karvetski, C., Toffanin, C., De Nicolao, G., Del Favero, S., Breton, M., Man, C.D., Renard, E., Zisser, H., Doyle, F.J., Cobelli, C. and Kovatchev, B.P. (2012). Modular closed-loop control of diabetes. *IEEE Transactions on Biomedical Engineering*, 59(11), 2986–2999.

Patterson, C.C., Dahlquist, G.G., Gyürüs, E., Green, A.M., Soltész, G. and the E.S. Group (2009). Incidence trends for childhood type 1 diabetes in Europe during 1989–2003 and predicted new cases 2005–20: a multicentre prospective registration study. *The Lancet*, 373(9680), 2027–2033.

Percival, M.W., Y. Wang, B. Grosman, E. Dassau, H. Zisser, L. Jovanovič, and F.J. Doyle III (2010). Development of a multi-parametric model predictive control algorithm for insulin delivery in type 1 diabetes mellitus using clinical parameters. *Journal of Process Control*, 1–14.

Pistikopoulos E.N. (2009). Perspectives in multiparametric programming and explicit model predictive control. *AIChE Journal*, 55(8), 1918–1925.

Pistikopoulos, E.N., Dua, V., Bozinis, N.A., Bemporad, A. and Morari, M. (2002). On-line optimization via off-line parametric optimization tools. *Computers & Chemical Engineering*, 26(2), 175–185.

Pistikopoulos, E.N., Faisca, N.P., Kouramas, K.I. and Panos, C. (2009). Explicit robust model predictive control. In: *Proceedings of the International Symposium on Advanced Control of Chemical Process (ADCHEM).*

Pistikopoulos, E.N., Georgiadis, C.M. and Dua, V. (2007). *Multi-parametric programming theory: theory and applications*, vol. 1.

PSE (2011a) *gPROMS ModelBuilder Guide*. London: PSE. http://www.psenterprise.com

PSE (2011b) *gPROMS Model Validation Guide*. London: PSE. http://www.psenterprise.com

PSE (2011c) *gPROMS Optimisation Guide*. London: PSE. http://www.psenterprise.com

Rave, K., Nosek, L., Posner, J., Heise, T. and Roggen, K. (2006). Renal glucose excretion as a function of blood glucose concentration in subjects with type 2 diabetes: results of a hyperglycaemic glucose clamp study. *Nephrology Dialysis Transplantation*, 21, 2166–2171.

Regittnig, W., Ellmerer, M., Fauler, G., Sendlhofer, G., Leis, H., Schaupp, L., Wach, P. and Pieber, T.R. (2003). Assessment of transcapillary glucose exchange in human skeletal muscle and adipose tissue. *American Journal of Physiology Endocrinology and Metabolism*, 285, E241–E251.

Roden, M. and Bernroider, E. (2003). Hepatic glucose metabolism in humans: its role in health and disease. *Best Practice & Research Clinical Endocrinology & Metabolism*, 17(3), 365–383.

Russell, S.J., El-Khatib, F.H. Nathan, D.M., Magyar, K.L., Jiang, J. and Damiano, E.R. (2012) Blood glucose control in type 1 diabetes with a bihormonal bionic endocrine pancreas. *Diabetes Care*, 35, 2148–2155.

Sakizlis, V., Dua, V., Perkins, J.D. and Pistikopoulos, E.N. (2004). Robust model-based tracking control using parametric programming. *Computers & Chemical Engineering*, 28 (1–2), 195–207.

Sorensen, J.T. (1978). *A Physiological Model of Glucose Metabolism in Man and Its Use to Design and Assess Improved Insulin Therapies for Diabetes*. Cambridge, MA: MIT Press.

Soru, P., De Nicolao, G., Toffanin, C., Dalla Man, C., Cobelli, C. and Magni, L. (2012). MPC based artificial pancreas: strategies for individualization and meal compensation. *Annual Reviews in Control*, 36(1), 118–128.

Tarín, C., Teufel, E., Picó, J., Bondia, J. and Pfleiderer, H.-J. (2005). Comprehensive pharmacokinetic model of insulin glargine and other insulin formulations. *IEEE Transactions on Biomedical Engineering*, 52(12), 1994–2005.

Thabit, H. and Hovorka, R. (2013). Closed-loop insulin delivery in type 1 diabetes. *Endocrinology and Metabolism Clinics of North America*, 41(1), 105–117.

Thabit, H. and Hovorka, R. (2014). Bringing closed-loop home: recent advances in closed-loop insulin delivery. *Current Opinion in Endocrinology, Diabetes, and Obesity*, 21(2), 95–101.

Tolić, I.M., Mosekilde, E. and Sturis, J. (2000). Modeling the insulin-glucose feedback system: the significance of pulsatile insulin secretion. *Journal of Theoretical Biology*, 207(3), 361–375.

Van Heusden, K., Dassau, E., Zisser, H.C., Seborg, D.E. and Doyle, F.J. (2012). Control-relevant models for glucose control using a priori patient characteristics. *IEEE Transactions on Biomedical Engineering*, 59(7), 1839–1849.

Wennesland, R., Brown, E., Hopper, J., Hodges, J., Guttentag, O.E., Scott, K.G., Tucker, I.N. and Bradley, B. (1959). Red cell, plasma and blood volume in healthy men measured by radiochromium (Cr51) cell tagging and hematocrit:

influence of age, somatotype and habits of physical activityon the variance after regression of volumes to height and weight combined. *Journal of Clinical Investigation*, 38(7), 1065–1077.

Wilinska, M.E., Chassin, L.J., Acerini, C.L., Allen, J.M., Dunger, D.B. and Hovorka, R. (2010). Simulation environment to evaluate closed-loop insulin delivery systems in type 1 diabetes. *Journal of Diabetes Science and Technology*, 4(1), 132–144.

Wilinska, M.E., Chassin, L.J., Schaller, H.C., Schaupp, L., Pieber, T.R. and Hovorka, R. (2005). Insulin kinetics in type-I diabetes: continuous and bolus delivery of rapid acting insulin. *IEEE Transactions on Biomedical Engineering*, 52(1), 3–12.

Ziehn, T. and Tomlin, A.S. (2009). GUI–HDMR: a software tool for global sensitivity analysis of complex models. *Environmental Modelling & Software*, 24(7), 775–785.

Zavitsanou, S., Mantalaris, A., Georgiadis, M.C. and Pistikopoulos, E.N. (2014). Optimization of insulin dosing in patients with type 1 diabetes mellitus. *Computer Aided Chemical Engineering*, 33, 1459–1464.

Part III

6

An Integrated Platform for the Study of Leukaemia

Eirini G. Velliou[1], Maria Fuentes-Gari[2], Ruth Misener[3], Eleni Pefani[4], Nicki Panoskaltsis[5], Athanasios Mantalaris[6], Michael C. Georgiadis[7], and Efstratios N. Pistikopoulos[8]

[1] *Department of Chemical and Process Engineering, Faculty of Engineering and Physical Sciences, University of Surrey, UK*
[2] *Process Systems Enterprise (PSE), London, UK*
[3] *Department of Computing, Imperial College London, UK*
[4] *Clinical Pharmacology Modelling and Simulation, GSK, UK*
[5] *Department of Medicine, Imperial College London, UK*
[6] *Department of Chemical Engineering, Imperial College London, UK*
[7] *Laboratory of Process Systems Engineering, School of Chemical Engineering, Aristotle University of Thesaloniki, Greece*
[8] *Texas A&M Energy Institute, Artie McFerrin Department of Chemical Engineering, Texas A&M University, USA*

6.1 Towards a Personalised Treatment for Leukaemia: From *in vivo* to *in vitro* and *in silico*

Leukaemia is a severe cancer of the blood, currently affecting more than 8000 individuals in the UK on an annual basis (for more details, see Chapter 7). Traditional clinical diagnosis and further treatment focus on each patient's clinical symptoms and signs, characteristics (e.g. sex and family history) and laboratory imaging evaluation. This process is a reactive approach to the disease, initiating after the disease symptoms appear. Moreover, in the past, drug development by pharmaceutical industries was based on empirical observations. However, nowadays, with the significant progress constantly taking place in the areas of genomics, proteomics and metabolomics, it is believed that specific information related to the genetic characteristics and proteomic and metabolomic profiles of an individual patient could be used to tailor medical care (so-called *personalised medicine*). In this context, personalised healthcare is expected to deliver a *step change* in (a) quality and value of care, through more precise and personalised diagnostics as well as cost-effective and targeted therapies; (b) the pharmaceutical industry through a more efficient drug development process based on improved disease and drug discovery platforms and modelling of patient-specific and disease-specific biomarker endpoints; and (c) the diagnostics industry through the advent of diagnostic tests. Some of the challenges in delivery of personalised medicine lie in (a) the *fidelity and validity of current experimental systems* (i.e. *in vitro*) used to investigate human disease, (b) *the integration of patient-specific and disease-specific datasets* (i.e. *in silico*) and (c) the application

Modelling Optimization and Control of Biomedical Systems, First Edition.
Edited by Efstratios N. Pistikopoulos, Ioana Naşcu, and Eirini G. Velliou.

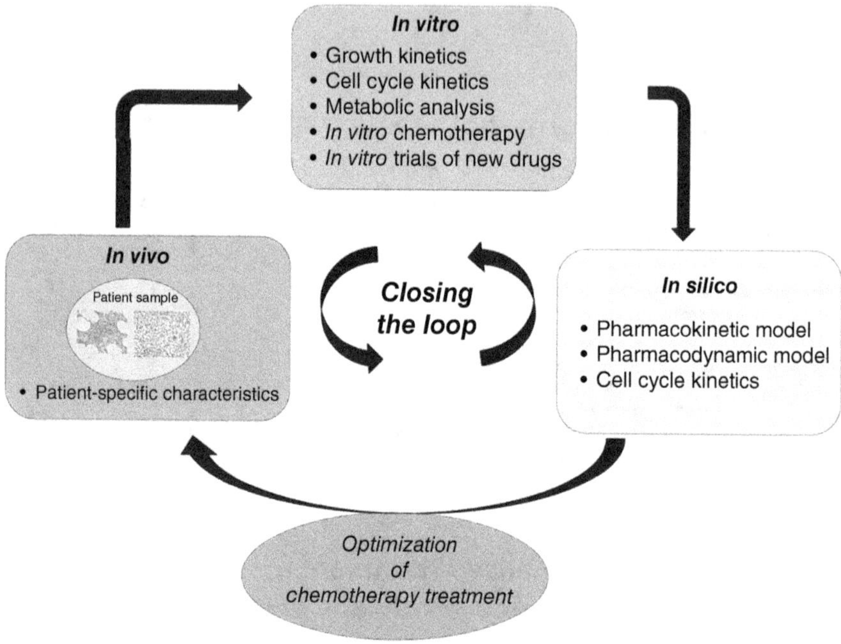

Figure 6.1 Towards optimisation/personalisation of chemotherapy for leukaemia treatment. *Source*: Velliou *et al.* (2014). Reproduced with permission of Elsevier.

of these models into practice to identify *simple targets* and *more efficient, yet less toxic, therapies* for a specific condition (i.e. *in vivo*). ***Therefore, "closing the loop" from* in vivo *to* in vitro *and* in silico *is a first step towards optimisation and, consequently, personalisation of chemotherapy treatment*.**

Figure 6.1 describes the different parts of the integrated platform for optimisation of chemotherapy for leukaemia treatment, which is currently being designed in the Centre for Process Systems Engineering at Imperial College London (Velliou *et al.*, 2014).

This platform includes (a) an *in vitro* and (b) an *in silico* part. Development and optimisation of (a) and (b) will eventually bridge the gap between laboratory experimentation/mathematical modelling and *in vivo* optimal chemotherapy treatment for a specific individual.

The following sections include a more detailed description of each individual block of the platform described in Figure 6.1.

6.2 *In vitro* Block of the Integrated Platform for the Study of Leukaemia

An appropriate *in vitro* platform (i.e. a 3D structure that mimicked the bone marrow) was developed by Mortera *et al.* (2010, 2011), consisting of highly porous polyurethane (PU; pore size = approx. 100 μm) of dimensions 5 × 5 × 5 mm

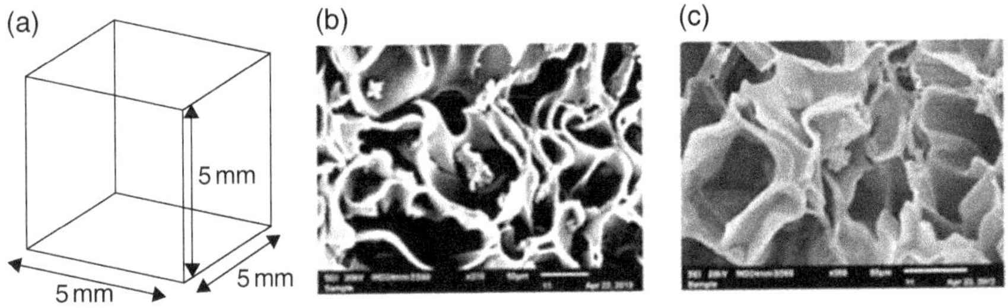

Figure 6.2 (a): Geometry of the 3D scaffolds; (b–c) scanning electron microscopy (SEM) images of the highly porous 3D scaffolds, including seeded leukaemic cells. *Source*: Velliou *et al.* (2014). Reproduced with permission of Elsevier.

(Figure 6.2), which allows perfusion of nutrients and oxygen within the matrix. The adhesion signals of the extracellular matrix (ECM) are recapitulated by coating this PU cube with collagen type 1. This 3D scaffolding system successfully supported the long-term expansion of leukaemic cell lines for over 6 weeks. Moreover, it successfully supported expansion and differentiation of umbilical cord blood cells (blood cells with high proliferation/differentiation potential that are extracted from the cord which arises from the navel that connects the foetus with the placenta) without any exogenous cytokines for a timeframe of 4 weeks, in contrast to traditional 2D culture systems that allowed umbilical cord blood cells expansion only for a few days in the absence of exogenous growth factors. Therefore, this 3D scaffolding system provides an ideal laboratory high-throughput technical platform for screening several environmental factors and identifying those that are crucial for the successful *ex vivo* expansion of normal and leukaemic blood.

More details on the *in vitro* studies of leukaemia are described and analysed in Chapter 7.

6.3 *In silico* Block of the Integrated Platform for the Study of Leukaemia

The most common treatment for acute myeloid leukaemia (AML), a severe and aggressive type of leukaemia, is intensive chemotherapy (see also Chapter 7). The latter involves exposure of the patient to cytotoxic drugs which interact with highly proliferative cells. More specifically, only cells that are in specific phases of the cell cycle (e.g., the process by which cells duplicate; see also Chapter 7) will be eliminated. Since healthy cells also proliferate in order to renew the cellular material, they will equally be affected. It is, therefore, crucial to maintain a balance between the number of cancer cells killed and the loss of healthy cells. However, clinical treatment protocols ignore the mechanisms behind drug action on the normal and abnormal populations, which can lead to over- or under-treatment. A more rational approach for the design of

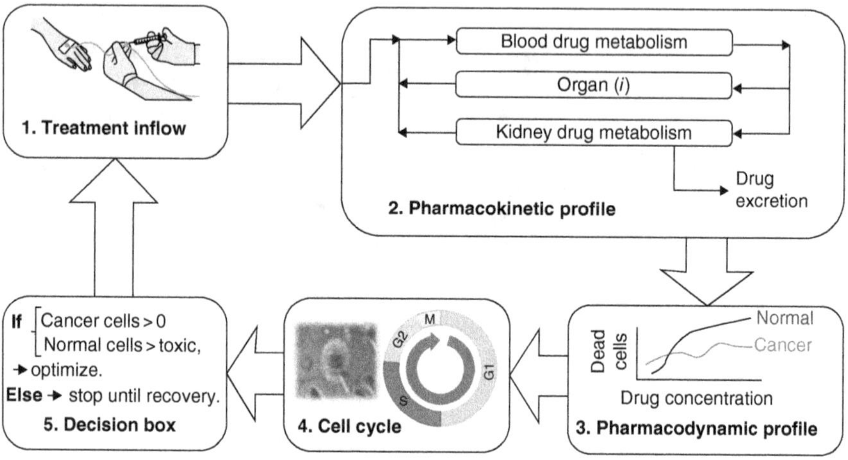

Figure 6.3 Mathematical optimisation of chemotherapy treatment for AML.
Source: Fuentes-Garí *et al.* (2015). Reproduced with permission of Elsevier.

clinical treatment protocols based on the personalisation of the chemotherapy schedule for each individual patient was developed by Pefani *et al.* (2013, 2014; for more details, please see Chapter 8). A schematic overview of the modelling approach is presented in Figure 6.3.

The model is composed of two main sections: pharmacokinetics (PK), which describes the elimination of the drug by the organs, and pharmacodynamics (PD), which accounts for the effects of the drug on the cells in the bone marrow, which is the location of the tumour generation. The main input to the system is the treatment inflow, the value of which is calculated based on the administration route and the injection rate of the drug. The resulting drug concentration reaches the body through the bloodstream, delivering it to the organs, which absorb it at different rates. Mass balances are performed in each of these organs, giving the drug concentration profiles. The drug concentration profiles calculated in the PK model are the main input for the PD model, in which the effect of the drug on the normal and the cancer cell populations is computed according to cell cycle kinetics of each population. Two separate models are used for each of them.

Because most of the cancer cells are proliferating, the cell cycle model in this case incorporates three compartments in which the cells are non-resting. Each of them is described by the mass balance between body compartments (including cell death by drug action, if applicable). The transition rates are dependent on cell cycle times and natural apoptosis rates in each of the phases (Basse *et al.*, 2003). The normal cell population model considers a proliferative population and a resting population that can move into a proliferative state. In both cases, the cell cycle kinetics are modelled through a set of ordinary differential equations (ODEs) (one per compartment y):

$$\frac{dP_y}{dt} = k_{y-1}\left(T_{y-1}\right)\cdot P_{y-1} - k_y\left(T_y\right)\cdot P_y - effect_j \cdot P_y \tag{6.1}$$

where P_y and P_{y-1} are the number of cells in compartments y and $y-1$; $k_y(T_y)$ and $k_{y-1}(T_{y-1})$ are the transition rates from compartment y and $y-1$, respectively (dependent on the duration of the corresponding phases, T_y and T_{y-1}); and $effect_j$ is the effect of drug j in the compartment.

More details on the *in silico* studies of leukaemia are described and analysed in Chapter 8.

6.4 Bridging the Gap Between *in vitro* and *in silico*

As mentioned in Section 6.2, an appropriate *in vitro* platform (see also Chapter 7) will enable the *ex vivo* cultivation of leukaemia derived from patients and, consequently, the identification and quantification of factors that crucially affect the cancer's evolution. Data derived from the *in vitro* platform will serve as an input for the *in silico* platform (see also Chapter 8), leading to the construction of advanced models and ultimately to the prediction of the optimal chemotherapy dosage for a specific individual (Figure 6.1).

Chapters 7 and 8 focus on a detailed description of the *in vitro* and *in silico* studies of leukaemia.

References

Basse, B., Baguley, B.C., Marshall, E.S., Joseph, W.R., Brunt, B.V., & Wake, G. (2003). A mathematical model for analysis of the cell cycle in cell lines derived from human tumours. *Mathematical Biology*, **47**, 295–312.

Fuentes-Garí, M., Velliou, E., Misener, R., Pefani, E., Rende, M., Panoskaltsis, N., Mantalaris, A., & Pistikopoulos, E.N. (2015). A systematic framework for the design, simulation and optimization of personalized healthcare: making and healing blood. *Computers and Chemical Engineering*, **81**, 80–93.

Mortera-Blanco, T., Mantalaris, A., Bismarck, A., Aqel, N., & Panoskaltsis, N. (2011). Long-term cytokine-free expansion of cord blood mononuclear cells in three-dimensional scaffolds. *Biomaterials*, **32**, 9263–9270.

Mortera-Blanco, T., Mantalaris, A., Bismarck, A., & Panoskaltsis, N. (2010). Development of a three-dimensional biomimicry of human acute myeloid leukemia *ex vivo*. *Biomaterials*, **31**, 2243–2251.

Pefani, E., Panoskaltsis N., Mantalaris, A., Georgiadis, M.C., & Pistikopoulos, E.N. (2013). Design of optimal patient-specific chemotherapy protocols for the treatment of patients with acute myeloid leukaemia. *Computers & Chemical Engineering*, **57**, 187–195.

Pefani, E., Panoskaltsis N., Mantalaris, A., Georgiadis, M.C., & Pistikopoulos, E.N. (2014). Chemotherapy drug scheduling for the induction treatment of patients diagnosed with acute myeloid leukaemia. *Transactions in Biomedical Engineering*, **61**(7), 2059–2056.

Velliou, E., Fuentes-Garí, M., Misener, R., Pefani, E., Rende, M., Panoskaltsis, N., Pistikopoulos, E.N., & Mantalaris A. (2014). A framework for the design, modeling and optimization of biomedical systems. *Computer Aided Chemical Engineering*, **34**, 225–236.

7

In vitro Studies: Acute Myeloid Leukaemia

Eirini G. Velliou[1], Eleni Pefani[2], Susana Brito dos Santos[3], Maria Fuentes-Gari[4], Ruth Misener[5], Nicki Panoskaltsis[6], Athanasios Mantalaris[3], Michael C. Georgiadis[7], and Efstratios N. Pistikopoulos[8]

[1] *Department of Chemical and Process Engineering, Faculty of Engineering and Physical Sciences, University of Surrey, UK*
[2] *Clinical Pharmacology Modelling and Simulation, GSK, UK*
[3] *Department of Chemical Engineering, Imperial College London, UK*
[4] *Process Systems Enterprise (PSE), London, UK*
[5] *Department of Computing, Imperial College London, UK*
[6] *Department of Medicine, Imperial College London, UK*
[7] *Laboratory of Process Systems Engineering, School of Chemical Engineering, Aristotle University of Thesaloniki, Greece*
[8] *Texas A&M Energy Institute, Artie McFerrin Department of Chemical Engineering, Texas A&M University, USA*

7.1 Description of Biomedical System

This chapter describes human normal and abnormal (i.e., so-called leukaemia) haematopoiesis, the latter being a severe cancer of the blood. The process of haematopoiesis – normal and abnormal – as well as the bone marrow (BM) architecture and the cell cycle procedure are analysed. Moreover, current experimental platforms for studying leukaemia *ex vivo* and parameters that have to be taken into consideration when designing a treatment protocol (i.e., chemotherapy) are discussed and analysed.

7.1.1 The Human Haematopoietic System

Haematopoiesis (or haemopoiesis) is the process of blood cell production from haematopoietic stem cells (HSCs). The word is derived from the Greek word *αἷμα*, which means *blood*, and the verb *ποιεῖν*, which means *to make*. In the human body, this process takes place in a three-dimensional (3D) microenvironment, the BM (see also Section 7.1.2).

HSCs are unique mainly due to their self-renewal ability (i.e., the ability to divide and produce at least one daughter cell identical to the mother one), as well as their ability to differentiate into either the myeloid or the lymphoid blood lineage, therefore leading to the formation of a variety of mature blood cells such as megakaryocytes, macrophages, erythroid cells (erythrocytes), platelets, dendritic cells (Quesenberry and Colvin, 2001; Buttery and Shakesheff, 2008), lymphocytes (T and B cells), natural killer cells and plasma cells (Figure 7.1) (Quesenberry and Colvin, 2001; Ryan *et al.*, 2001). HSCs can

Modelling Optimization and Control of Biomedical Systems, First Edition.
Edited by Efstratios N. Pistikopoulos, Ioana Naşcu, and Eirini G. Velliou.

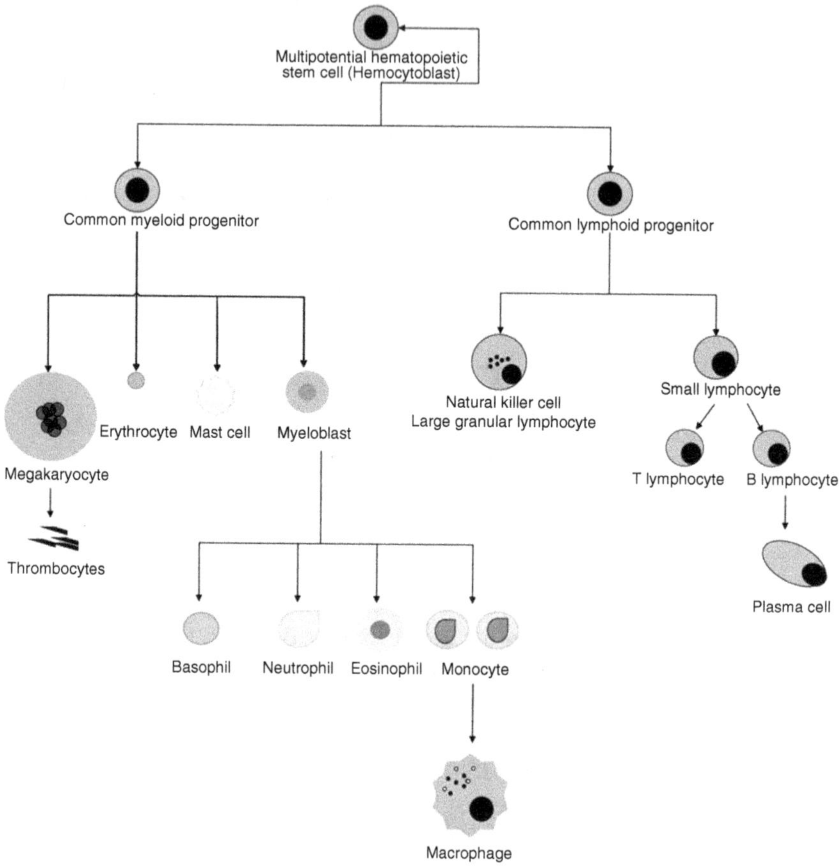

Figure 7.1 The human haematopoietic system.

be divided into long-term stem cells (LT-HSCs) and short-term progenitor cells (ST-HSCs). LT-HSCs have a self-renewal capability which may last up to several months, and they may differentiate towards progenitor cells. ST-HSCs have shorter self-renewal capability compared to LT-HSCs; they can differentiate immediately towards a variety of mature blood cells (Marshak *et al.*, 2001).

The HSC differentiation process is regulated by cytokines and growth factors. Cytokines are small molecules of protein or glycoprotein nature that act as signalling molecules (Gilman *et al.*, 2001). They act in short distances and concentrations by binding in specific cell membrane receptors, and their role is crucial for the progression of haematopoiesis, inflammation and immunity. Many cytokines can participate in a variety of actions at different stages and lineages of haematopoiesis. Examples of cytokines that regulate the haematopoietic system are the interleukins (cytokines with pleiotropic actions), thrombopoietin (regulator of megakaryocytopoiesis and platelet synthesis), erythropoietin (crucial for erythrocyte production) and the inhibitors (which mainly inhibit the differentiation of progenitor to mature blood cells, therefore blocking the haematopoietic process) (Quesenberry and Colvin, 2001).

7.1.2 General Structure of the Bone Marrow Microenvironment

In adult humans, the process of haematopoiesis takes place in the BM, which is a 3D organ of high complexity. In the BM of an adult human, approximately 6 billion cells are produced per kilogram of body weight on a daily basis, from which 2.5 billion cells are red blood cells, 2.5 billion are platelets and 1 billion are granulocytes (Abboud and Lichtman, 2001; Beutler *et al.*, 2001). The BM lies within the trabecular bone which, together with the BM stroma, is the physical support for the maintenance of haematopoiesis (Panoskaltsis *et al.*, 2005). The stroma consists of a variety of cells: endothelial cells (they cover the inner part of the sinus and control the inflow and outflow of molecules), reticular cells (they form part of the adventitial coating of the vascular sinus), adipocytes (they develop by lipogenesis of fibroblast-like cells, and they promote haematopoiesis), bone cells (osteoblasts and osteoclasts), macrophages and lymphocytes (they enhance cell–cell interaction in the BM and produce growth factors essential for haematopoiesis). In addition to the stroma cells that nurture the haematopoietic process in microenvironmental niches, an important feature of the BM is the extracellular matrix (ECM) (Abboud and Lichtman, 2001; Beutler *et al.*, 2001; Charbord, 2001). The ECM is a supportive structure which consists of proteins: the so-called extracellular matrix proteins (EMPs), such as proteoglycans and glycoaminoglycans (structural macromolecules); laminin, hemonectin and fibronectin, which are adhesive proteins; as well as collagen types I, III and IV (Abboud and Lichtman, 2001; Beutler *et al.*, 2001; Panoskaltsis *et al.*, 2005).

The ECM together with the BM stroma and the trabecular bone form a 3D structure which in combination with the blood vessels and sinusoids serve as an appropriate microenvironment for the haematopoietic process to take place (i.e., the haematopoietic inductive microenvironment [HIM]) (Trentin, 1970; Naito *et al.*, 1992). In the HIM, growth, proliferation, differentiation and self-renewal of HSCs take place via the interactions between HSCs, the ECM, the BM stroma and cytokines, leading to the production of a variety of mature blood cells. In haematopoiesis, one HSC is able to divide at least 50 times and generate 10^{15} cells (Dorshkind, 1990; Kollet *et al.*, 1993; Panoskaltsis *et al.*, 2005).

For the efficient reproduction of *ex vivo* haematopoiesis, it is essential to maintain, on the one hand, the self-renewal potential and, on the other hand, the differentiation capability of HSCs towards progenitors and eventually mature blood cells. For this maintenance to be achieved, it is critical to provide an *in vitro* HIM which should mimic the structure and properties of the *in vivo* HIM (see also Section 7.2).

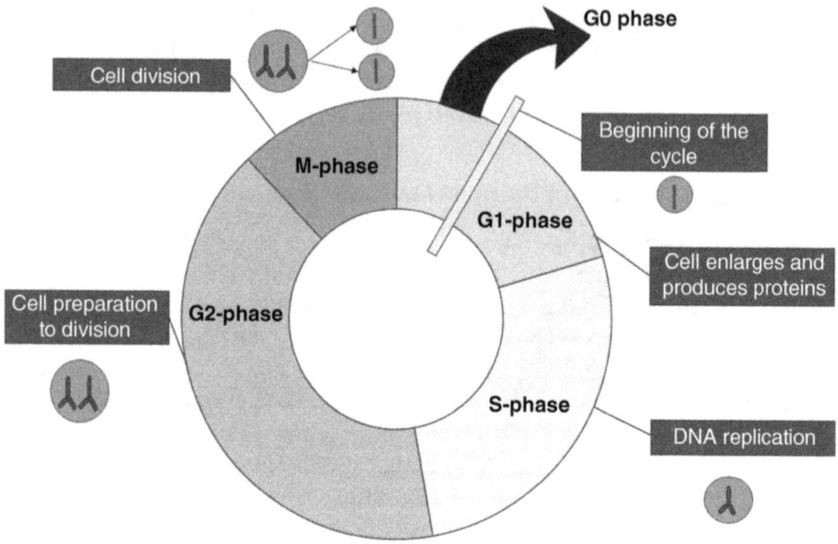

Figure 7.2 The cell cycle.

Table 7.1 The phases of the cell cycle.

State	Phase	Process taking place in each phase
Quiescent/resting	G_0	The cell either has left the cycle or has not entered it yet; the cell does not divide.
Interphase	G_1	Cells increase in size.
	S	The genetic material (DNA) of the cell is duplicated.
	G_2	At this gap phase between DNA synthesis and mitosis, cell growth takes place.
Cell division	M	Cell division (cytokinesis) takes place, leading to two daughter cells.

7.1.3 The Cell Cycle

Each living cell results from the division of a mother cell. This process of cell reproduction is essential for the maintenance of a living organism from the simplest single cell to advanced organisms. The cell cycle can be defined as a series of events which lead to the duplication and further division of a single cell to two daughter cells. The cell cycle can be divided in four stages, or phases: the G_1 phase (growth phase), the S phase (DNA-synthesis phase), the G_2 phase (pre-mitotic phase) and the M phase (mitosis phase) (Figure 7.2 and Table 7.1). G_1 and G_2 are considered gap phases as they are phases at which more time is

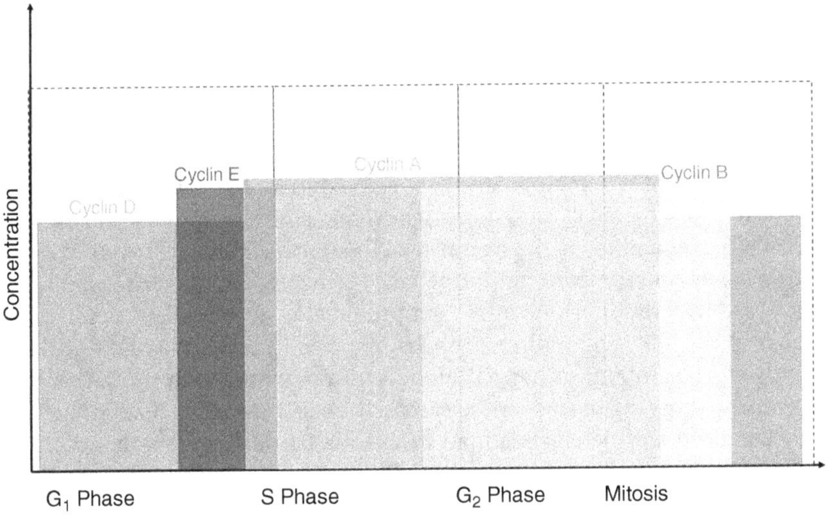

Figure 7.3 Cyclin expression throughout the cell cycle.

given for the cells to grow (a process that is much more time-consuming than duplication and chromosome segregation). An additional phase is the G_0, which describes a phase at which the cells are not in any of the cell cycle phases, and therefore are non-dividing (Morgan, 2007).

Control of the cell cycle can be defined as a series of tightly regulated molecular switches which trigger the initiation of specific cell cycle events at a certain regulatory checkpoint (Morgan, 2007). Control of the cell cycle and consequently transition from one phase to another are regulated by enzymes, the cyclin-dependent kinases (CDKs). CDKs are activated at specific stages of the cell cycle by binding to cyclins (i.e., regulatory proteins). Cyclins function as positive regulatory molecules that help the transition from one stage of the cell cycle to the next, depending on their concentration which fluctuates throughout the cell cycle. More specifically, the initiation of the cell cycle starts in mid- to late G_1 phase, when cyclin E reaches a maximum concentration/expression. Cyclin E binds to its CDKs, giving the signal for the G_1/S switch. Similarly, cyclin A regulates the S/G_2 transition and cyclin B the entrance to mitosis (i.e., the G_2/M transition) (Hartwell *et al.*, 1989; Murray and Krischner, 1989; Morgan, 1997).

Extensive research has indicated that cell cycle progression is crucial for the evolution of leukaemia since the main characteristic of a cancer cell is an uncontrolled high proliferation (Figure 7.3). Consequently, most available chemotherapy drugs target stopping this high proliferation by interfering with the cell cycle and blocking cells that are in the process of replication (see, e.g., Balgobind *et al.*, 2010; Kaiser *et al.*, 2011). Different chemotherapy drugs act in different phases of the cell cycle (see also Section 7.1.5).

7.1.4 Leukaemia: The Disease

Leukaemia is a severe cancer of the haematopoietic system characterized by the inability of blood progenitors to mature normally, leading to the accumulation of immature white blood cells (i.e., the so-called blasts) in the BM (Beutler *et al.*, 2001). Alternatively, this disease can be viewed as the formation of an abnormal haematopoietic tissue, the initiation of which is the result of the function of a small amount of leukaemic stem cells (LSCs) (Passegué *et al.*, 2003). Depending on the decrease of the normal blood cell populations, the disease symptoms can consist of fatigue, haemorrhage, infections and fever. Moreover, dyspnoea or other symptoms may occur as leukaemic cells circulate in the body and infiltrate tissues. The health condition of the patient depends on the amount of normal blood cells compared with those that are leukaemic. According to Cancer Research UK in 2011, 8616 people in the UK were diagnosed with leukaemia, and in 2010 4503 deaths from that disease occurred. Moreover, 82,300 new cases of leukaemia were diagnosed in 2012 within the European Union.

Leukaemia can be divided in different types, depending on the haematopoietic lineage at which the proliferation disorder occurs. More specifically, *myeloid leukaemia* occurs in the myeloid lineage, and *lymphocytic leukaemia* occurs in the lymphoid lineage. Depending on the speed of the disease, the evolution of leukaemia can be divided into *acute* (i.e., the number of the blasts increases rapidly, leading to a faster disease evolution) and *chronic* (i.e., the progress of the disease is slower, as there is a production of partly mature but not functional white blood cells). Based on this categorization, the following four general types of leukaemia can be found: acute myeloid leukaemia, chronic myeloid leukaemia, acute lymphocytic leukaemia and chronic lymphocytic leukaemia.

Acute myeloid leukaemia (AML) is one of the most aggressive types of leukaemia. According to Cancer Research UK, approximately 2921 cases of AML occurred in 2011 in the UK. AML is a type of leukaemia which is characterized by an increased proliferation of immature blasts in the myeloid lineage and as a consequence insufficient red blood cell production. These blasts have low proliferation capability. However, a small population of cells with higher proliferation capacity and self-renewal potential (i.e., the LSCs) are the key component for the maintenance of the disease (Bonnet and Dick, 1997). AML is usually the result of somatic mutations in either a pluripotent HSC or a slightly differentiated progenitor cell, and it leads to a deregulation and/or inhibition of normal haematopoiesis due to space restrictions and inhibitory and clonal factors specific to the disease (Lowenberg *et al.*, 1999; Lichtman, 2001; Panoskaltsis *et al.*, 2003).

Chronic myeloid leukaemia (CML) (alternatively known as chronic granulocytic leukaemia [CGL]) is a stem cell disease of the myeloid lineage that is characterized by an exaggerated granulocytosis, anaemia, granulocytic immaturity, basophilia and splenomegaly. There is an extreme cellular accumulation in the BM, and from a genetic point of view in 90% of the cases a reciprocal translocation between chromosomes 9 and 22 is observed (i.e., the so-called *Philadelphia (Ph) chromosome* or *translocation*). In many cases, the disease can progress into a very high-speed phase, resembling AML (Lichtman, 2001). CML is a rather rare cancer of the blood. According to Cancer Research UK, 675 patients were diagnosed with CML in 2011 in the UK.

Acute lymphocytic leukaemia (ALL) (alternatively known as acute lymphoblastic leukaemia) is a type of leukaemia which mostly affects children. Approximately 4000 cases of ALL are reported on an annual basis in the USA, of which two-thirds are children (Pui and Evans, 2006). According to Cancer Research UK, 654 cases of ALL were diagnosed in 2011 in the UK ALL is a leukaemia in which the cellular deregulation occurs in the lymphoid lineage of the haematopoietic system. More specifically, an abnormality in the lymphocytes takes place, leading to an accumulation of non-functional white blood cells (blasts) in the BM, as a result of abnormal cellular proliferation, blocking of cellular differentiation and increased resistance to apoptosis (i.e., cell death) (Pui, 2009).

Chronic lymphocytic leukaemia (CLL) is a type of leukaemia which is characterized by an accumulation of immature lymphocytes (i.e., of the B-cell lineage) in the human BM, peripheral blood and lymphoid tissues (Kipps, 2001). The progress of CLL is much slower than that of ALL, and it affects older adults (i.e., people above 51 years of age). According to the National Health Service (NHS), about 2400 people are diagnosed in the UK with CLL on an annual basis, and the disease affects 2.7 persons per 100,000 in the USA (Kipps, 2001). According to Cancer Research UK, in 2011, 3233 people were diagnosed with CLL in the UK. According to the American Cancer Society, 15,490 people in the USA were diagnosed with CLL in 2009.

7.1.5 Current Medical Treatment

The most common treatment for all types of leukaemia is *chemotherapy*. For more accelerated types (such as AML), a first round of *induction chemotherapy* starts directly after diagnosis, followed by a *remission period* before the *consolidation*, which normally consists of additional chemotherapy and in some cases *transplantation* from a donor or from the patient's HSCs – the latter being rarer (Cancer Research UK, 2011).

A variety of chemotherapy drugs are generally used, depending on the type of leukaemia. The most commonly used drugs for treatment of leukaemia are *cytarabine* (cytosine arabinose [ara-C]), which is an antimetabolite targeting to block the DNA/RNA replication by attacking the cells that are in the S phase of the cell cycle, and the *anthracycline* drugs (such as fludarabine), which attach cells that are in the G_1 phase of the cell cycle (American Cancer Society, 2013). Current chemotherapy treatment protocols are designed based on pre-clinical animal experiments, empirical clinical trials as well as the acquired experience of subspecialist physicians. The design parameters for these protocols consist of the patient BM aspirate examination (blasts percentage, immunophenotype, cytogenetic and molecular analysis) and patient physiological characteristics (height and weight) for the normalization of the dose applied on the body surface area (BSA).

In the remainder of this chapter, the available experimental platforms for *in vitro* studies of haematopoietic normal and abnormal (i.e., cancerous cells) are described in detail. Moreover, the effect of a variety of environmental factors on the cellular evolution – on a macroscopic kinetics, a cell cycle progression and a metabolic level – is pointed out. These environmental factors should be taken into consideration when designing experiments, as their impact is of great importance for the leukaemic evolution *in vitro* and, therefore, for understanding the mechanisms of the blood cancer progression.

7.2 Experimental Part

7.2.1 Experimental Platforms

A first step towards understanding and further optimising chemotherapy for AML treatment is the conduction of *in vitro* studies that aim at the reproduction of the *in vivo* environment (Mayani *et al.*, 2009). A variety of *ex vivo* research in both normal and abnormal haematopoiesis is conducted in a two-dimensional (2D) environment (e.g., laboratory T-flasks or micro-plates). For example, Koller *et al.* (1993) achieved expansion of BM mononuclear cells (MNCs) *ex vivo* in stirred 2D laboratory flasks. Collins *et al.* (1998) successfully expanded peripheral and cord blood cells in serum-free spinner cultures. Giarratana *et al.* (2005) successfully grew and differentiated cord blood cells *ex vivo*. Despite the fact that 2D cultures enable the understanding of growth and differentiation of HSCs, (a) they require a high amount of exogenous growth factors that differ from the *in vivo* environment, therefore creating an artefact in the culture and consequently in the cell kinetics; and (b) they are unable to recapitulate the *in vivo* BM microenvironment (Rabinowitz *et al.*, 1993; Engelhardt *et al.*, 2001; Kirito *et al.*, 2003; Levac *et al.*, 2005; Panoskaltsis *et al.*, 2005).

For the latter to take place, several researchers focused on the development and further study of normal and leukaemic haematopoiesis in 3D culturing systems. In particular, these cultures recapitulate the architectural characteristics of the BM microenvironment by using a variety of polymer-based biomaterials and/or the BM ECM by using specific proteins and/or the BM stromal support by co-culturing HSCs with appropriate cellular systems (Lee-Thedieck and Spatz, 2012). For example, Tun *et al.* (2002) observed an increased expansion and proliferation of murine BM cells in a 3D polyvinyl formal (PVF) porous scaffolding system compared to suspension cultures. Feng *et al.* (2006) managed to successfully expand HSCs in 3D polyethylene terephthalate (PET) scaffolds coated with fibronectin to mimic the ECM of the BM. The expansion in 3D was much higher than that observed in 2D controls. Lutolf *et al.* (2009) observed increased proliferation of murine HSCs in polyethylene glycol (PEG) hydrogels. Moreover, in the presence of specific proteins such as Wtn3 and N-cadherin, proliferation decreased in the PEG hydrogels similarly to what was observed *in vivo*, pointing out that the hydrogels mimicked efficiently the BM niche. Lee and Kotov (2009) manufactured inverted colloidal crystal (ICC) scaffolds using a layer-by-layer (LBL) assembly technique. The created microfilm functioned as a 3D *ex vivo* mimicry of the thymic microenvironment as successful differentiation of HSCs towards T cells was observed. Nichols *et al.* (2009) fabricated ICC polystyrene and silica gel scaffolds that successfully supported maintenance and differentiation of murine HSCs towards B cells for a time frame of 2 weeks. Mortera-Blanco *et al.* (2010, 2011) developed 3D collagen-coated polyurethane (PU) scaffolds that supported the long-term (up to 6 weeks) growth of AML and the long-term cytokine-free expansion of HSCs derived from umbilical cord blood, respectively. Leisten *et al.* (2012) proposed a 3D collagen-based gel system as an efficient *in vitro* bone marrow niche model in which haematopoietic progenitor cells (HPCs) proliferate and expand highly when co-cultured with MSCs. Ferreira *et al.* (2012) constructed 3D fibrin scaffolds in which successful expansion of HSCs derived from umbilical cord blood took place in the presence of exogenous cytokines and stromal support. Alijtawi *et al.* (2014) observed an increased resistance of the HL-60 and Kashumi leukaemic cell lines to drug (cytarabine)-induced apoptosis in 3D polyglycolic acid/poly-L-lactide (PGA/PLLA) copolymer disks, compared to cell co-cultures in 2D. Raic *et al.* (2014) developed macroporous PEG diacrylate (PEGDA) hydrogels which were co-polymerized with the minimal integrin RGD in order to enable cell adhesion. Within these hydrogels when co-cultured with human BM mesenchymal cells and/or osteoblast-like cells, expansion, differentiation and maintenance of HSC stemness were achieved.

7.2.2 Crucial Environmental Factors in an *in vitro* System

In order to achieve an efficient *ex vivo* recapitulation of the BM niche, next to the 3D architectural characteristics (e.g., structure, presence of adhesion proteins or appropriate cellular stromal support), particular consideration needs to be given to environmental conditions such as oxygen, nutrient and growth factor concentrations. In Section 7.2.2.1, a detailed description of the impact of environmental parameters on normal and/or abnormal haematopoiesis is given.

7.2.2.1 Environmental Stress Factors and Haematopoiesis

Fluctuations in the microenvironmental conditions (e.g., oxygen concentration, and composition and concentration of nutrients such as glucose, cytokines or other growth factors) may be experienced as an environmental stress and, as a consequence, can highly affect the normal and abnormal haematopoietic proliferation, metabolic activity, resistance and further evolution. For example, a variety of researchers have shown that oxidative stress (i.e., increase in the concentration of reactive oxygen species [ROS]) leads to activation of survival pathways, and it is a key factor that promotes progression of cancerous stem cells as well as resistance to chemotherapy (see as examples Fruehauf and Meyskens, 2007; Lyu *et al.*, 2008; Liu *et al.*, 2009; Adbal Dayem *et al.*, 2010).

Especially in the case of the abnormal haematopoietic situation of AML, alterations of the oxygen and glucose concentration, on the one hand, in the different body compartments (e.g., in the BM and the peripheral blood or the liver) and, on the other hand, between patients (e.g., individual cases of hypoglycaemia or hyperglycaemia) may lead to a different stress adaptation of the leukaemic population. The latter will most likely affect the cancer growth, the inactivation kinetics and the response to a chemotherapeutic drug *in vivo*.

A variety of research has revealed the strong relation between resistance, that is, longer survival and increased proliferation of haematopoietic and/or leukaemic cells and/or resistance to chemotherapy, with (1) oxygen or (2) starvation stress *in vitro* and/or *in vivo*.

For example, Fecteau *et al.* (2013) observed an increased *in vitro* survival of cells from BM aspirates of patients with CLL in 5% O_2 compared to 20% O_2. This increased survival under hypoxic conditions was a result of the MSC increased proliferation and production of soluble pro-survival factors (i.e., CXCL12). Interactions between CLL and (increased) MSCs lead to enhanced CLL resistance.

Wilkinson *et al.* (2012) state that chronic oxidative stress may contribute to increased resistance of lymphoma patients to chemotherapy. This is due to the fact that oxidative stress leads to alteration of the mitochondria, that is, release of intermembrane proteins which lead to an increased permeabilization of the outer mitochondrial membrane and further resistance to apoptosis.

Zhou *et al.* (2010) pointed the possible relation between the relapse of AML and increased oxidative stress *in vivo*. Specifically, parameters related to oxidative stress – such as activities of adenosine deaminase and xanthine oxidase, antioxidant capacity (T-AOC), and levels of human thioredoxin (TRX) and indoleamine 2,3-dioxygenase – as well as expression of specific genes related to oxidative stress were monitored in patients with AML for a time period from a primary to a relapsed status. Low T-AOC and upregulated TRX expression lead to a relapse of the disease, indicating a strong correlation between oxidative stress and AML development and relapse.

Lodi *et al.* (2011) observed that hypoxia is a key factor that affects the metabolic activity, that is, the adaptation of phospholytic and glycolytic metabolism, and evolution of KG1a and K562 leukaemic cell lines. Mitochondrial respiration remained unaltered for both cell lines, indicating the ability of these leukaemic cell lines to increase their resistance under oxidative stress.

Giuntoli *et al.* (2011) studied the effect of the level of glucose on the growth and proliferation of K562 cell lines, U937 cell lines or primary chronic myeloid leukaemia cells under hypoxic conditions (i.e., 0.1% O_2) as well as under normoxia (i.e., 21% O_2). Although in general, lower growth was observed for lower glucose concentrations in hypoxia as well as in normoxia, glucose shortage in hypoxia led to increased size of the leukaemic (compared to the normal) haematopoietic population.

Herst *et al.* (2011) has pointed a possible relation between the level of glycolytic metabolism of AML blasts and resistance to chemotherapy. Analysis of 26 BM aspirates showed that AML cells with higher glucose consumption were more tolerant to *in vitro* apoptosis caused by all-trans retinoic acid (ATRA) and/or arsenic trioxide (ATO).

Deorosan and Nauman (2011) studied the effect of the glucose level on the metabolic state of MSCs cultivated *in vitro* in a 3D collagen-based matrix as well as in a 2D system. They showed that the initial glucose level highly affects the metabolic state of the cells, throughout 6 days of culture. More specifically, they characterized the metabolic state of the MSCs via the change in the level of lactate (i.e., the more lactate produced, the more the cellular metabolism was on the anaerobic site) and via the change in the level of pyruvate (i.e., in the case of glycolysis, pyruvate is the main metabolic product). It was shown that in the 2D culture when starting with higher glucose levels, higher lactate was produced, indicating that the relative cellular metabolic state was anaerobic, compared to lower initial levels of glucose for which the lactate-to-pyruvate ratio was lower. However, in the 3D system, although the glucose level clearly affects the cellular viability (i.e., it is higher for higher glucose), no systematic trend of the lactate-to-pyruvate ratio was observed, pointing out that the 3D system needs more in-depth investigation.

In general, from all the studies discussed here, it is clear that a systematic monitoring of environmental factors (Section 7.2.1), such as oxygen and glucose concentration, are essential, on the one hand, for a better understanding of the (normal and abnormal) haematopoietic cellular proliferation and metabolic activity, and, on the other hand, for more efficient *in vitro* mimicry of *in vivo* systems (i.e., in which fluctuations of those factors occur). For the latter to take place efficiently, progressive replacement of the 2D cultures with 3D scaffolds is essential (Section 7.2.2). In the 3D systems, in addition to the more efficient reproduction of the actual characteristics of the BM, such as the desired niche and an environment suitable for HSC expansion and differentiation (Panoskaltsis *et al.*, 2005), oxygen and glucose form gradients within the matrix, therefore mimicking more realistically the human BM. Efficient control and monitoring of oxygen and glucose gradients should take place throughout the conduction of experiments in 3D systems (Provin *et al.*, 2008; Xu *et al.*, 2008; Streeter and Cheema, 2011).

7.2.3 Growth and Metabolism of an AML Model System as Influenced by Oxidative and Starvation Stress: A Comparison Between 2D and 3D Cultures

This section describes in detail an experimental study which deals with the systematic monitoring and comparison of the proliferation and metabolic evolution of an AML model system in 3D and 2D cultures for different oxygen and glucose conditions close to physiological (*in vivo*) levels (Velliou *et al.*, 2013, 2015). More specifically, K-562 cell lines were cultivated in 3D PU porous scaffolds coated with collagen as well as in 2D suspension cultures in 5% (hypoxia) as well as 20% (normoxia) oxygen and for three different glucose levels: 4.3 g/L (optimal level generally applied in laboratory growth media), 1.3 g/L (highest human physiological level *in vivo*) and 0.6 g/L (lowest human physiological level *in vivo*) for 2 weeks. This time span was selected as the total chemotherapy cycle, for AML treatment lasts approximately 10 days (NHS Trust, UK), and understanding the effect of environmental factors on a leukaemic population within that time frame would be appropriate for comparison and/or future combination with *in vitro* chemotherapy experiments. The cellular proliferation and metabolic evolution were monitored.

7.2.3.1 Materials and Methods

Fabrication, Coating and Sterilization of 3D Scaffolds

3D scaffolds were fabricated and coated as previously described in Mortera Blanco *et al.* (2010).

Scaffold Fabrication For the scaffold fabrication, 3 g of PU (Noveon, Belgium) was added in 60 mL of dioxan (99.8% pure; Sigma Aldrich, UK). The foams were fabricated by thermally induced phase separation (TIPS) of polymer solution, followed by solvent sublimation as described in Safinia *et al.* (2005). The PU solution was frozen at −186 °C and maintained there for 2 h. Removal of the solvent took place by freeze-drying in an ethylene glycol bath at −15 °C for approximately 3 days. Prior to coating, scaffolds were cut into $5 \times 5 \times 5\,mm^3$ cubes.

Scaffold Coating After cutting, the scaffolds were coated with collagen type I from calf skin (Sigma Aldrich). The cubes were first dipped in ethanol (70% v/v) for 1 min, then transferred in phosphate buffered saline (PBS; Gibco, UK) for 10 min, and then centrifuged at 2500 rpm for 10 min. Hereafter, the scaffolds were transferred in 62.5 μg/mL of collagen type I solution – 0.1 M soluble acetic acid, dissolved in deionised water – of pH 7 (re-adjusted with addition of 0.1 M NaOH) and centrifuged at 2000 rpm for 20 min. At the end, scaffolds were washed once more with PBS at 1500 rpm for 10 min.

Scaffold Sterilisation Prior to cell seeding, sterilisation of the scaffolds took place by combinatory exposure to ethanol (70% v/v) and UV (230 v, 50 Hz, 0.14 A; Kedro Laboratory Products, UK). More specifically, the scaffolds were immersed in ethanol for 2 h and then washed twice with PBS – in which they were kept for 15 min for each washing step – before exposure for 8 min under UV. After sterilisation, the scaffolds were added in growth medium (see Section 2.3) in a humidified incubator for 3 days at 37 °C and 5% CO_2 in order to be examined for possible contaminations.

Inoculum Preparation

The human leukaemic cell line K-562 (human erythromyelo-blastoid leukaemic cell line; product no. CCL-243, ATCC, UK) which was originally derived from a patient with a blast crisis of CML (i.e., therefore resembling AML) was chosen as an AML model system. It has been previously reported by our group that this cell line can successfully expand in 3D PU collagen-coated scaffolds (Mortera-Blanco *et al.*, 2010). After thawing, as recommended by the cell line supplier, the cells were expanded by cultivation on standard polystyrene non-pyrogenic tissue culture flasks (Fisher Scientific, USA) and in a humidified incubator at 37 °C in an atmosphere of 5% CO_2. The cultivation medium was IMDM medium (Invitrogen Ltd, UK), supplemented with 10% foetal bovine serum (FBS, heat inactivated; Invitrogen Ltd) and 1% penicillin/streptomycin (Invitrogen Ltd). Feeding and appropriate re-suspension of the cells in fresh medium took place every 48 h, as indicated by the supplier, until a sufficient cell number for the experimental purposes was reached.

Cell-culturing Conditions

During the experiments, cells were cultured in DMEM, no-glucose medium (Invitrogen Ltd) supplemented with 10% foetal bovine serum (FBS, heat inactivated; Invitrogen Ltd), and 1% penicillin/streptomycin (Invitrogen Ltd) and either 4.3 (optimal glucose concentration as suggested by the cell line supplier and, therefore, considered as control [CTR]), 1.3 (highest concentration in the human blood, resembling hyperglycaemia [HIGH]) or 0.6 (lowest concentration in the human blood, resembling hypoglycaemia [LOW]) g/L of glucose.

2D Cultures For culturing the cells in suspension (i.e., 2D), 100 μL of cell suspension (5×10^5 cells/well) was added in a high-throughput system (i.e., a six-well tissue culture plate) that contained 1.5 mL of exhausted medium at 37 °C and 5% CO_2 and in either normoxia (20% O_2) or hypoxia (5% O_2) with appropriate glucose concentration (i.e., [LOW], [HIGH] or [CTR]). The plates were placed in either a hypoxic or a normoxic incubator. Medium was changed approximately every 48 h, and the volume was re-adjusted when needed – depending on the cell counts – in order to maintain the cell density suggested by the supplier, that is, between 10^5 and 10^6 cells per millilitre for a time period of 2 weeks.

3D Cultures For culturing the cells in the scaffolds (i.e., 3D), 100 μL of cell suspension (6×10^5 cells/well) was used in order to have approximately the same initial inoculum level as the 2D system, since we have observed that the seeding efficiency of the K-562 cell lines was approximately 70–80% (see also Mortera Blanco *et al.*, 2010). They were seeded onto the sterilized collagen scaffolds, placed in 24-well tissue culture plates and incubated for 15 min at 37 °C, 5% CO_2 and either 5 or 20% O_2 (depending on the oxygen condition of the experiment) in order to allow the cells to settle into the scaffolds. Hereafter, 15 mL of appropriate medium (as described above for the 2D cultures) was added in each well. The plates were placed in either a hypoxic or a normoxic incubator, and medium was changed approximately every 48 h for 2 weeks.

Determination of Growth in Two Dimensions

Determination of growth took place almost on a daily basis for all the conditions under study. For growth determination in the 2D system, the whole medium of three different wells was removed (triplicate) and a cell count was performed by using the haemocytometer. 20 μL of cell suspension was added to 20 μL of erythrosin B (ATCC, USA). 10 μL of this solution was used to fill the two chambers of the haemocytometer. The number of viable (unstained) and non-viable (stained red) cells were counted in five 4×4 squares, and this number was extrapolated to the total volume of the equivalent well.

Determination of Growth in Three Dimensions

Quantitative assessment of the cell proliferation in the 3D scaffolds took place by means of changes in the number of metabolically active cells using the tetrazolium compound (3-(4,5-dimetylthiazol-2-yl)-5-(3-carboxylmethox-yphenyl)-2-(4sulfophenyl)-2H-tetrazolium [MTS]; CellTiter96® AQ_{ueous} Solution Cell Proliferation Assay, Promega, USA) (Yang *et al.*, 2002). MTS was added and incubated with the scaffolds seeded with cells over 3 h at 37 °C, 5% CO_2 and either 5 or 20% O_2 (depending on the oxygen condition of the experiment). Absorbance was measured at 490 nm. In order to correlate absorbance with actual cell numbers, a calibration curve was constructed, correlating the cell number in known 2D suspension with the absorbance at 490 nm after applying the MTS assay. This calibration curve was used for the calculation of cell numbers, enabling the comparison with the 2D system.

Analysis of Metabolites

Analysis of nutrients and metabolites took place with the use of the Nova Bioprofiler 400 (Nova Biomedical, UK). 1.5 ml of (waste) medium was pipetted and transferred in an Eppendorf tube. Centrifugation took place at 2000 rpm for 5 min for removal of cells that could be present in the medium. Supernatants were stored at −20 °C prior to analysis. Samples for nutrient and metabolite analysis were collected approximately on a daily basis for all the conditions under study, and from two different wells/conditions in order to account for possible variability between different wells. Nova Bioprofiler can detect glucose of levels up to 0.1 mM. For glucose levels lower than that, a glucose assay kit (abcam, UK) was used, which is able to determine glucose levels lower than 1 mM.

Statistical Analysis

The experiments were performed in duplicate ($n = 2$), and cell proliferation was determined in triplicate ($n = 3$) and bioprofiler in duplicate ($n = 2$) for each occasion. Analysis of variance was performed with a level of significance of $p < 0.05$ in order to evaluate the statistical difference/indifference between different conditions.

7.2.3.2 Results and Discussion

Cellular Proliferation

In general, the cellular proliferation was higher in normoxia (20% O_2) compared to hypoxia (5% O_2) for both the 3D and 2D system (Figure 7.4) (Velliou *et al.*, 2015), apart from the [CTR] (i.e., 4.3 g/L glucose) in the 3D system at which no significant ($p < 0.05$) difference in the cell proliferation was observed for the two different oxygen conditions (Figure 7.4c and 7.5d).

In normoxia (i.e., 20% O_2), for both the 2D and the 3D systems, no effect of the glucose concentration on the cellular proliferation is observed until day 9 (Figure 7.4a and 7.5c), after which significant differences ($p < 0.05$) can be seen for different glucose levels. More specifically, in the 2D culture system, significantly lower ($p < 0.05$) cell numbers are observed for the [LOW] level of glucose (i.e., 0.6 g/L), compared to [HIGH] (i.e., 1.3 g/L) and [CTR] (i.e., 4.3 g/L), from day 9 up to day 12. For this time period, no significant differences were observed in the cell numbers between [CTR] and [HIGH], and only at the end of the culturing period (i.e., day 14), the [CTR] level leads to significantly higher ($p < 0.05$) cell numbers compared to the lower glucose levels (Figure 7.4a). Similar trends can be observed for the 3D system: from day 9 on, a [LOW] level of glucose results in significantly lower cell proliferation compared to [CTR] and [HIGH]. Moreover, no significant difference in the proliferation between the two latter conditions is observed (Figure 7.4c).

In contrast, in hypoxia (i.e., 5% O_2), for both the 2D and 3D systems, there is a higher impact of the glucose level on the cell proliferation (Figure 7.4b and 7.5d). More specifically, for the 2D system, the effect of glucose on the K-562 growth becomes significant ($p < 0.05$) from day 9 on. From that time point, lower glucose (i.e., [LOW]) leads to lower proliferation, and the highest proliferation is observed for the [CTR] (Figure 7.4b). The difference in the cell number for [CTR], [HIGH] and [LOW] is more evident compared to normoxia (Figure 7.4a and 7.4b). In the scaffolds, the impact of the level of glucose on the K-562 expansion becomes significant ($p < 0.05$) between the [CTR] and the two

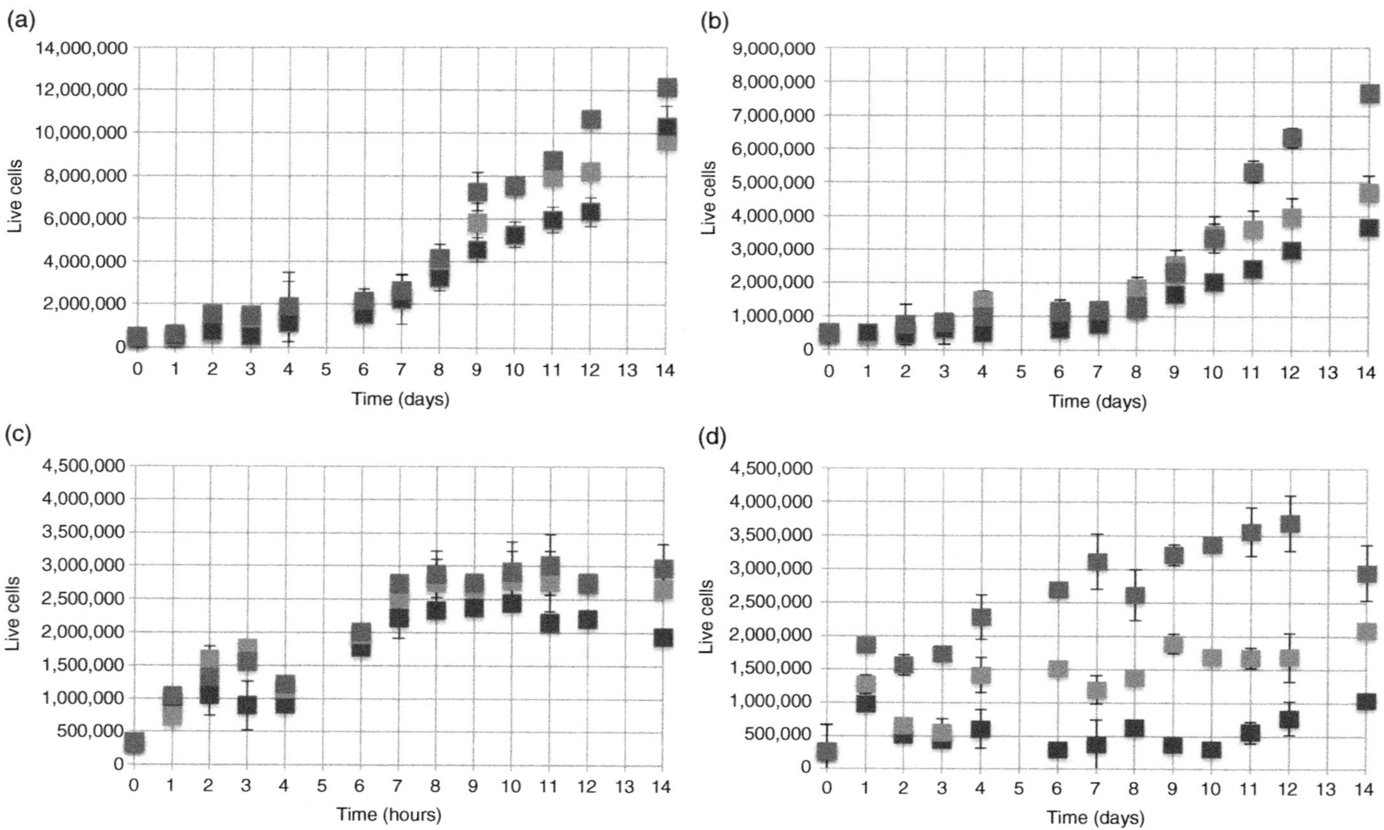

Figure 7.4 K-562 growth (a,b) the 2D and (c,d) the 3D systems, at different oxygen levels: (a,c) 20% O_2, and (b,d) 5% O_2. Different colours represent different glucose levels: (□) [CTR], (□) [HIGH] and (□) [LOW]. *Source*: This figure was originally published in (Velliou *et al.*, 2015), and has been re-used with permissions.

lower glucose concentrations ([HIGH] and [LOW]) on day 2 and among all three glucose levels from day 4 on. Especially, the lowest glucose level (i.e., [LOW]) seems to be limiting for growth in the 3D scaffolds in 5% O_2, as the cell expansion throughout 2 weeks is very low (Figure 7.4d).

Currently, there is very limited literature available, especially in 3D scaffolding systems, studying the proliferation of haematopoietic cells under oxygen and glucose concentrations close to physiological levels. However, several of the findings of Velliou *et al.* (2013, 2015) are in agreement with the available literature. For example, lower growth of K-562 for lower glucose levels and in hypoxic compared to normoxic conditions was also observed by Giuntoli *et al.* (2011) in a 2D culture system. More specifically, K-562 as well as chronic leukaemia primary cells had lower proliferation in hypoxia (0.1% O_2) compared to normoxia (20% O_2), and low glucose levels in the growth medium (1.1 g/L) lead to lower proliferation for both oxygen conditions compared to higher glucose (4.3 g/L). A reversed effect of the impact of oxygen on cell survival, compared to our findings, was observed by Fecteau *et al.* (2013) for MSCs derived from the BM of patients with CLL 2D culture; that is, higher survival and proliferation of the primary cells were observed in hypoxia (5%) compared to normoxia (20%).

Metabolic Evolution

Next to the cell growth, the metabolic activity of the cells (i.e., production of glutamate and lactate) for all the conditions under study was monitored and compared for different environmental conditions ([LOW], [HIGH] and [CTR] glucose, and 5 and 20% O_2) and systems (2D vs. 3D). On the one hand, monitoring the cell metabolism could better elucidate the differences among the culturing systems and conditions, and, on the other hand, possible alterations of the metabolic activity could affect AML progression as well as drug response (e.g., Herst *et al.*, 2011).

From a glutamate production point of view, for [CTR] glucose for almost all days, no significant ($p < 0.05$) difference between 2D and 3D can be seen in either hypoxia or normoxia (Figure 7.5a and 7.5b). This was most likely due to the fact that glucose was present in excess and the cells did not prioritise the glutamine (-glutamate) pathway as a glucose supplement in cellular metabolism (Deberardinis and Cheng, 2010). In contrast, under glucose limitations (i.e., [LOW] and [HIGH] glucose), significantly higher ($p < 0.05$) glutamate production took place in the 2D compared to the 3D system (Figure 7.5c, 7.6d, 7.5e and 7.6f), most likely due to glucose limitations: glutamine acts supplementary to glucose metabolism for the macromolecule production and cellular energy demands (Deberardinis and Cheng, 2010; Wise and Thompson, 2010). In general, higher glutamate synthesis took place in hypoxia, compared to normoxia both for the 2D and 3D systems (Figure 7.5). This is not surprisingly, as several studies have shown that under hypoxia, a switch from oxidative

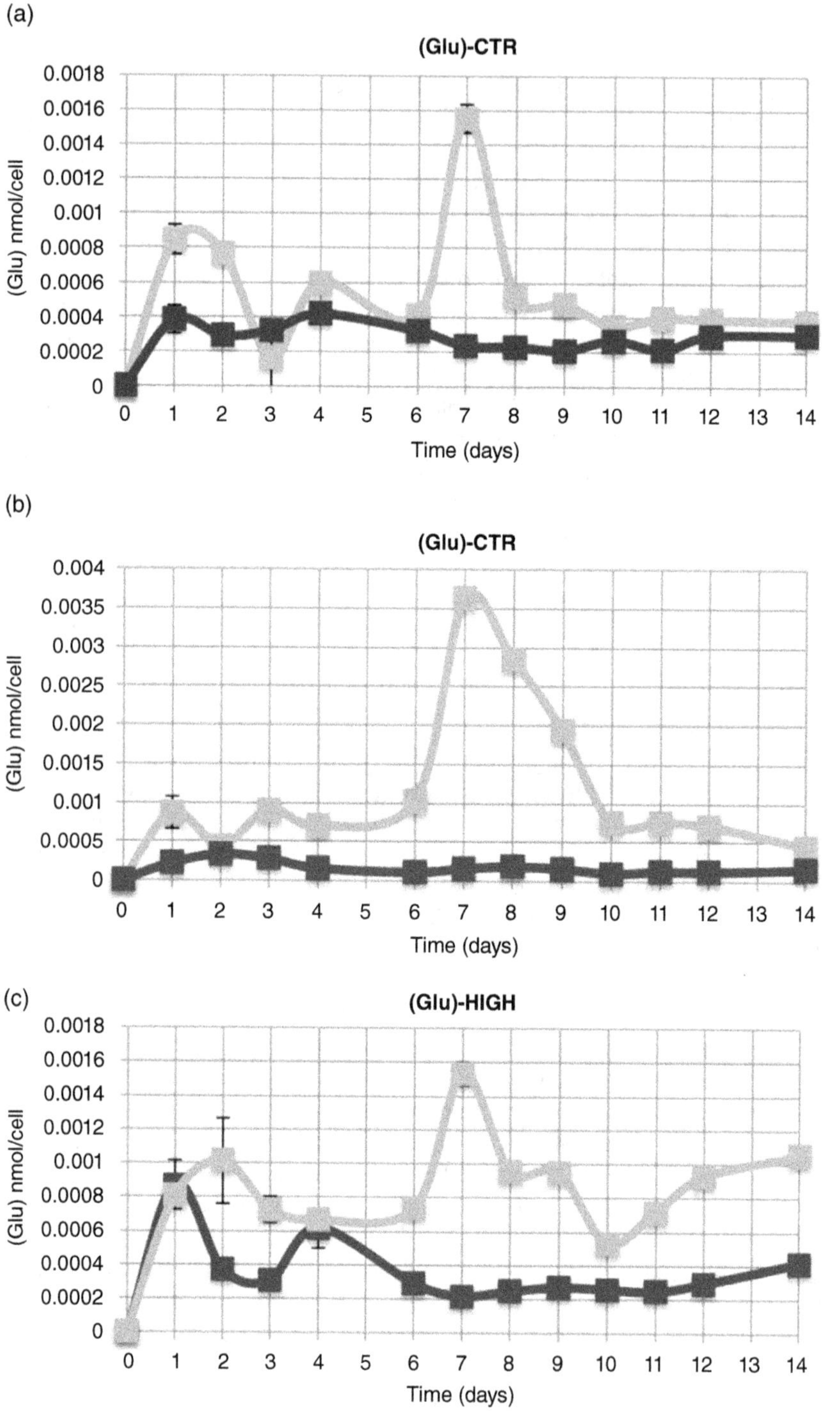

Figure 7.5 Evolution of glutamate [Glu] for all the environmental conditions under study. Different colours represent the two different culturing systems: (□) 2D cultures and (□) 3D scaffolds. *Source*: This figure was originally published in (Velliou *et al*., 2015), and has been re-used with permissions.

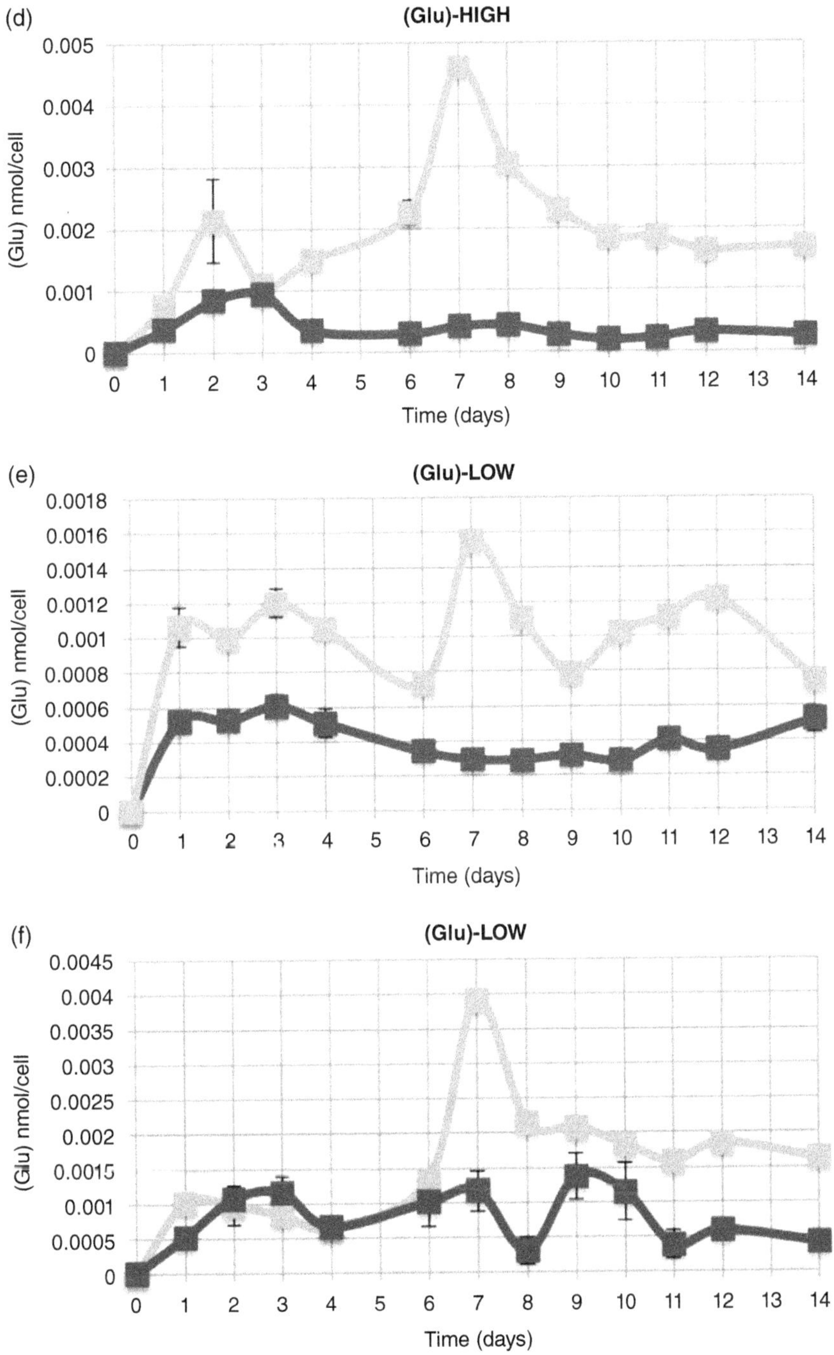

Figure 7.5 (Cont'd)

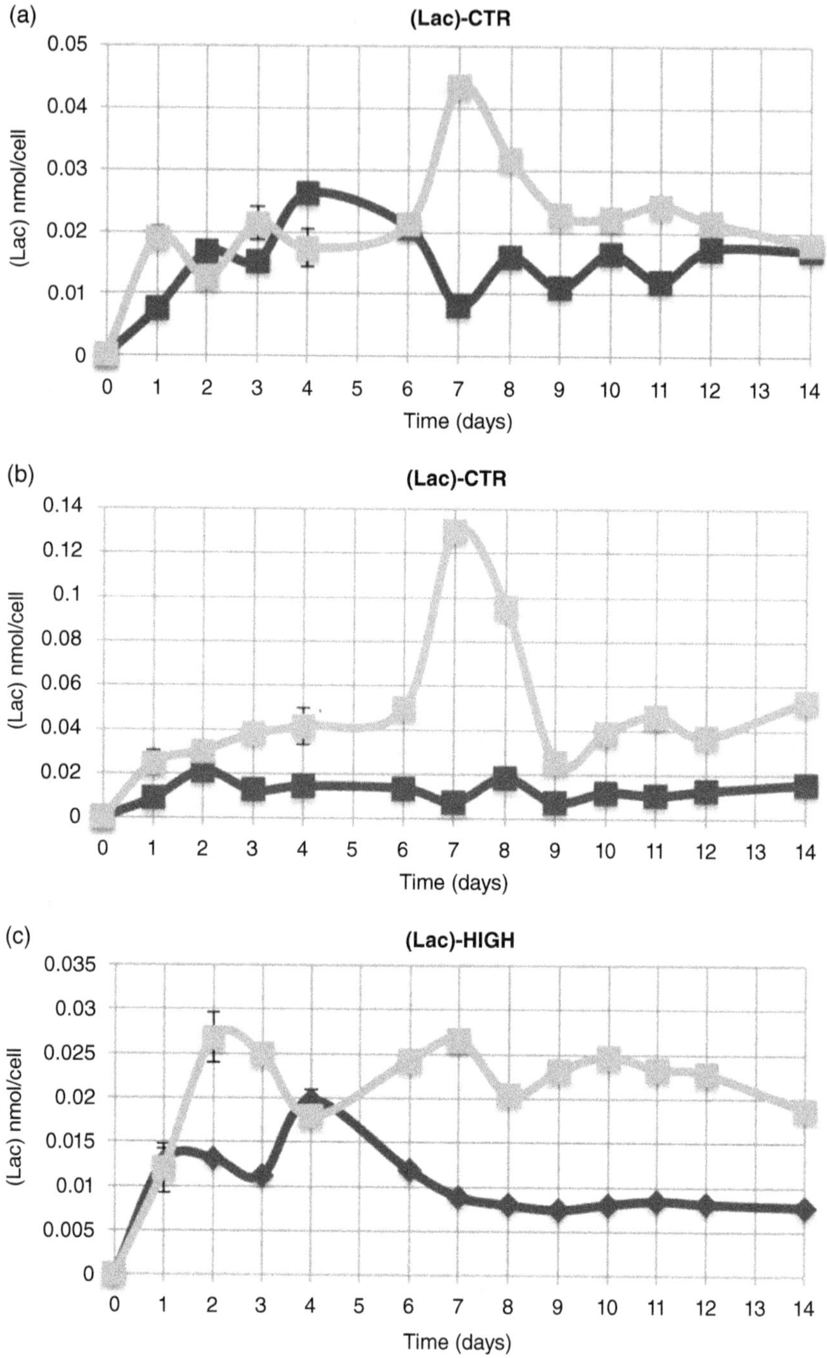

Figure 7.6 Evolution of lactate [Lac] for all the environmental conditions under study. Different colours represent the two different culturing systems: (□) 2D cultures and (□) 3D scaffolds. *Source*: This figure was originally published in (Velliou *et al.*, 2015), and has been re-used with permissions.

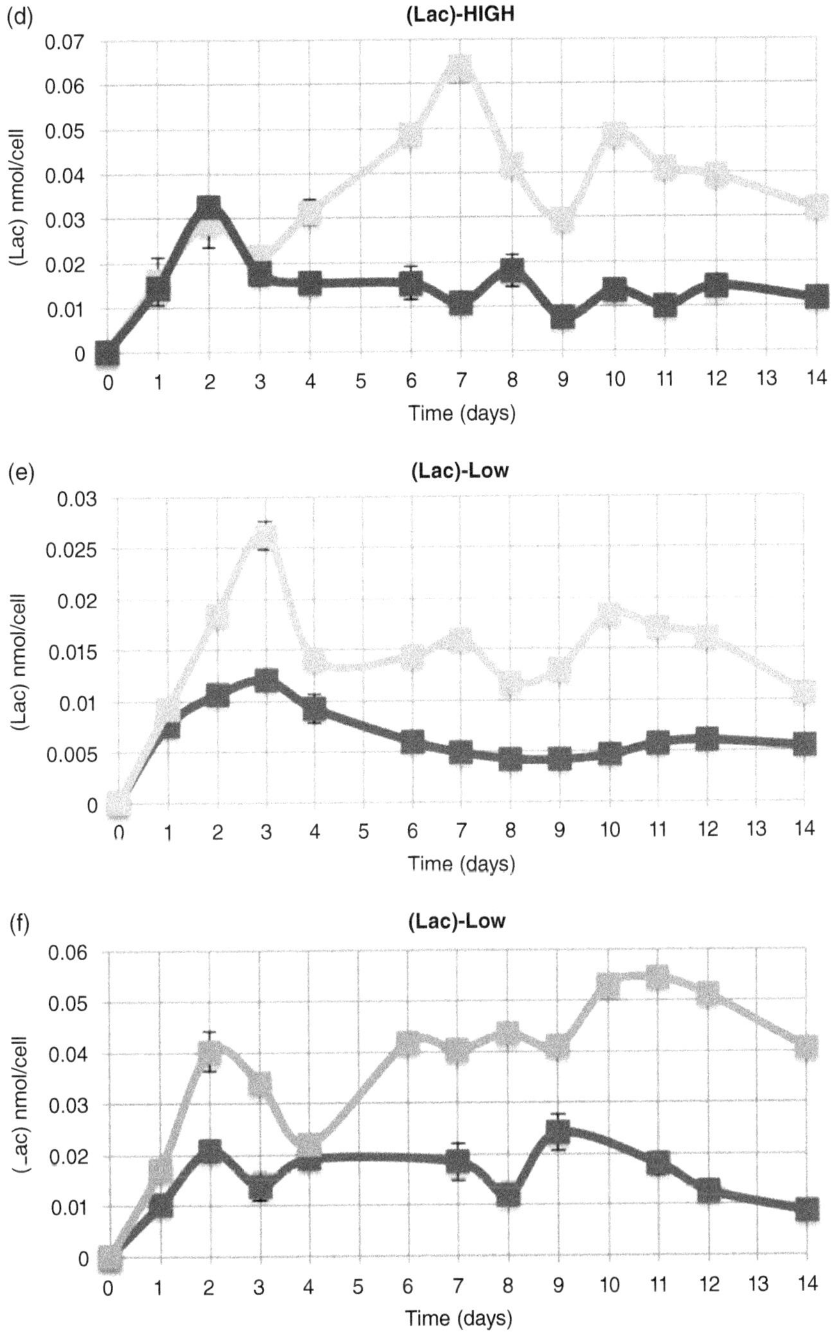

Figure 7.6 (Cont'd)

metabolism (alteration of acetyl-CoA to citrate ratio) to reductive glutamine metabolism (increase in a-ketoglutarate to citrate) takes place (Fedt *et al.*, 2013). This switch leads consequently to an enhanced glutamate production.

For lactate synthesis, apart from the [CTR] level of glucose in normoxia at which the lactate production was the same for 2D and 3D systems (Figure 7.6a), for all other conditions significantly higher ($p < 0.05$) lactate was produced in 2D compared to 3D, as cellular proliferation is much higher in the 2D system (Figure 7.6b–f). For all glucose concentrations under study, as expected, higher lactate was produced in hypoxia, confirming the anaerobic cell metabolism (Figure 7.6). Nevertheless, a significant amount of lactate was also produced in normoxia as well, possibly as a result of the Warburg effect. The latter is suggested to be a characteristic of highly proliferative and/or cancerous cells that – due to alterations or damage at the mitochondria – independently of the presence of oxygen do not enter the tricarboxylic acid (TCA) cycle, therefore following the anaerobic pathway (Warburg 1956; Vander Heiden *et al.*, 2009, 2010). The K-562 cell lines are cancerous, highly proliferative cells, and therefore most likely following partly the anaerobic pathway under normoxic conditions in both the 2D and the 3D systems (Figure 7.6).

7.2.3.3 Conclusions

From the findings of Velliou *et al.* (2013, 2015), it can be concluded that the K-562 cell lines can proliferate and maintain for 2 weeks in medium with glucose close to *in vivo* physiological levels (much lower than generally applied in laboratory systems) in both normoxia (20% O_2) and hypoxia (5% O_2). Generally, the cells proliferate higher and are metabolically more active in suspension (i.e., 2D), compared to the PU collagen-coated 3D scaffolds, following a similar trend. Higher glutamate production was observed in hypoxia for all glucose conditions under study, and in hypoxia, as expected, the anaerobic metabolic pathway of lactate synthesis is followed. Lactate is also produced in normoxia, possibly as a result of the Warburg effect.

The differences that we report on the K-562 proliferation under different oxygen and glucose levels as well as between the 2D and 3D cultures can be of significant importance when applying chemotherapy *in vitro*. Most chemotherapy agents that are generally applied for the treatment of patients with AML are targeting highly proliferative cells, such as anthracyclines such as DNR which are targeting cells of the G_1 phase of the cell cycle (Cancer Research UK, 2011). Therefore, under oxidative or glucose stress or in a 3D microenvironment, cells proliferating slower may be less susceptible to chemotherapeutics. Future work should focus on unravelling the mechanisms of interaction of the cells with the 3D microenvironment in order to give more light to the observed differences in proliferation. Moreover, studying the combined effect of the environmental stresses and chemotherapy in both the 2D and 3D system would be of high importance towards chemotherapy optimisation.

7.3 Cellular Biomarkers for Monitoring Leukaemia *in vitro*

The possible oxidative and starvation cellular stress, as shown in Sections 7.2.2 and 7.2.3, should be systematically followed and/or quantified through *in vitro* studies. Efficient monitoring of the level of oxidative and metabolic stress and further adaptation can take place via the selection and further quantification of specific biomarker(s) (i.e., intracellular molecules), the expression and/or concentration of which may alter depending on the fluctuation of oxygen and glucose in the system. In order to select 'stress' biomarker(s), an in-depth investigation of possible mechanisms of oxidative and starvation stress cellular responses is needed. Hereafter, autophagy, a crucial mechanism which is activated in the absence of nutrients and at low oxygen levels, is described, and potential biomarker molecules are summarized (Velliou *et al.*, 2014).

7.3.1 (Macro-)autophagy: The Cellular Response to Metabolic Stress and Hypoxia

Autophagy is a cellular mechanism which aims at the maintenance of homeostasis of a normal cell, via degradation of organelles and cellular components by the lysosomes (Levine and Kroemer, 2008; Banerji *et al.*, 2012; Kongara and Karantza, 2012; Lozy and Karantza, 2012; Choi *et al.*, 2013). Autophagy is activated as a result of exposure to a stress factor and most probably in the absence of nutrients (i.e., glucose starvation), as well as under hypoxic conditions (Lum *et al.*, 2005; Scherz-Shoural *et al.*, 2007). Degradation of damaged mitochondria, aggregated proteins and other cellular organelles via the autophagic mechanism protects the cells from apoptotic signalling (Moore *et al.*, 2006; Jin and White, 2007). Autophagy may be important in the regulation of cancer development and in the determination of the response of cancer cells to chemotherapy (Degenhardt *et al.*, 2006; Hippest *et al.*, 2006; Wilkinson *et al.*, 2012).

Several researchers have shown that autophagy plays a crucial role in maintenance of normal haematopoiesis and function of HSCs (Kundu *et al.*, 2008; Warr *et al.*, 2013), and others have shown that autophagy may lead to an increased resistance (to chemotherapy) and survival of several cancers, including haematological malignancies. For example, Mortensen *et al.* (2011b) observed that loss of autophagy resulted in loss of normal function of murine HSCs, leading to the expansion of a progenitor cell population in the BM which has as a consequence a severe myeloproliferation. This myeloproliferation strongly resembled human AML, indicating a possible link between maintenance of autophagy and avoidance of malignancies such as AML. Wallington-Beddoe *et al.* (2011) showed that activation of autophagy (induced by the FTY720 drug) leads to increased survival of ALL cells.

7.3.2 Biomarker Candidates

7.3.2.1 (Autophagic) Biomarker Candidates

For 'switching on' and maintenance of the autophagic response, a variety of genes are overexpressed, and proteins – mainly kinases – are activated and/or de-activated (i.e., depending on whether they have a positive or negative regulatory role in autophagy). Therefore, in order to monitor autophagy in an *in vitro* system, many different biomarkers of the genomic and/or protein levels can be considered (see, as an overview, Table 7.2).

A possible candidate-biomarker is the serine threonine kinase ULK1, which is a key initiator of autophagy; its activation is essential for clearance of cellular mitochondria and ribosomes (Kundu *et al.*, 2008). This kinase is activated under growth factor deprivation and leads to activation of the glycogen synthesis kinase-3 (i.e., GSK3) that phosphorylates the acetyltransferase TIP60, which in turn acetylates and activates ULK1 (Lin *et al.*, 2012). Another crucial kinase which is directly related to autophagy is the *mTOR kinase*, which under normal nutrient concentrations binds and phosphorylates the ULK1, therefore suppressing autophagy. Under nutrient deprivation, this kinase is a result of the activation of the *P53* gene, therefore enabling activation of ULK1 and autophagy (Feng *et al.*, 2005). Another possible candidate biomarker is the *Atg7* gene, which is an essential gene for activation of autophagy and further regulation of HSC maintenance (Kundu *et al.*, 2008; Mortensen *et al.*, 2010, 2011a). It has been shown that deleting this gene in murine HSCs leads to death as a result of an accumulation of mitochondria and ROS, increased proliferation and DNA damage (Mortensen *et al.*, 2010, 2011a). FOXO3A also can be a possible biomarker. It has been found that it has a critical role in autophagy induced in mice in a cytokine-free environment (Warr *et al.*, 2013).

Table 7.2 Stress biomarkers for normal and abnormal HSCs.

Biomarker	Genomic	Protein
LKBI-AMPK		+
AMPK		+
ULK1		+
P53	+	
PTEN	+	
Atg7	+	
HIF		+
mTOR kinase		+
FOXO3A	+	
FUMH		+

Several researchers have reported a tumour suppressor role of autophagy. More specifically, it has been shown that activation of the AMPK pathway has a suppressor role in AML. AMPK is a protein kinase which regulates protein and energy homeostasis at an intracellular level via autophagic recycling of intracellular components. Practically, AMPK acts as a metabolic sensor of alteration of the intracellular lipid composition, and it restores energy by maintaining the balance of adenosine triphosphate (ATP) versus adenosine monophosphate (AMP), through the LKBI-AMPK activation. The latter (i.e., LKBI-AMPK) is a tumour suppressor in AML (Green *et al.*, 2010). Activation of the *P53* and *PTEN* genes has shown to have a tumour-suppressive role as they allow initiation of autophagy via inhibition of the activity of the mTOR kinase (Feng *et al.*, 2005).

7.3.2.2 (Non-autophagic) Stress Biomarker Candidates

A biomarker related to oxidative stress is the hypoxia-induced factor (HIF), which is the central regulator of oxygen homeostasis. More specifically, the HIF1-a regulator is overexpressed in many cancer types, and HIF proteins *mediate* cell adaptation to hypoxia (Zhong *et al.*, 1999; Birner *et al.*, 2000; Talks *et al.*, 2000; Warr and Passegue, 2013).

A possible starvation stress biomarker is the metabolic enzyme *fumarate hydratase* (FUMH) which converts fumarate to malate and is essential for glycolysis. This enzyme highly controls the intracellular levels of fumarate, and silencing of the expression of FUMH leads to fumarate intracellular accumulation (Ratcliffe, 2007).

7.4 From *in vitro* to *in silico*

As described in this chapter, appropriate *ex vivo* culturing systems can be of substantial importance, elucidating key mechanisms and factors that affect the evolution of leukaemia. Quantitative information on these key mechanisms is an appropriate input for the construction of predictive models for the *in silico* description of leukaemia. These models can serve as tools for the optimisation of chemotherapy protocols leading to a personalised treatment for each individual patient (see also Chapter 8). Moreover, quantification of appropriate intra-cellular biomarkers that are related to the chemotherapy cell resistance can enable the *combination of macroscopic kinetics with microscopic information* (specific for individual patients), leading to the construction of more detailed models of grey or white box nature. Due to their nature, these models will be more accurate and precise (in comparison to available macroscopic black box models) in the prediction of the response of individual patients to chemotherapy, as they will incorporate microscopic genetic and/or metabolic information which is patient specific.

References

Abboud, C.N. & Lichtman, M. (2001) Structure of the marrow and the hematopoietic microenvironment. In: E. Beutler, B.S. Coller, M.A. Lichtman, T.J. Kipps & U. Seligsohn (eds.), *Williams Hematology*. McGraw-Hill, New York.

Alijtawi, O.S., Li, D., Xiao, Y., Zhang, D., Ramachandran, K., Stehno-Bittel, L., Van Veldhuizen, P., Lin, T.L., Kambhampati, S. & Garimella, R. (2014) A novel three-dimensional stromal-based model for *in vitro* chemotherapy sensitivity testing of leukemia cells. *Leukemia & Lymphoma*, **55(2)**, 378–391.

Adbal Dayem, A., Choi, H.-Y., Kim, J.-H. & Cho, S.-G. (2010) Role of oxidative stress in stem, cancer, and cancer stem cells. *Cancers*, **2**, 859–884.

Balgobind, B.V., Zwaan, C.M., Reinhardt, D., Arentsen-Peters, T.J.C.M., Hollink, I.H.I.M., de Haas, V., Kaspers, G.J.L., de Bont, E.S.J.M., Baruchel, A., Stary, J., Meyer, C., Marschalek, R., Creutzig, U., den Boer, M.L., Pieters, R. & van den Heuvel-Eibrink (2010) High BRE expression in pediatric MLL-rearranged AML is associated with favorable outcome. *Leukemia*, **24(12)**, 2048–2055.

Banerji, V. & Gibson, S.B. (2012) Targeting metabolism and autophagy in the context of haematologic malignancies. *International Journal of Cell Biology*, **2012**, 1–9.

Beutler, E., Coller, B.S., Lichtman, M.A., Kipps, T.J. & Seligsohn, U. (eds.) (2001) *Williams Hematology*. McGraw-Hill, New York.

Birner, P., Schind, M., Obermair, A., Plank, C., Breitenecker, G. & Oberhuber, G. (2000) Overexpression of hypoxia-inducible factor 1a is a marker for an unfavorable prognosis in early-stage invasive cervical cancer. *Cancer Research*, **60**, 4693–4696.

Bonnet, D. & Dick, J. (1997) Human acute myeloid leukemia is organized as a hierarchy that originates from a primitive hematopoietic cell. *Nature Medicine*, **3(7)**, 730–737.

Buttery, L. & Shakesheff, K.M. (2008) A brief introduction to different cell types. In J. Polak, S. Mantalaris & S.E. Harding (ed.), *Advances in Tissue Engineering*, 39–41. Imperial College Press, London.

Cancer Research UK (2011) Leukaemia. http://www.cancerresearchuk.org/health-professional/cancer-statistics/statistics-by-cancer-type/leukaemia

Charbord, P. (2001) Microenvironmental cell populations essential for the support of hematopoietic stem cells. In L.I. Zon (ed.), *Hematopoiesis: A Developmental Approach*, 691–701. Oxford University Press, New York.

Choi, A.M.K., Ryter, S.W. & Levine, B. (2013) Autophagy in human health and disease. *New England Journal of Medicine*, **368**, 651–552.

Collins, P.C., Miller, W.M. & Papoutsakis, E.T. (1998) Stirred culture of peripheral and cord blood hematopoietic cells offers advantages over traditional static systems for clinically relevant applications. *Biotechnology and Bioengineering*, **59(5)**, 534–543.

DeBerardinis, R.J. & Cheng, T. (2010) Q's next: the diverse function of glutamine in metabolism, cell biology and cancer. *Oncogene*, **29(3)**, 313–324.

Degenhardt, K., Mathew, R., Beaudoin, B., Bray, K., Anderson, D., Chen, G., Mukherjee, C., Shi, Y., Gelinas, C., Fan, Y., Nelson, D.A. & White, E. (2006) Autophagy promotes tumor cell survival and restricts necrosis, inflammation and turmirogenesis. *Cancer Cell*, **10(1)**, 51–54.

Deorosan, B. & Nauman, E.A. (2011) The role of glucose, serum, and three-dimensional cell culture on the metabolism of bone marrow-derived mesenchymal stem cells. *Stem Cells International*, **2011**, 1–12.

Dorshkind, K. (1990) Regulation of hemopoiesis by bone marrow stromal cells and their products. *Annual Reviews of Immunology*, **8**, 111–137.

Engelhardt, M., Douville, J., Behringer, D., Jahne, A., Smith, A., Henschler, A. & Lange, W. (2001) Hematopoietic recovery of *ex vivo* perfusion culture expanded bone marrow and unexpanded peripheral blood progenitors after myeloblastive chemotherapy. *Bone Marrow Transplant*, **27(3)**, 249–259.

Fedt, S.M., Bell, L.M., Keibler, M.A., Olenchock, B.A., Mayers, J.R., Wasylenko, T.M., Vokes, N.I., Guarente, L., Vander Heiden, M.G. & Stephanopoulos, G. (2013) Reductive glutamine metabolism is a function of a-ketoglutarate to citrate ratio in cells. *Nature Communications*, **4(2236)**. doi:10.1038/ncomms3236

Fecteau, J.-F., Messmer, D., Zhang, S., Cui, B., Chen, L. & Kipps, T.J. (2013) Impact of oxygen concentration on growth of mesenchymal stromal cells from the marrow of patients with chronic lymphocytic leukemia. *Blood*, **121**, 971–974.

Feng, Q., Chai, C., Jiang, X.S., Leong, K.W. & Mao, H.Q. (2006) Expansion of engrafting human hematopoietic stem/progenitor cells in three-dimensional scaffolds with surface-immobilized fibronectin. *Journal of Biomedical Material Research PART A*, **78**, 781–791.

Feng, Z., Zhang, H., Levine, A.J. & Jin, S. (2005) The coordinate regulation of the p53 and mTOR pathways in cells. *PNAS*, **102(23)**, 8204–8209.

Ferreira, M.S., Jahnen-Dechent, W., Labude, N., Bovi, M., Hieronymus, T., Zenke, M., Schneider, R.K. & Neuss, S. (2012) Cord blood-hematopoietic stem cell expansion in 3D fibrin scaffolds with stromal support. *Biomaterials*, **33(29)**, 6987–6997.

Fruehauf, J.-P. & Meyskens, F.-L. Jr. (2007) Reactive oxygen species: a breath of life or death? *Clinical Cancer Research*, **13**, 89–794.

Giarratana, M.C., Kobari, L., Lapillonen, H., Chalmers, D., Kiger, L., Cynober, T. *et al.* (2005) Ex vivo generation of fully mature human red blood cells from hematopoietic stem cells. *Nature Biotechnology*, **23**, 69–74.

Gilman, A., Goodman, L.S., Hardman, J.G. & Limbird, L.E. (2001) *Goodman & Gilman's: The Pharmacological Basis of Therapeutics*, 9th ed. McGraw-Hill, New York.

Giuntoli, S., Tanturli, M., Gesualdo, F.D., Barbetti, V., Rovida, E. & Sbarba, P.D. (2011) Glucose availability in hypoxia regulates the selection of chronic myeloid leukemia progenitor subsets with different resistance to imatinib-mesylate. *Haematologica*, **96(2)**, 204–212.

Green, A.S., Chapuis, N., Maciel, T.T., Willems, L., Lambert, M., Arnoult, C., Boyer, O., Bardet, V., Park, S., Foretz, M., Viollet, B., Ifrah, N., Dreyfus, F., Hermine, O., Moura, I.C., Lacombe, C., Mayeux, P., Bouscary, D. & Tamburini, J. (2010) The LKB1/AMPK signaling pathway has tumor suppressor activity in acute myeloid leukemia through the repression of mTOR-dependent oncogenic mRNA translation. *Blood*, **116**, 4262–4273.

Herst, P.M., Howman, R.A., Neeson, P.J., Berridge, M.V. & Ritchie D.S. (2011) The level of glycolytic metabolism in acute myeloid leukemia blasts at diagnosis is prognostic for clinical outcome. *Journal of Leukocyte Biology*, **89**, 51–55.

Hippest, M.M., O'Toole, P.S. & Thorburn, A. (2006) Autophagy in cancer: goof, bad or both? *Cancer Research*, **66**, 9349–9351.

Jin, S. & White, E. (2007) Role of autophagy in cancer: management of metabolic stress. *Autophagy*, **3(1)**, 28–31.

Kaiser, M., Kuhnl, A., Reins, J., Fischer, S., Ortiz-Tanchez, J., Schlee, C., Mochmann, L.H., Heesch, S., Benlasfer, O., Hofmann, W.K., Thiel, E. & Baldus, C.D. (2011) Antileukemic activity of the HSP70 inhibitor pifithrin-mu in acute leukemia. *Blood Cancer Journal*, **1(7)**, e28.

Kipps, T.J. (2001) Classification of malignant lymphoid disorders In: E. Beutler, B.S. Coller, M.A. Lichtman, T.J. Kipps & U. Seligsohn (eds.), *Williams Hematology*. McGraw-Hill, New York.

Kirito, K., Fox, N. & Kaushansky, K. (2003) Thrombopoietin stimulates Hoxb4 expression: an explanation for the favourable effects of TPO on hematopoietic stem cells. *Blood*, **102(9)**, 3172–3178.

Koller, M.R. & Palsson, B.O. (1993) Tissue engineering: reconstitution of human hematopoiesis *ex vivo*. *Biotechnology and Bioengineering*, **42**, 909–930.

Kongara, S. & Karantza, V. (2012) The interplay between autophagy and ROS in tumorigenesis. *Frontiers in Oncology*, **2(175)**, 1–13.

Kundu, M., Lindsten, T., Yang, C.-Y., Wu, J., Zhao, F., Zhang, J., Selak, M.A., Ney, P.A. & Thompson, C.B. (2008) Ulk1 plays a critical role in the autophagic clearance of mitochondria and ribosomes during reticulocyte maturation. *Blood*, **112**, 1493–1502.

Lee, J. & Kotov, N.A. (2009) Notch ligand presenting acellular 3D microenvironments for ex vivo human hematopoietic stem-cell culture made by layer-by-layer assembly. *Small*, **5**, 1008–1013.

Lee-Thedieck, C. & Spatz, J.P (2012) Artificial niches: biomimetic materials for hematopoietic stem cell culture. *Macromolecular Rapid Communications*, **33**, 1432–1438.

Leisten, I., Kramann, R., Ventura-Ferreira, M.S.V., Bovi, M., Neuss, S., Ziegler, P., Wagner, W., Knuchel, R. & Schneider, R.K. (2012) 3D co-culture of hematopoietic stem and progenitor cells and mesenchymal stem cells in collagen scaffolds as a model of the hematopoietic niche. *Biomaterials*, **33**, 1736–1747.

Levac, K., Karabu, F. & Bhatia, M. (2005) Identification of growth factor conditions that reduce *ex vivo* cord blood progenitor expansion but not alter human repopulating function *in vivo*. *Haematologica*, **90(2)**, 166–172.

Levine, B. & Kroemer, G. (2008) Autophagy in the pathogenesis of disease. *Cell*, **132(1)**, 27–42.

Lichtman, M.A. (2001) Classification and clinical manifestations of the clonal myeloid disorders. In: E. Beutler, B.S. Coller, M.A. Lichtman, T.J. Kipps & U. Seligsohn (eds.), *Williams Hematology*. McGraw-Hill, New York.

Lin, S-Y., *et al.* (2012) GSK3-TIP60-ULK1 signaling pathway links growth factor deprivation to autophagy. *Science*, **336**, 477(2012).

Liu, R. Li, Z., Bai, S., Zhang, H., Tang, M., Lei, Y., Chen, L., Liang, S., Zhao, Y.-I., Wei, Y. & Huang, C. (2009) Mechanism of cancer cell adaptation to metabolic stress: proteomics identification of a novel thyroid hormone-mediated gastric carcinogenic signaling pathway. *Molecular and Cellular Proteomics*, **8(1)**, 70–84.

Lodi, A., Tiziani, S., Khanim, F.L., Drayson, M.T., Gunther, U.L., Bunce, C.M. & Viant, M.R. (2011) Hypoxia triggers major metabolic changes in AML cells without altering indomethacin-induced TCA cycle deregulation. *ACS Chemical Biology*, **6**, 169–175.

Lowenberg, B., Downing, J. & Burnett, A. (1999) Acute myeloid leukemia. *New England Journal of Medicine*, **341(14)**, 1051–1062.

Lozy, F. & Karantza, V. (2012) Autophagy and cancer cell metabolism. *Seminars in Cell and Developmental Biology*, **23**, 395–401.

Lum, J., Bauer, D.E., Kong, M., Harris, M.H., Li, C., Lindsten, T. & Thompson, C.B. (2005) Growth factor regulation of autophagy and cell survival in the absence of apoptosis. *Cell*, **120**, 237–248.

Lutolf, M.P., Doyonnas, R., Havenstrite, K., Koleckar, K. & Blau, H.M. (2009) Perturbation of single hematopoietic stem cell fates in artificial niches. *Integrative Biology (Cambridge)*, **1**, 59–69.

Lyu, B.N., Ismailov, S.B., Ismailov, B. & Lyu, M.B. (2008) Mitochondrial concept of leukemogenesis: key role of oxygen-peroxide effects. *Theoretical Biology and Medical Modelling*, **5(23)**, 1–11.

Marshak, D.R., Gardner, R.L. & Gottlieb. D. (eds.) (2001) *Stem Cell Biology*. Cold Spring Harbor Laboratory Press, Cold Spring Harbor, NY.

Mayani, H., Flores-Figueroa, E. & Chavez-Gonzalez, A. (2009) *In vitro* biology of human myeloid leukaemia. *Leukaemia Research*, **33**, 624–637.

Moore, M.N., Allen, J.I. & Somerfield, P.J. (2006) Autophagy: role in surviving environmental stress. *Marine Environmental Research*, **62**, S420–S425.

Morgan, D.O. (1997) Cyclin-dependent kinases: engines, clocks, and microprocessors. *Annual Review of Cell and Developmental Biology*, **13**, 261–291.

Morgan, D.O. (2007) *The Cell Cycle-Principles of Control*. New Sciences Press, London.

Mortensen, M. & Simon, A.K. (2010) Nonredundant role of Atg7 in mitochondrial clearance during erythroid development. *Autophagy.* **6**, 423–425.

Mortensen, M., Soilleux, E.J., Djordjevic, G., Tripp, R., Lutteropp, M., Sadighi-Akha, E., Stranks, A.J., Glanville, J., Knight, S., Jacobsen, S.-E.W., Kamil R. Kranc, K.R. & Simon, A.K. (2011a) The autophagy protein Atg7 is essential for hematopoietic stem cell maintenance. *The Journal of Experimental Medicine,* **208(3)**, 455–467.

Mortensen, M., Watchon A.S. & Simon, A.K. (2011b) Lack of autophagy in the hematopoietic system leads to loss of hematopoietic stem cell function and dysregulated myeloid proliferation. *Autophagy,* **7(9)**, 1069–1070.

Mortera-Blanco, T., Mantalaris, A., Bismarck, A., Aqel, N. & Panoskaltsis, N. (2011) Long-term cytokine-free expansion of cord blood mononuclear cells in three-dimensional scaffolds. *Biomaterials,* **32**, 9263–9270.

Mortera-Blanco, T., Mantalaris, A., Bismarck, A. & Panoskaltsis, N. (2010) Development of a three-dimensional biomimicry of human acute myeloid leukemia *ex vivo*. *Biomaterials,* **31**, 2243–2251.

Murray, A.W. & Krischner, M.W. (1989) Dominoes and clocks: the union of two views of the cell cycle. *Science,* **246**, 614–621.

Naito, K., Tamahashi, N., Chiba, T., Kaneda, K., Okuda, M., Endo, K., Yoshinaga, K. & Takahashi, T. (1992) The microvasculature of the human bone marrow correlated with the distribution of hematopoietic cells: a computer-assisted three-dimensional reconstruction study. *Tohoku Journal of Experimental Medicine,* **166**, 439–450.

Nichols, J.E., Cortiella, J., Lee, J., Niles, J.A., Cuddihy, M., Wang, S. *et al.* (2009) *In vitro* analog of human bone marrow from 3D scaffolds with biomimetic inverted colloidal crystal geometry. *Biomaterials,* **30**, 1071–1079.

Panoskaltsis, N., Mantalaris, A. & Wu, D.J.H. (2005) Engineering a mimicry of bone marrow tissue *ex vivo*. *Journal of Bioscience Engineering,* **100**, 28–35.

Panoskaltsis, N., Reid, C.D.L. & Knight, S.C. (2003) Quantification and cytokine production of circulating lymphoid and myeloid cells in acute myelogenous leukemia (AML). *Leukemia,* **17**, 716–730.

Passegué, E., Jamieson, C.H.M., Ailles, L.E. & Weissman, I.L. (2003) Normal and leukemic hematopoiesis: are leukemias a stem cell disorder or a reacquisition of stem cell characteristics? *PNAS,* **100**, 11842–11849.

Provin, C., Takano, K., Sakai, Y., Fujii, T. & Sirakashi, R. (2008) A method for the design of 3D scaffolds for high-density cell attachment and determination of optimum perfusion culture conditions. *Journal of Biomechanics,* **41**, 1436–1449.

Pui, C.H. (2009) Acute lymphoblastic leukaemia: introduction. *Seminars in Hematology,* **46(1)**, 1–2.

Pui, C.H. & Evans, W.E. (2006) Treatment of acute lymphoblastic leukemia. *The New England Journal of Medicine,* **354**, 166–178.

Quesenberry, P.J. & Colvin, G.A. (2001) Hematopoietic stem cells, progenitor cells, and cytokines. In: E. Beutler, B.S. Coller, M.A. Lichtman, T.J. Kipps & U. Seligsohn (eds.), *Williams Hematology.* McGraw-Hill, New York.

Rabinowitz, J., Petros, W., Stuart, A. & Peter, W. (1993) Characterization of endogenous cytokine concentrations after high-dose chemotherapy with autologous bone marrow support. *Blood*, **81(9)**, 2452–2459.

Ratcliffe, P.J. (2007) Fumarate hydratase deficiency and cancer: activation of hypoxia signaling? *Cancer Cell*, **11**, 303–305

Safinia, L., Mantalaris, A. & Bismarck A. (2005) Towards a methodology for the effective surface modification of porous polymer scaffolds. *Biomaterials*, **26**, 7537–7547.

Scherz-Shouval, E., Shvets, E., Fass, E., Shorer, H., Gil, L. & Elazar, Z. (2008) Reactive oxygen species are essential for autophagy and specifically regulate the activity of Atg4. *The EMBO Journal*, **26**, 1749–2760.

Streeter, I. & Cheema, U. (2011) Oxygen consumption rate of cells in 3D culture: the use of experiment and simulation to measure kinetic parameters and optimise culture conditions. *Analyst*, **136**, 4013–4019.

Talks, K.L., Turley, H., Gatter, K.C., Maxwell, P.H., Pugh, C.W., Ratcliffe, P.J. & Harris, A.L. (2000) The expression and distribution of the hypoxia-inducible factors HIF-1a and HIF-2a in normal human tissues, cancers, and tumor associated macrophages. *American Journal of Pathology*, **157**, 411–421.

Trentin, J.J. (1970) Influence of hematopoietic organ stroma (hematopoietic inductive microenvironments) on stem cell differentiation. In: A.S. Gordon (ed.), *Regulation of Hemopoiesis*, 161–186. Appleton-Century-Crofts, New York.

Tun, T., Miyoshi, H., Aung, T., Takahashi, S., Shimizu, S., Kuroha, T., Yamamoto M. & Ohshima, N. (2002) Effect of growth factors on ex vivo bone marrow cell expansion using three-dimensional matrix support. *Artificial Organs*, **26(4)**, 333–339.

Vander Heiden, M.G., Cantley, L.C. & Thompson, C.B. (2009) Understanding the Warburg effect: the metabolic requirements of cell proliferation. *Science*, **324**, 1029–1033.

Vander Heiden, M.G., Locasale, J.W., Swanson, K.D., Sharfi, H., Heffron, G.J, Amador-Noguez, D., Christofk, H.R., Wagner, G., Rabinowitz, J.D., Asara, J.M. & Cantley L.C. (2010) Evidence of an alternative glycolytic pathway in rapidly proliferating cells. *Science*, **329**, 1492–1499.

Velliou, E., Dos Santos, S.B., Fuentes-Gari, M., Misener, R., Pefani, E., Panoskaltsis, N., Mantalaris, A. & Pistikopoulos, E.N. (2014) Key environmental stress biomarker candidates for the optimisation of chemotherapy treatment of leukaemia. *Malta Journal of Health Science*, accepted. doi:10.14614/CHEMOLEUK.2.29

Velliou, E.G., Dos Santos, S.B., Papathanasiou, M.M., Fuentes-Gari, M., Misener, R., Panoskaltsis, N., Pistikopoulos, E.N. & Mantalaris, A. (2015) Towards unravelling the kinetics of an acute myeloid leukaemia model system under oxidative and starvation stress: a comparison between two- and three-dimensional cultures. *Bioprocess and Biosystems Engineering*, **38(8)**, 1589–1600.

Velliou, E., Fuentes Gari, M., Dos Santos, S., Misener, R., Panoskaltsis, N., Mantalaris, A. & Pistikopoulos, E.N. (2013) The effect of oxygen and glucose stress on the evolution of a leukemia model system in an in vitro bone marrow mimicry. Paper presented at the 13th AIChE annual meeting, San Francisco, CA.

Wallington-Beddoe, C.T., Hewson, J., Bradstock, K.F. & Bendall, L.J. (2011) FTY720 produces caspase-independent cell death of acute lymphoblastic leukaemia cells. *Autophagy*, **7(7)**, 707–715.

Warburg, O. (1956) On the origin of cancer cells. *Science*, **123**, 309–314.

Warr, M.R., Binnewies, M., Flach, J., Reynaud, D., Garg, T., Malhorta, R., Debnath, J. & Passegue, E. (2013) FOXO3A directs a protective autophagy program in hematopoietic stem cells. *Nature*, **494**, 323–327.

Warr, M.R. & Passegue, E. (2013) Metabolic makeover of HSC. *Cell Stem Cell*, **12**, 1–3.

Wilkinson, S.T., Tome, M.E. & Briehl, M.M. (2012) Mitochondrial adaptations to oxidative stress confer resistance to apoptosis in lymphoma cells. *International Journal of Molecular Sciences*, **13**, 10212–10228.

Wise, D.R. & Thompson, C.B. (2010) Glutamine addiction: a new therapeutic target. *Trends in Biochemical Science*, **35**, 427–433.

Xu, S., Li, D., Xie, Y., Lu, J. & Dai, K. (2008) The growth of stem cells within β-TCP scaffolds in a fluid-dynamic environment. *Materials Science and Engineering*, **28**, 164–170.

Zhong, H., Marzo, A.M., Laughner, E., Lim, M., Hilton, D.A., Zagzag, D., Buechler, P., Isaacs, W.B., Semenza, G.L. & Simons, J.W. (1999) Overexpression of hypoxia-inducible factor 1α in common human cancers and their metastases. *Cancer Research*, **59**, 5830–5835.

Zhou, F.-L., Zhang, W.-G., Wei, Y.-C., Meng, S., Bai, G.-G., Bai- Wang, B.-Y., Yang, H.-Y., Tian, W., Meng, X., Zhang, H. & Chen, S.-P. (2010) Involvement of oxidative stress in the relapse of acute myeloid leukemia. *The Journal of Biological Chemistry*, **285**, 15010–15015.

8

In silico Acute Myeloid Leukaemia

Eleni Pefani[1], *Eirini G. Velliou*[2], *Nicki Panoskaltsis*[3], *Athanasios Mantalaris*[4], *Michael C. Georgiadis*[5], *and Efstratios N. Pistikopoulos*[6]

[1] *Clinical Pharmacology Modelling and Simulation, GSK, UK*
[2] *Department of Chemical and Process Engineering, Faculty of Engineering and Physical Sciences, University of Surrey, UK*
[3] *Department of Medicine, Imperial College London, UK*
[4] *Department of Chemical Engineering, Imperial College London, UK*
[5] *Laboratory of Process Systems Engineering, School of Chemical Engineering, Aristotle University of Thesaloniki, Greece*
[6] *Texas A&M Energy Institute, Artie McFerrin Department of Chemical Engineering, Texas A&M University, USA*

8.1 Introduction

In the UK (Cancer Research UK, 2008), it is estimated that more than 1 in 3 people will be afflicted with cancer in their lifetime. For one such cancer, leukaemia – a neoplasm of the blood and bone marrow (BM) – 1 in 71 men and 1 in 105 women will be affected, with incidence sharply rising in adults over the age of 50. Approximately 40% of those affected with leukaemia will have acute myeloid leukaemia (AML).

Leukaemia is a cancer of the BM and blood wherein blood cells are unable to develop or function normally, are overproduced at an immature stage of development and overtake any normal elements remaining in the BM and blood (see also Chapter 7). This uncontrolled growth compounds the morbidity and mortality due to the disease by inhibiting development of healthy blood and immune cells through multiple mechanisms (Panoskaltsis *et al.*, 2003, 2005).

The most common treatment for most types of leukaemia is intensive chemotherapy given through the vein (intravenous [IV]). This therapy can be life-threatening since only relatively few patient-specific and leukaemia-specific factors are considered in current protocols; choice of chemotherapy, intensity and duration often depends on either the availability of a clinical trial, the treating physician's experience or the collective experience of the treating centre, with significant international protocol variability. Inter-patient and intra-leukaemia variability add complexity to these treatment decisions and are not yet adequately addressed, possibly accounting for the 30–45% long-term survival rates in young people with one type of BM cancer, AML. For

Modelling Optimization and Control of Biomedical Systems, First Edition.
Edited by Efstratios N. Pistikopoulos, Ioana Naşcu, and Eirini G. Velliou.

those who are cured, there is a lifetime of increased risks of secondary cancers, cardiovascular disease and diabetes due to the adverse effects of treatment.

In order to overcome these limitations, there is a need for personalised treatments that incorporate both the individual patient characteristics and features specific to the patient's leukaemia (different for every patient).

Mathematical modelling is undoubtedly a useful tool that can be used for the automation of chemotherapy treatment due to its advantages in systematically exploring extensive datasets in order to capture a system's dynamics and subsequently provide better insight for process enhancement. Towards this direction, various mathematical models have been developed for different biomedical systems (Parker and Doyle, 2001; Sherer *et al.*, 2006; Ledzewicz and Schättler, 2007; Dua *et al.*, 2008; Harrold and Parker, 2009; Krieger *et al.*, 2013) with the aim to describe the disease under chemotherapy and afterwards propose the optimal treatment design. For the most part, these models aim to describe the disease dynamics of a hypothesised average patient case study. Under this assumption, these models do not include patient and disease-specific characteristics as parameters in the model, but they use mean values derived from a number of patient-volunteer studies. However, there is a lack to our knowledge of models that include personalised patient and disease information and use optimisation methods in order to design optimal personalised chemotherapy protocols.

The need for more personalised treatment design has been discussed in various works (Undevia *et al.*, 2005; Garattini, 2007; Essers and Trumpp, 2010), and the main sources of inter- and intra-patient variability are in the cellular kinetics of the tumour and normal cell populations and the kinetics of the anticancer agents when they enter the human body. *Thus, the desired mathematical model for the simulation of patient behaviour and tumour response during chemotherapy should consist of three parts: (a) the cell cycle model, which is the target of drug action; and the (b) pharmacokinetic (PK) and (c) pharmacodynamic (PD) aspects that provide the complete description of drug diffusion and action after administration.*

8.1.1 Mathematical Modelling of the Cell Cycle

The *cell cycle* is the set of cell mechanisms that duplicate the cell's material and divide it into two daughter cells, with the purpose either to preserve or to expand the cell population (see also Chapter 7). Cancer is unavoidably connected to the cell cycle, as the origin of the disease is out-of-control cell growth due to an abnormality in the process of cell proliferation. The macro effect of this abnormal cell proliferation is the uncontrollable regulation of tissue growth leading to the creation of tumours, masses of malfunctioning cells that are harmful for the body. In that sense, a better insight of the cell cycle procedure will provide more information about cancer dynamics, which would unquestionably be useful for chemotherapy treatment design.

Chemotherapy drugs aim to stop this uncontrollable cell proliferation by interfering with the cell cycle and killing the cells in replication. Chemotherapeutic drugs are classified according to the cell cycle phase in which they are active. Some drugs react selectively with cells during a particular cell cycle phase (cell cycle–specific drugs), whereas others react with cells in all phases (cell cycle–nonspecific drugs). Some examples of commonly used cell-specific drugs are the anthracyclines, such as daunarubicin (DNR), doxorubicin and idarubicin, that inhibit DNA and RNA synthesis (i.e. during the S and G_1 phases); antimetabolites, such as methotrexate and Ara-C, that are S phase–specific drugs; and cell cycle–nonspecific drugs, such as platinum drugs (e.g. carboplatin), that react with all the phases of the cell cycle. A drug's specificity makes the cell cycle's role a critical factor for the efficacy of the treatment.

Advances in experimental tools and setups gave rise to more sophisticated experiments and research in the 1980s and 1990s, which have focused mainly on the characterisation of patient variability in cell cycle kinetics.

Mathematical models that describe the cell cycle have been developed since the very first experimental data were obtained. The most common type of modelling approach is the compartmental model, where compartments are used to describe the different cell phases or combination of phases into clusters. In that case, the mathematical model consists of the mass balances of cells in each compartment of the model. The simplest mathematical model assumes that the entire cell cycle forms one compartment (Swierniak et. al, 2009), whereas the most detailed model considers each phase as one compartment (Sherer *et al.*, 2006).

However, compartmental models fail to account for system heterogeneity, and a rigid population is assumed with common characteristics of size, age, mass and so on. Population-based models (PBMs) are another modelling type which describes the effect of the cell heterogeneity on the cell culture dynamics. In these models, the cell cycle is organised into population balances differentiated over time and another property that evolves in parallel with the cell cycle progression (i.e. over mass for the mass-structured cell cycle) (Sidoli *et al.*, 2006).

Although the PBMs are more robust and accurate than compartmental models, they introduce a considerable number of unknown parameters, some of which are difficult to experimentally measure. In general, for both types of models (compartmental and PBM), the required parameters for the cell cycle mathematical model consist of the transition rates of cells between cell cycle phases, the proliferation rates of the cells, the distribution of cell populations (normal and abnormal) into the cell cycle phases and the natural apoptotic rate of each cell cycle phase. For the calculation of these parameters, the experimental measurement available is the duration of the cell-cycling phases. If the time-history profile of each cell cycle phase is known thereafter, the distribution of the cell population into phases, proliferation rates and transition rates can be estimated (Basse *et al.*, 2003). As far as the cell natural apoptotic rate is concerned, since the model purpose is chemotherapy action on cells, a valid assumption often used is that there is a minor probability that the cell will follow a natural apoptosis path and the apoptotic rate equals zero (Basse *et al.*, 2003). In that sense, the prevailing system measurement required is the duration of each phase in the mitotic cycle.

Models in the literature assume constant cell-cycling times and transition rates between the phases which make the model purely deterministic (Fister and Paneta, 2000; Ledzewicz and Schättler, 2002; Dua *et al.*, 2008; Swierniak *et al.*, 2009). These models fail to capture the intra-patient invariability in the duration of the cycling phases and are adequate only to capture the behaviour of a mean cell cycle function. Works with more accurate approach are those which consider the cycle duration as a stochastic distribution between a minimum and maximum range (Basse *et al.*, 2003; Kimmel and Swierniak, 2003; Sherer *et al.*, 2006). In this approach, cells of the same population have different distribution characteristics. However, a large window for modelling amelioration still remains for the improvement of the compartmental models to account for intra-patient variability, let alone the PBMs that should include this variability as functions over both time and another system characteristic.

8.1.2 Pharmacokinetic and Pharmacodynamic Mathematical Models in Cancer Chemotherapy

The most crucial branches of pharmacology essential for both drug development and management of drug information are PK and PD. PK generally aims to give the time-concentration history of the drug throughout the body, while PD aims to describe the drug effects on the body. Those two are intimately connected, as the effects of the drug on the body depend on the drug concentration at the molecular site of action.

The combination of PK and PD consists of the complete action of the drug on the human body (i.e. the time-dependent procedure for a drug to reach and act on a cell). The steps of the drug action in the body are as follows, and they are described in Figure 8.1 (Ratain and Plunkett, 2003):

1) Drug administration
2) Drug absorption and metabolism through gastrointestinal tract in case of oral administration of a drug
3) Metabolism of the drug in the liver
4) Drug delivery in the cell environment, and protein binding to act on the cell
5) Drug action (PD)
6) Drug returned either to the liver or to the kidney, and excreted by biliary or urinary excretion.

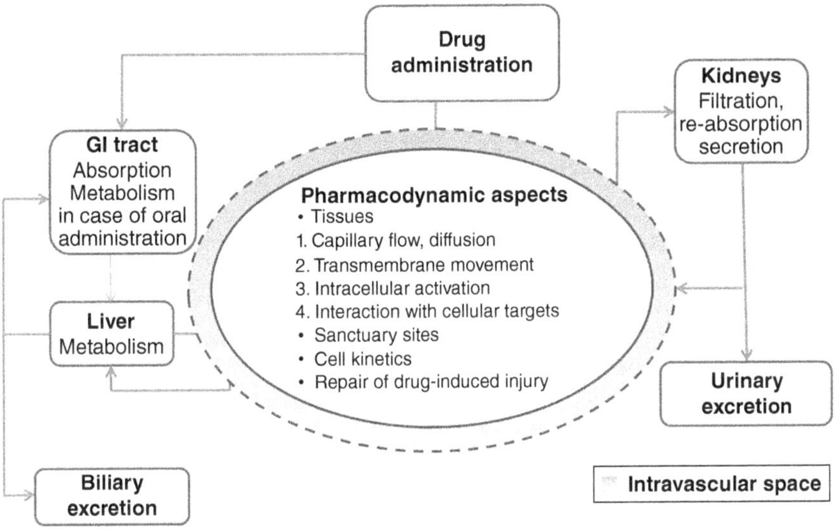

Figure 8.1 Schematic diagram of PK and PD: blue boxes are for the PK model, and they are connected to the red cycle that represents the PD part of drug action. *Source*: Ratain and Plunkett (2003). Reproduced with permission of Elsevier.

8.1.2.1 PK Mathematical Models

The major mechanisms of the *PK drug action* in the body consist of the drug absorption, distribution, metabolism and excretion (Figure 8.2). Drug absorption is considered in cases of non-IV dose administration (subcutaneous [SC], oral etc.) where the drug inflow reaches the systemic circulation with a certain time delay (absorption rate) and in a decreased amount (bioavailability) as some of the initial drug given is being bounded during the absorption. Afterwards, the drug is distributed throughout the fluids and tissues of the body and is then metabolised in the liver and the kidneys. Finally, the drug is

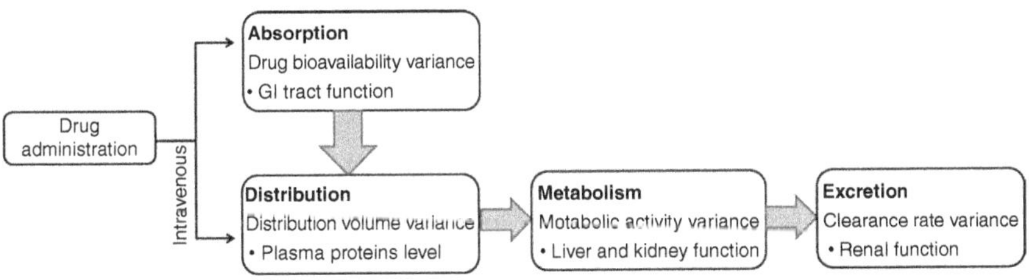

Figure 8.2 The process of drug delivery. Drug delivery is governed by four mechanisms: absorption, distribution, metabolism and excretion. Each of these mechanisms is deprived of further mechanisms. Inter- and intra-patient variability in these mechanisms is the probable source of PK variability.

eliminated and excreted via the kidneys by either the urinary or biliary route (Saltzman, 2001).

High inter-patient variability exists in the amount of drug concentration produced by the same dose administered. This inter-patient variability is certainly correlated to the PK mechanisms described here, although the exact source of variability is yet to be defined. Extensive works (Undevia *et al.*, 2005; Garattini, 2007) review the probable patient information that could correlate with the drug concentration in the tumour site and successively in the treatment outcome.

To begin with, absorption depends on the drug absorption rate and bioavailability. The *absorption rate* is the time delay from the time of administration until the drug reaches the systemic circulation, when it is given non-intravenously. This term has meaning mainly for the case of oral administration, where the drug is first inserted to the gastrointestinal (GI) tract; for SC dosage, this rate represents a drug leak from the SC injection site to the blood. *Bioavailability* refers to the final amount of drug reaching the blood compartment. Again, this term is mostly applied to oral administration, where drug losses occur in the GI tract. Both bioavailability and absorption rate vary between different patients and depend on factors such as the absorptive area, the transition time of the drug to the blood, the blood flow and the GI environment, which are all probable sources of patient variability.

Moreover, drug distribution involves the drug transition from the intravascular to the extravascular space. The amount of the distributed drug will define the distribution volume of the drug and depends on the level of binding proteins where a proportion of the drug is bounded and the amount of free drug is reduced. The drug metabolism takes place on the remaining free drug that reaches the liver and in some cases the kidney. The metabolism kinetics depends on the patient hepatic blood uptake and the enzymatic activity of the patients. Lastly, excretion is related to the kidney action to eliminate and remove the inactive drug and will again differ between patients, resulting in varying drug clearance rates.

In summary, there is patient variability in all four mechanisms of drug delivery described here. This patient variability definitely contributes to the different treatment outcome between patients. Especially for the anticancer agents, the variability in the amount of active drug that reaches the tumour location (without the drug bounded during absorption and distribution) will finally affect the concentration of the active metabolite produced and the drug intra-cellular activation.

Currently, two types of PK model are used: the compartmental and the physiological models. In compartmental modelling, body organs are grouped into compartments, and drug is assumed to be absorbed, distributed and eliminated in these compartments. These are standardised models in frequent use by pharmaceutical companies, and commercial tools exist for the development of this type of models for a variety of drugs (see e.g. http://www.iconplc.com/technology/products/nonmem/, http://www.pharsight.com/products/prod_winnonlin_home.php and http://www.mathworks.co.uk/discovery/pharmacokinetic.html). However, the ability of these models to give a valid estimation of the drug profile of a newly studied patient is rather questionable. The major source of model uncertainty is due to the fact that the values of the variables are based on the interpretation of the mean concentration profile of a group of patients. This *mean-concentration profile* in most of the cases is not representative of the behaviour of patients in the group studied, let alone the whole patient population. Furthermore, since these models are empirical and include the concentration profile in the body, as a totality they do not allow for the model extension to account for more detailed phenomena in the tumour location, such as by linking the concentration profile to a detailed PD model of the drug mechanism when it is intracellularly activated and acts.

These drawbacks are satisfied to a certain extent by the physiologically based pharmacokinetic (PBPK) models. A *physiological model* is a highly compartmental model that considers all the organs reacting with the drug. The model is derived from the equilibrium balances in these organs. PBPK models depend on two types of information: the patient physiological and the drug biochemical information. Physiological parameters consist of the body organ volume (Vi) and the blood flow rate in the body organs (Q_i). These parameters have been extensively measured during the past decade and are correlated to patient characteristics such as sex, age, body mass index and cardiac output (Brody, 1945; Wennesland *et al.*, 1959; Brown *et al.*, 1962; Chouker *et al.*, 2004; Pichardo *et al.*, 2007). Moreover, the biochemical parameters are the parameters for the calculation of the drug metabolism rate. One common assumption in PBPK models is that the metabolism follows Michaelis–Menten kinetics, and the required parameters are the drug maximal velocity (V_{max}) and the Michaelis–Menten affinity constant (k_m).

In recent decades, remarkable progress has been noted regarding the experimental design for the parameters of PBPK models. Initially for the PK model metabolism, information was only available from animal experiments, and scaling was used afterwards for the approximation of the equivalent human values. However, there is a level of uncertainty in this method, as the human organism is much more complex than other species and such a relation can only succeed to give a rough guess of a human value and not the accuracy required. Nowadays, established methods exist to correlate PBPK parameters to *in vitro* and allometric data (Chaturvedi *et al.*, 2001; Jones *et al.*, 2009), and commercial tools exist for the calculation of these parameters for given drugs (http://www.cyprotex.com/home/, http://www.simulations-plus.com and http://www.simcyp.com).

A level of detail can be added to the PBPK models by further separating each compartment (i.e. organ) into the vascular, interstitial and intracellular parts.

The vascular is the part of the organ with the blood vessels from which the drug is inserted within the organ (interstitial), and then the drug reaches the intracellular part of the cell environment. This modification would introduce a wale of parameters that includes the physiochemical characteristics of drug chemicals within the interstitial, vascular components (Schmitt *et al.*, 2008; Peyret and Krishnan, 2011; Yun and Edginton, 2013). More elegant future models would combine this level of information with cellular information to provide cellular-level PBPK models that could give insight of probable correlation between patient cell characteristics and the different PK profile (Caldwell *et al.*, 2012). Table 8.1 lists some paradigms of the two types for PK models for anticancer agents.

Table 8.1 PK models of cancer drugs.

Compartmental models	
Eight-compartmental model for methotrexate (MTX)	(Reich *et al.*, 1977)
Multicompartmental model for doxorubicin	(Reich *et al.*, 1977)
Three-compartmental model for idarubicinol	(Looby *et al.*, 1997)
Two-compartmental model for methotraxate	(Hijiya *et al.*, 2006)
Two-compartmental model for etoposide	(Hijiya *et al.*, 2006)
Two-compartmental model for teniposide	(Hijiya *et al.*, 2006)
One-compartmental model for Ara-C	(Hijiya *et al.*, 2006)
Two-compartment model for etoposide	(Relling *et al.*, 1998)
Two-compartmental pharmacokinetic model for etoposide	(Panetta *et al.*, 2002)
Compartmental modelling of cyclosporine, etoposide and mitoxantrone	(Lacayo *et al.*, 2002)
Two-compartment model of idarubicin	(Gillies *et al.*, 1987)
Three-compartment model of mitoxantrone	(Richard *et al.*, 1992)
Physiological models	
Physiological model for Ara-C	(Morrison *et al.*, 1975)
Physiological model for Ara-C (Ara-C)	(Dedrick *et al.*, 1972)
Physiological model for thriopental	(Bischoff and Dedrick, 1968)
Physiological model for methotrexate	(Himmelstein and Lutz, 1979)
Physiologican model for adriamycin/doxorubicin	(Himmelstein and Lutz, 1979)
Physiological model for actinomycin-d	(Lutz *et al.*, 1977)
Physiological model for adriamycin	(Chen and Gross, 1979)
Physiological model for cis-dichlorodiammine-platinum	(Chen and Gross, 1979)
Physiological model for cyclotidine	(Chen and Gross, 1979)

Table 8.2 Formulas of PD models.

Model	Model equations	Description
Linear model	$E = S \cdot C + E_o$	*E*: drug effect *C*: drug concentration *S*: slope parameter E_o: initial drug effect
Log-linear model	$E = S \cdot \log C + I$	*E*: drug effect *C*: drug concentration *S*: slope parameter *I*: constant
E_{max} model	$E = E_o - \frac{E_{max} \cdot C}{EC_{50} + C}$	*E*: drug effect *C*: drug concentration E_{max}: maximum drug effect E_o: initial drug effect from previous application EC_{50}: concentration producing half of the maximum drug effect
Sigmoid E_{max} model	$E = \frac{E_{max} \cdot C^n}{EC_{50}^n + C^n}$	*n*: constant affecting the shape of the drug effect–concentration curve

Source: Holford and Sheiner (1982). Reproduced with permission of Elsevier.

8.1.2.2 PD Mathematical Models

The PD model describes the effect of the drug action to enter the cell and cease its function. Because of the complexity of the drug mechanism of action, detailed PD models are not in use. Expressions relating the drug concentration to the drug effect are preferable. In that way, the accuracy of the PD model is highly dependent on the precision of the PK model.

The most widely used models are the drug effect models. These models relate the drug concentration to the drug effect, which in our case is the number of dead cells. These models contain estimated parameters from real-life data, and drug effect depends only on the drug concentration. Table 8.2 summarises the expressions of these models.

8.2 Chemotherapy Treatment as a Process Systems Application

Models aiming to describe the actions of chemotherapy should consist of mathematical expressions for all steps of drug treatment, from administration to intracellular action. All parts are described in Figure 8.3, where the general

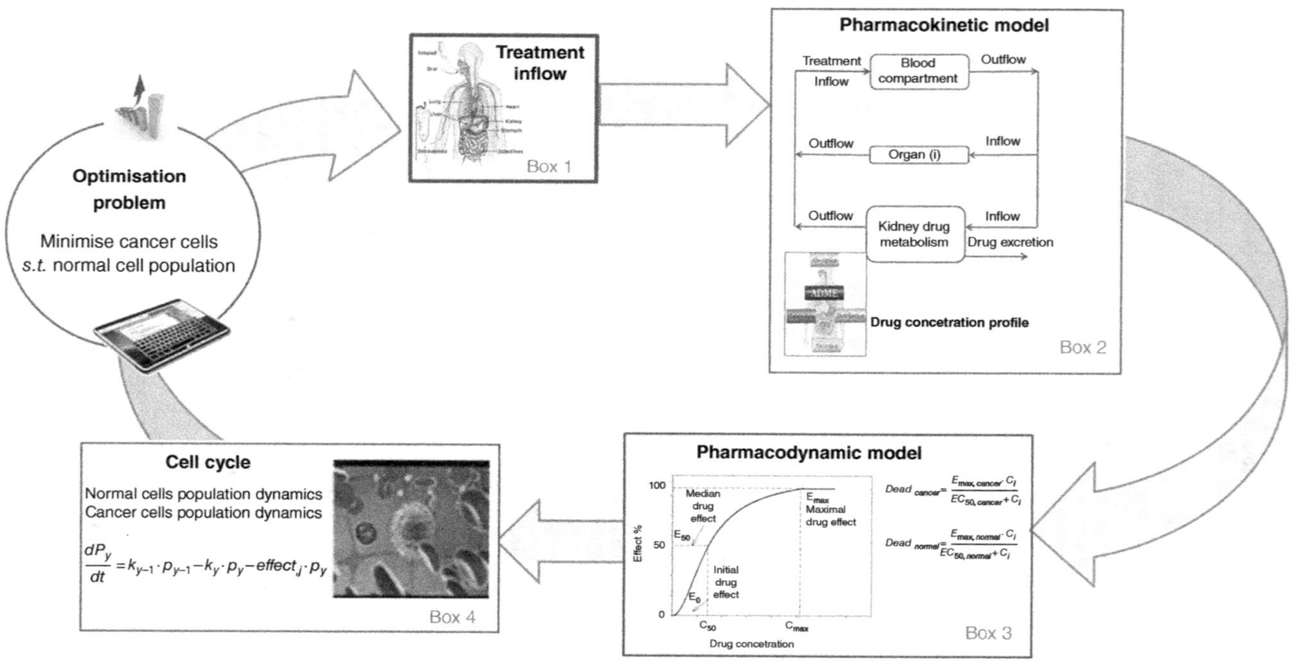

Figure 8.3 Framework for the derivation of an optimal personalised chemotherapy protocol.

framework for the derivation and function of such a mathematical model is described.

To start, the initial dose load given to the patient in combination with the administration route and injection rate will be used for the calculation of treatment inflow (Figure 8.3, box 1), the main input for the PK model. The PK model (Figure 8.3, box 2) depends on patient-specific characteristics and is composed of the set of drug mass balances in patient organs for the calculation of the drug concentration profile. This profile is the main input for the PD model. The PD model (Figure 8.3, box 3) calculates the number of both normal and cancer cells which have died due to drug administration; these are then successively subtracted from the starting number of cells (Figure 8.3, box 4) in order to calculate the number of each cell type which remain following the chemotherapy cycle. Thereafter, a new optimisation problem will be introduced and solved only if there are tumour cells still present (in this model), and normal cells are in sufficient number such that the patient can tolerate another chemotherapy cycle. Within this framework, optimal chemotherapy cycles will be designed which aim to effectively control the treatment schedule (inflow, dose load) in order to eradicate the maximum possible disease while maintaining normal cells within predefined limits. The optimal treatment will be different on a case-by-case basis and will be based on physiological characteristics of the patient that determine the drug kinetics and cell characteristics that determine the diseased and normal population dynamics.

8.2.1 Physiologically Based Patient Model for the Treatment of AML With DNR and Ara-C

In this section, a mathematical model is formulated to simulate the chemotherapeutic action of two anti-leukaemic drugs, DNR and Ara-C, commonly used in clinical practice for the treatment of AML. The model describes the dynamic interactions of leukaemic and normal cells exposed to chemotherapeutic drugs by a system of ordinary differential equations (Pefani *et al.*, 2013, 2014).

A quick guide to the developed model is presented in Table 8.3, while Appendix 8A includes the model in full detail. Initially, a drug dose ($u_{,j}$) of anti-leukaemic agent j is injected into the patient intravenously over duration$_{,j}$ (8.1). The inflow rate of drug j is then transmitted by direct injection into the blood and is circulated to the whole body (8.2). This inflow is the main input for the calculation of the drug concentration in the blood ($C_{B,j}$), taking into account patient-specific parameters such as the total patient blood volume (V_B), the blood flow in organs (Q_i, with i: heart [H], liver [Li], bone marrow [M], lean [Le] and kidneys [K]) and $C_{i,j}$ is the concentration of drug j in organs i. The drug is transmitted via the blood to the organs, and the general mass balance in the organs is the one in Equation (8.3) that includes the elimination

Table 8.3 Brief guide to model equations.

Model part	Equations	
PK (IV)	$inf\,low_{,j} = \dfrac{u_{,j}}{duration_{,j}}$	(8.1a)
	$V_B \cdot \dfrac{dC_{B,j}}{dt} = \sum_{i:H,Li,M,Le,K} Q_i \cdot C_{i,j} - Q_B \cdot C_{B,j} + \inf low_{,j}$	(8.1b)
	$V_i \cdot \dfrac{dC_{i,j}}{dt} = Q_i \cdot C_{B,j} - Q_i \cdot C_{i,j} - k_{k,j} \cdot C_{B,j} - k_i \cdot C_{i,j} \cdot V_{iT}$	(8.1c)
PK (SC)	$\dfrac{dS}{dt} = Inflow - k_a \cdot k_b \cdot S$	(8.2a)
	$V_B \cdot \dfrac{dC_B}{dt} = \sum_{i:H,Li,M,Le,K} Q_i \cdot C_{i,j} - Q_B \cdot C_{B,j} + k_a \cdot k_b \cdot S$	(8.2b)
PD	$effect_{,j} = \dfrac{E_{\max,j} \cdot C_{M,j}^{slope,j}}{E_{50,j} + C_{M,j}^{slope,j}}$	(8.3)
Cell cycle	$\dfrac{dP_y}{dt} = k_{y-1} \cdot P_{y-1} - k_y \cdot P_y - effect_{,j} \cdot P_y$	(8.4)

Source: Pefani *et al.* (2014). Reproduced with permission of IEEE.

rate of the drug in the kidneys and the liver ($k_{,i,j}$). After drug elimination and action, the drug is excreted through the urine with a clearance rate ($k_{k,j}$) from the kidneys.

The PD model is used for the calculation of drug effect, which is the percentage of dead cells due to drug action. The main input for PD is the drug concentration in the tumour location, which for AML is the concentration of drug in the BM ($C_{M,j}$) as is calculated by the PK model; and $E_{max,j}$, $E_{50,j}$ and slope are the PD parameters that depend on the drug j. The effect$_{,j}$ calculated by the PD model is the percentage of cells which react with the drug j and are killed. This effect is multiplied by the number of cells reacting with the drug in order to calculate the number of cells that died due to treatment and the remaining cells after drug action. The mass balance is in Equation (8.5), where y is the cell cycle phase (y: G_1, S, G_2 or M), P_y is the cell population in phase y and k is the transition rate of cells from one phase to the next (i.e. k_{y-1} is the transition rate from phase $y-1$ to phase y).

The SC route is an alternative dose route where the drug is injected in the individual's derma (S). In this type of drug administration, the drug inflow reaches the systemic circulation with a certain time delay (absorption rate) and in a decreased amount (bioavailability), as some of the initial drug given is being bounded during the drug absorption from the SC site to the blood compartment. For the model of the SC route, Equation (8.2) will be replaced by Equations (8.2a) and (8.2b), which account for drug bioavailability (k_b) and absorption delay (k_a).

8.2.2 Design of an Optimal Treatment Protocol for Chemotherapy Treatment

Treatment design will be mainly based on the control of four schedule parameters:

- Drug use
- Dose load
- Dose duration
- Number of dose applications.

The optimisation algorithm is presented in Table 8.4.

The objective function is the minimisation of the leukaemic ($Cells_{leuk}$) cells subject to the treatment schedule, which is defined by the drug use (j), the dose load ($u_{n,j}$), the dose duration ($t_{n,j}$), the number of applications (NA) and the interval period between two succeeding dose applications (τ_n). The four first parameters are the optimisation schedule variables, whereas the interval period between two doses is a design variable calculated by the frequency

Table 8.4 Chemotherapy process optimisation algorithm.

Objective function	$\min_{j,u_{n,j},t_{n,j},NA} Cells_{leuk}$
Equality constraints	$Cells_{nor,n,j} = f\left(effect_{n,j}\right)$ $Cells_{leuk,n,j} = f\left(effect_{n,j}\right)$ $effect_{n,j} = f\left(C_{M,n,j}\right)$ $C_{M,n,j} = f\left(\text{inflow}_{n,j}\right)$ $Inflow_{n,j} = \sum_{n=1}^{n=NA} \frac{u_{n,j}}{t_{n,j}} \cdot appl, \quad where \begin{cases} appl = 1, \ \tau_{n-1} < time < \tau_{n+1,j} \\ appl = 0, \tau_{n+1,j} < time < \tau_{n+2,j} \end{cases}$
Inequality constraints	$Cells_{nor}\big\vert_{n=NA} \geq Cells_{leuk}\big\vert_{n=NA}$ $Cells_{nor} > 3 - \log reduction$

*n is the number of dose applications; j is the drug; $t_{n,j}$ is the duration of each dose application; $u_{n,j}$ is the dose load of each application; $Cells_{leuk,n,j}$ is the number of leukaemic cells; $Cells_{nor,n,j}$ is the number of normal cells; $effect_n$ is the the PD effect of drug j over application n; $C_{Mn,j}$ is the BM concentration; $Inflow_{n,j}$ is the inflow of drug j during application n; τ_n is the duration between two succeeding dose applications; NA is the total number of applications.

Source: Pefani *et al.* (2014). Reproduced with permission of IEEE.

of doses (i.e. if two or four doses will be applied daily), as defined by clinicians. The control parameters define the drug inflow that has physical meaning only for the periods of chemotherapy treatment, whereas the value of the inflow is set to 0 for the periods between two succeeding chemotherapy cycles.

The feasible optimisation solutions are defined by the set of the equality and inequality constraints. Equality constraints consist of the expressions used to calculate the number of leukaemic ($Cells_{leuk}$) and normal ($Cells_{nor}$) cells throughout the treatment. Both cell populations are functions of the drug PD effect ($effect_n$), which is defined by the drug concentration profile in the tumour location (i.e. the BM) ($C_{M,n,j}$). The drug concentration profile is determined by the treatment inflow, a variable calculated by the schedule and the design parameters. Moreover, the inequality constraints consist of constraints on the number of normal cells that will have to be higher than a 3-log reduction throughout the treatment (path constraint), and by treatment completion they will have to be higher than the number of leukaemic cells (endpoint constraint).

8.2.3 Mathematical Model Analysis Using Patient Data

8.2.3.1 Model Sensitivity Analysis

To gain a further understanding of the model and the crucial parameters that highly affect the treatment outcome (i.e. the level of leukaemic cells), a global sensitivity analysis and quasi–Monte Carlo based high-dimensional model representation using Sobol's indices were performed using the graphical user interface/high-dimensional model representation (GUI-HDMR) software (Ziehn and Tomlin, 2009). The output of interest is the number of leukaemic cells, and the parameters checked are the cell cycle times and the PK and PD parameters, which are listed in Table 8.5.

Specifically, the drug elimination rates in the liver were included for the studied drugs, as inter-patient variability has been indicated and reported in the ULCH (2009) and BC Cancer Agency (2007) works. The same works also report patient variability for the DNR kidney clearance rate; however, there is no measured variability of the kidney clearance rate for Ara-C, and this parameter is not included in the sensitivity analysis. For the inter-patient variability of the PD parameters, the work of Quartino *et al.* (2007) has been used, which includes analysis of PD action of DNR and Ara-C on BM samples of 179 patients with AML. Moreover, the cell cycle parameters' ranges are as calculated and reported in the work of Raza *et al.* (1990) with experiments on the cell kinetics characterisation of 54 patients diagnosed with ALM.

For calculation of the parameters sensitivity index (SI), 40,000 simulations were run of all the possible combinations of the tested parameters within their assigned ranges. The SI results are presented in Table 8.5.

Table 8.5 PK, PD and cell cycle parameters and inter-individual ranges used for model sensitivity analysis and sensitivity index results.

	Symbol	Default value	Deviation	Reference	Sensitivity Index
PK	$k_{l,Ara\text{-}C}$	0.069	0.067–0.07	(UCLH, 2009)	0.0007
	$k_{k,DNR}$	1.5	0.036–1.7	(BC Cancer Agency, 2007)	0.017
	$k_{l,DNR}$	0.015	0.014–0.017	(BC Cancer Agency, 2007)	0.000085
PD	$E_{max,Ara\text{-}C}$	0.83	0.79–0.86	(Quartino *et al.*, 2007)	0.0003
	$E_{50,Ara\text{-}C}$	0.29	0.25–0.33	(Quartino *et al.*, 2007)	0.0049
	$E_{max,DNR}$	0. 91	0.88–0.93	(Quartino *et al.*, 2007)	0.00925
	$E_{50,DNR}$	0.09	0.076–0.1	(Quartino *et al.*, 2007)	0.0928
	$slope_{DNR}$	1.23	1.06–1.4	(Quartino *et al.*, 2007)	0.000468
Cell cycle	T_s	15	6–43	(Raza *et al.*, 1990)	0.2705
	T_C	60	18–211	(Raza *et al.*, 1990)	0.604

* $k_{l,Ara\text{-}C}$ is the Ara-C liver elimination rate (min^{-1}); $k_{k,DNR}$ is the DNR clearance rate by the kidneys $\left(\frac{lt}{min}\right)$; $k_{l,DNR}$ is the DNR elimination rate in the liver (min^{-1}); $E_{max,Ara\text{-}C}$ is the Ara-C maximum drug effect; $E_{50,Ara\text{-}C}$ is the Ara-C concentration at half drug effect $\left(\frac{mg}{lt}\right)$; $E_{max,DNR}$ is the DNR maximum drug effect; $E_{50,DNR}$ is the DNR concentration at half drug effect $\left(\frac{mg}{lt}\right)$; $slope_{DNR}$ is the slope scaling factor for DNR kinetics; T_s is the S phase duration (h); and T_C is the total cell cycle duration (h).

Source: Pefani *et al.* (2014). Reproduced with permission of IEEE.

The SA results clearly indicate cell cycle phases' duration as the most crucial parameters, where *Tc* has an effect of 60.4% on the treatment outcome and *Ts* has 27.05% effect. Of note, the limit for a parameter to be accounted as crucial for the measured variable is at least 10%.

8.2.3.2 Patient Data

The project is submitted and approved by the North West London Hospitals Trust for the provision of health records of patients diagnosed with AML and treated within Northwick Park Hospital using DNR and Ara-C anti-leukaemic agents under either IV or SC doses applied. The clinical data of the six patients are under two clinical applied treatment protocols, one intensive and the other non-intensive: (a) daunarubicin (DNR) and cytosine arabinoside (Ara-C) used in standard IV doses (DA 3 + 10), and (b) low-dose Ara-C (LDAC) administered subcutaneously (SC). The clinical data comprised the required patient physiological characteristics, the blast percentage in BM aspirate at disease diagnosis, the chemotherapy treatment protocol and the blast percentage together with marrow cellularity of the marrow examinations after the applied chemotherapy protocol. For conversion purposes, the measured number of cells in the BM of an average AML patient is 1 trillion cells (Williams *et al.*, 1983). This average number of cells is then multiplied by the cellularity factor given for each patient; this is 0.2 for a hypo-cellular BM, 0.9 for a normo-cellular BM of a patient younger than 30 years, 0.5 for a normo-cellular BM of a patient between 30 and 65 years old, 0.4 for a normo-cellular BM of a patient older than 65 years and 0.95 for a hyper-cellular BM (Williams *et al.*, 1983). These conversion rules used data for the current presented analysis that are provided in Appendix 8B.

8.2.3.3 Estimation of Patient-Specific Cell Cycle Parameters

In this section, data presented in Appendix 8B of the health records of six AML patients are used for the estimation of leukaemic cell cycle parameters that are the critical model parameters as indicated by the sensitivity analysis results. One important assumption on which the estimation is based is the cell cycle times during the interval period between two succeeding chemotherapy cycles. As mentioned, this period is a recovery period lasting 20–30 days during which the patient receives no treatment.

Under this assumption, two problems are formulated and solved for the estimation of the cell cycle parameters. The first problem concerns the interval period between completion of one cycle and the BM aspirate prior to its successive cycle. For this period, the leukaemic cell cycle parameters are set to $Ts = 40$ h and $Tc = 211$ h. The provided leukaemic population measurement at the end of this interval period together with the duration of this period are then used for the estimation of the leukaemic population at the beginning of the recovery period, which is the leukaemic population at completion of the last applied chemotherapy cycle.

The second parameter estimation problem uses this calculated leukaemic population at the end of the chemotherapy cycle together with the provided initial tumour burden at the beginning of each chemotherapy cycle and the treatment schedule in order to fit and estimate the leukaemic cell cycle parameters (Ts, Tc) under chemotherapy.

This parameter estimation problem is solved using gPROMs (gPROMs, 2003), and the fitted cell cycles for the six patients are listed in Table 8.6.

The results show variability of the cycling times that are different between patients and between the chemotherapy cycles of the same patient. The mean calculated time for Ts is 15 h, with a range between 9 and 21 h, and for Tc the mean value is 47.5 h with variability within 33–68 h.

Another observation from the fitted cell cycle results is that the longer Tc times were indicative of disease relapse (P001 fourth cycle, P006 third cycle). This relation between Tc and disease increase has a scientific explanation, as the longer cycling times indicate a longer G_0/G_1 phase. It is well reported (Komarova and Wodarz, 2005; Lewin *et al.*, 2007; Michor, 2008) that the G_0/G_1 phase is a factor related to disease resistance and relapse, since cells in this

Table 8.6 Cell cycle times fitted for the clinical data of 6 patients under LD and DA protocol (Appendix 8B).

	Patient number	*Ts* (h)*	*Tc* (h)*
	Patients under LD protocol		
	001 (First cycle)	13	45
	001 (Second cycle)	16	40
	001 (Third cycle)	11	45
	001 (Fourth cycle)	18	65
	002 (First cycle)	21	45
	006 (First cycle)	20	33
	006 (Second cycle)	14	46
	006 (Third cycle)	14	68
	006 (Fourth cycle)	20	40
	016 (Second cycle)	14	45
	Patients under DA protocol		
	011 (First cycle)	9	53
	026 (First cycle)	15	47
	026 (Second cycle)	15	40
	016 (First cycle)	10	54
Mean		15	47.5
Range		(9–21)	(33–68)

phase are not detected by the drugs and they form residual disease after treatment completion. The reverse relation was observed for *Ts* time, where the longer *Ts* indicated lower leukaemic cell populations. The slower S phase duration is linked to a higher percentage of cell population in this phase that respectively increases the probability of the leukaemic cells to be detected and eradicated by anti-leukaemic S phase–specific drugs such as DNR and Ara-C.

Moreover, a very interesting point in the resulting cell cycle distributions is that patients successfully treated under the LDAC protocol are characterised by lower *Tc* duration compared to patients under the DA protocol. An interesting fact in clinical practice is that patients who receive a low dose of SC treatment present as good treatment results as patients who receive much higher doses of DNR and Ara-C administered IV. In order to capture this fact, the model uses lower duration of the non-proliferating phase for the cases of patients with successful results of low-dose Ara-C treatment. Physically, this means that for a patient to be successfully treated by a low-dose treatment, an explanatory scenario is that the majority of his or her cells will be in proliferation, and thus susceptible to the drug.

8.3 Analysis of a Patient Case Study

Patient 016 is a patient case study treated under the DA protocol for the first chemotherapy cycle and under the LDAC protocol for the second cycle. As shown in the simulation results (Table 8.7), this patient presents leukaemic population reduction from the first chemotherapy cycle, and normal cells are higher than leukaemic cells. However, by the completion of the second cycle, residual disease exists and BM hypoplasia is not achieved. For this reason, the optimisation problem is solved for both chemotherapy cycles.

8.3.1 First Chemotherapy Cycle

For the first chemotherapy cycle, Ara-C is suggested to be continuously administered over daily infusions. The total dose of Ara-C is kept constant over the simulation protocols (i.e. 200 mg/m^2 daily dose load). For DNR, the same schedule is followed with a dose increase to 90 mg/m^2 (Table 8.8).

Under this chemotherapy protocol, the leukaemic population is further minimised, and by completion of the first cycle, the leukaemic population is 2.43×10^8 cells less with a cost of 2.3×10^8 normal cells (Figure 8.4).

8.3.2 Second Chemotherapy Cycle

For the second chemotherapy protocol, the schedule suggested includes daily doses of 40 mg of Ara-C applied as daily continuous infusions for 10 days (Table 8.9).

Table 8.7 Leukaemic population of patient P016 based on simulation model results.

Date	Leukaemic population (model)
First cycle start date: 03/07/2010	8.55×10^{11} cells
First cycle end date: 13/07/2010	3.29×10^8 cells
BM aspirate after first cycle: 17/08/2010	9.46×10^9 cells
Second cycle start date: 07/09/2010	7.62×10^{10} cells
Second cycle end date: 17/09/2010	3.96×10^8 cells
Bone marrow aspirate after second cycle: 12/10/2010	4×10^9 cells

Table 8.8 Optimal schedule of the first chemotherapy cycle for Patient P016.

Protocol	Dose load	Dose duration	Application route	Application schedule
DA protocol				
DNR	90 mg/m^2	1 min	IV	One daily applications on days 1, 3 and 5
Ara-C	200 mg/m^2	Daily	IV	1 daily application, for days 1–10

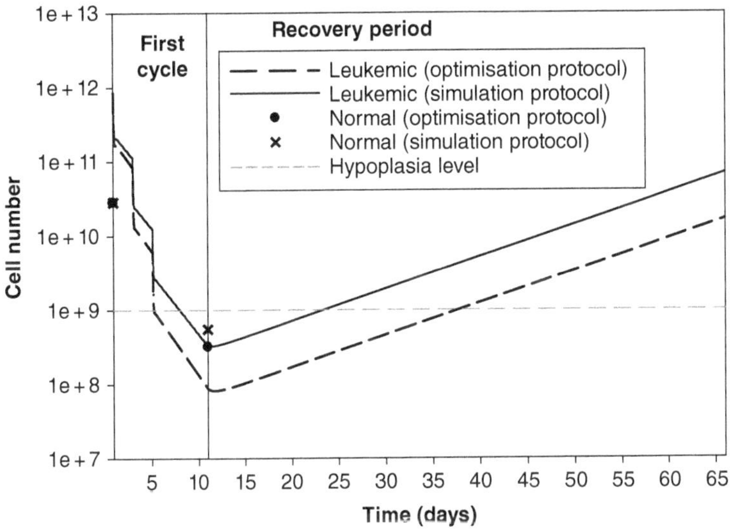

Figure 8.4 Patient P016 behaviour over the first chemotherapy cycle (days 1–11) and the recovery period prior to the second chemotherapy cycle (days 11–67). The dashed line is for the leukaemic cell population over the optimised protocol; the straight black line is for the leukaemic cell over the simulation of the clinical applied protocol; the cycle signs are for the normal population at the start and end dates of the optimisation protocol; the *x* signs are for the normal population at the start and end dates of the simulation protocol; and the grey line represents BM hypoplasia objective.

Table 8.9 Optimal LDAC induction treatment protocol for Patient P016.

Protocol	Dose load	Dose duration	Application route	Application schedule
SC Ara-C				
First cycle	40 mg	Daily	SC	One daily application for days 1–10

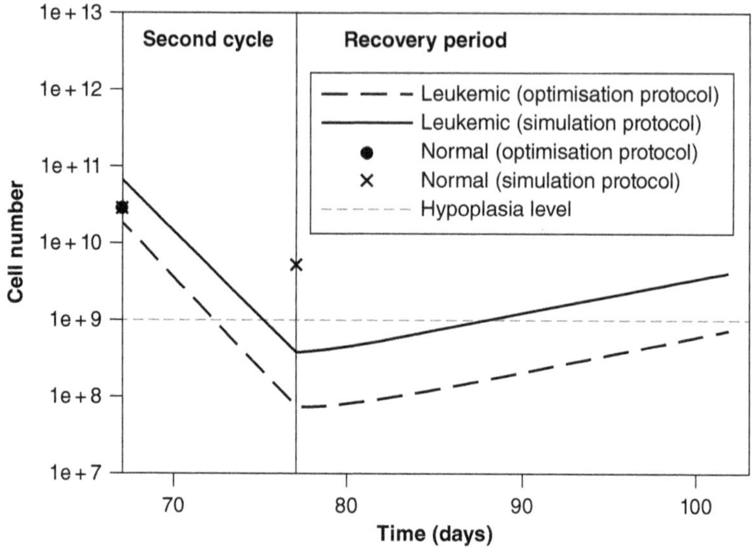

Figure 8.5 Patient P016 behaviour over the second chemotherapy cycle (days 67–77) and the recovery period prior to the BM aspirate at treatment completion (days 77–100). The dashed line is for the leukaemic cell population over the optimised protocol; the straight black line is for the leukaemic cell over the simulation of the clinical applied protocol; the cycle signs are for the normal population at the start and end dates of the optimisation protocol; the *x* signs are for the normal population at the start and end dates of the simulation protocol; and the grey line represents BM hypoplasia objective.

Figure 8.5 presents the normal and leukaemic cell dynamics. The leukaemic population has a further decrease of 3.2×10^8 cells, and the normal population is kept in the same order of magnitude. This is expected if we consider that the normal population consists of proliferating cells susceptible to the treatment and quiescent cells serving as backup cells in times of BM depletion. Since the transition rate of quiescent cells depends on the population depletion, the population will be adjusted to the loss and the transition rate will be adapted to keep the population constant. For the optimal protocol, since dose injection rate is lower and constant over the optimal treatment protocol, it will allow a constant transition of quiescent cells to proliferation, which will result in a more rigid normal cell population recovery over this protocol. Moreover, by treatment completion, the leukaemic population is reduced 3.3×10^9 cells, resulting in BM hypoplasia, as the final population is lower to the limit of leukaemic population (reduced 1×10^9 cells) (Table 8.10).

Table 8.10 Leukaemic and normal cell populations for P016, over the simulation and optimisation induction treatment protocols.

Date	Leukaemic population over simulation	Normal population over simulation	Leukaemic population over optimisation	Normal population over optimisation
Beginning of first cycle	8.55×10^{11}	2.83×10^{10}	8.55×10^{11}	2.83×10^{10}
End of first cycle	3.29×10^{8}	5.5×10^{8}	8.6×10^{7}	3.2×10^{8}
Beginning of second cycle	7.62×10^{10}	2.83×10^{10}	1.82×10^{10}	2.83×10^{10}
End of second cycle	3.96×10^{8}	5.39×10^{9}	7.6×10^{7}	5.35×10^{9}
BM aspirate after second cycle	4×10^{9}		7×10^{8}	

8.4 Conclusions

Treatment for AML with chemotherapy may result in acute and long-term life-threatening complications due to drug toxicity. Only relatively few patient- and leukaemia-specific factors are taken into consideration in current protocols, and choice of treatment often depends on the treating physician's experience. With the advent of novel treatments and large amounts of patient- and leukaemia-specific genomic data, there is a clear need for a systematic approach to the design and execution of chemotherapy regimens. We address these challenges in AML treatment by deriving a mathematical model that combines the leukaemia-specific actions on the cell cycle (i.e. drug target) with patient-specific pharmacology of the drugs (pharmacokinetics). Mathematical modelling is a tool that can be used for the automation of chemotherapy treatment due to its advantages in systematically exploring extensive datasets in order to capture a system's dynamics and subsequently provide better insight for process enhancement. We have developed a model for the simulation of patients with AML undergoing treatment with two standard chemotherapy protocols, one intensive and the other non-intensive: (a) daunarubicin (DNR) and cytosine arabinoside (Ara-C) used in standard IV doses (DA = 3 + 7 or 3 + 10), and (b) low-dose Ara-C (LDAC) administered SC. Sensitivity analysis of the developed model identifies cell cycle times as the critical parameters that control treatment outcome. For model analysis, clinical data of six patients who underwent chemotherapy are used for the estimation of cell cycle time distribution. The patient data comprise disease characteristics (tumour burden, cell cycle times and normal cell population) as well as patient-specific characteristics (gender, age, weight and height). The estimated mean S phase duration (*Ts*) is 15 h (range: 9–21 h), and mean whole cell cycle duration (*Tc*) is 47.5 h (range: 33–68 h). The estimated data reveal a clear relationship of cell cycle times to treatment outcomes. Specifically, low *Ts* duration combined with high *Tc* duration indicate worse treatment outcomes, whereas the reverse combination is indicative of a good response to treatment. In order to improve effectiveness of AML therapy and reduction of toxicity, treatment with chemotherapy is presented as an optimal control problem with the main aim of obtaining a treatment schedule which could maximise leukaemic cell kill, yet minimise death of the normal cell population in the bone marrow. By the end of treatment, the leukaemic population should be reduced to a level of approximately 10^9 cells, at which point BM hypoplasia is achieved. Both the mathematical modelling and optimisation algorithm are illustrated through the analysis of a patient case study treated under the two analysed protocols, and the results clearly demonstrate the potential amelioration of treatment design through optimisation.

In summary, this chapter presents the potential for improved treatment design in AML therapy, dependent on disease and patient characteristics, defined on a case-by-case basis. This design would provide the opportunity to personalise treatment protocols for gold standard intensive and non-intensive therapies as well as for novel drugs through the use of optimisation methods.

Appendix 8A Mathematical Model

8A.1 Treatment Inflow

$$\inf low_{,j} = \frac{dose_{,j}}{duration_{,j}} \tag{8A.1}$$

$$dose_{,j} = u_{,j} \cdot bsa \tag{8A.2}$$

$$bsa = \sqrt{\frac{height \cdot weight}{3600}} \tag{8A.3}$$

The inflow rate is the rate of the administered dose applied over the dosage duration. The dose is adjusted to the patient by its multiplication with the body surface area, calculated by the Mosteller empirical equation as is currently done in clinical practice. These equations are used for the calculation of the inflow rate given the treatment schedule characteristics (i.e. the dose load and duration of administration). Moreover, these two characteristics comprise the control variables for the optimisation problem.

8A.2 Pharmacokinetic Model

For both drugs, DNR and Ara-C, physiologically based PK (PBPK) models are used to calculate drug concentration of the active metabolite in specific human organs at each time point. Initially, the drug is injected into the blood and circulates to the whole body. The mass balance for the blood compartment is:

$$V_B \cdot \frac{dC_{B,j}}{dt} = \sum_{\substack{i:H,Li,M,Le \\ j:ara-C,DNr}} Q_i \cdot C_{i,j} + Q_K \cdot C_{K,j} - Q_B \cdot C_{B,j} + inflow_{,j} \tag{8A.4}$$

where $C_{B,j}$ is the concentration of drug j in the blood compartment; V_B is the total patient blood volume; Q_i is the blood flow in organs i: heart (H), liver (Li), bone marrow (M), lean (Le) and kidney (K); $C_{i,j}$ is the concentration of drug j in organs i; and *inflow* is the treatment inflow as calculated in Equation (8A.1).

The metabolic action takes place in the liver, and then the active metabolite is circulated in the body via the blood. The mass balance in the body organs is as follows:

$$V_i \cdot \frac{dC_{i,j}}{dt} = Q_i \cdot C_{B,j} - Q_i \cdot C_{i,j} - k_{i,j} \cdot C_{i,j} \cdot V_{iT} \tag{8A.5}$$

The drug is transmitted via the blood to the organs, and the general mass balance in the organs is the one in Equation (8A.5). The term $k_{,i,j}$ is the elimination rate of the drugs in the body organs and has physical meaning only for the liver. After drug elimination and action, the drug is excreted through urine with clearance rate ($k_{k,j}$) from the kidneys. An extra factor is introduced in the mass balance of the kidneys (8A.6) to account for the drug clearance ($k_{k,j}$). After its metabolism and action, the drug j is excreted through urine, and the cumulative excretion is calculated by Equation (8A.7):

$$V_i \cdot \frac{dC_{i,j}}{dt} = Q_i \cdot C_{B,j} - Q_i \cdot C_{i,j} - k_{k,j} \cdot C_{B,j} \cdot V_{iT} \tag{8A.6}$$

$$U_{,j} = \int_0^t k_{k,j} \cdot C_{B,j} dt \tag{8A.7}$$

8A.3 Pharmacodynamic Model

The PD model is used for the calculation of the drug effect, which is the percentage of dead cells due to drug action. The PD model is derived from one equation (Equation 8A.8) where the main input is drug concentration in the location of the tumour, which for AML is the concentration in BM ($C_{M,j}$) and is calculated by the PK model.

$$effect_{,j} = \frac{E_{\max,j} \cdot C_{M,j}^{slope,j}}{E_{50,j} + C_{M,j}^{slope,j}} \tag{8A.8}$$

where $E_{max,j}$, $E_{50,j}$ and slope are the PD parameters that depend on the drug j and are validated using clinical data.

8A.4 Cancer Cell Cycle Model

A dynamic model is used for the description of the cell cycle through chemotherapy treatment. The selected compartments are the cells in the G_1, S, G_2 and M phases. G_1 is the first compartment after the starting point of the cell cycle and lasts T_{G1} hours. Afterwards, the cell proceeds to the S phase (DNA replication). The S phase lasts T_S hours, and the cell is transferred to the last compartment, G_2 and M (G_2M), that last T_{G2M} hours and result in two newborn cells. The mathematical model consists of the mass balances between these compartments and is described by the following equations:

$$\frac{dG_1}{dt} = 2 \cdot k_3 \cdot G_2M - k_1 \cdot G_1 - effect_{,j} \cdot G_1 \tag{8A.9}$$

$$\frac{dS}{dt} = k_1 \cdot G_1 - k_2 \cdot S - effect_{,j} \cdot S \tag{8A.10}$$

$$\frac{dG_2M}{dt} = k_2 \cdot S - k_3 \cdot G_2M - effect_{,j} \cdot G_2M \tag{8A.11}$$

where G_1, S and G_2M represent the cell population in cell cycle compartments; k_1, k_2 and k_3 are the transition rates between cell phases; and $effect_{,j}$ is calculated by the PD model (8A.8) and is the percentage of each cell cycle population killed by the anticancer drug. This parameter has physical meaning only if a drug acts on a particular cell phase, that is, for drug Ara-C the effect will be 0 for phases G_1 and G_2M, whereas for DNR the effect will be 0 only for phase G_2M. The transition rates are functions of the duration of the cell cycle phases and are calculated by Equations (8A.12) through (8A.14):

$$k_1 = \frac{1}{T_{G1} + \mu_{G1}} \tag{8A.12}$$

$$k_2 = \frac{1}{T_S + \mu_S} \tag{8A.13}$$

$$k_3 = \frac{1}{T_{G2M} + \mu_{G2M}} \tag{8A.14}$$

where μ_{G1}, μ_S and μ_{G2M} are the natural apoptosis rates for each cell cycle phase.

As the cell cycle is a dynamic model, it depends on the initialisation state. The initial distribution of the cell population in the cell phases is difficult to measure and will be estimated by Equations (8A.15) through (8A.17):

$$G_1\big|_{t=0} = \frac{T_{G1}}{Tc} \cdot N(0) \tag{8A.15}$$

$$S\big|_{t=0} = \frac{T_S}{Tc} \cdot N(0) \tag{8A.16}$$

$$G_2M\big|_{t=0} = \frac{T_{G2M}}{Tc} \cdot N(0) \tag{8A.17}$$

where T_C is the total cell cycle time and $N(0)$ is the initial number of cancer cells in the modelled cell cycle population.

8A.5 Normal Cell Cycle Model

The normal stem cell reserve contains cells which can replicate, differentiate or die. These cells are grouped into two compartments, proliferating (P) and non-proliferating (Q) cells. Non-proliferating cells are G_1 phase cells grouped together with quiescent cells. These cells are activated and transmitted to the proliferating compartment at a rate ($\beta(Q)$) that is reciprocal to the number of quiescent cells (8A.20), that is, when the number of cells is low, more cells will be activated in order to preserve the stem cell population. The set of mathematical equations expressing the behaviour of normal cells is as follows:

$$\frac{dQ}{dt} = -\delta \cdot Q - \beta(Q) \cdot Q + 2 \cdot e^{-\gamma \cdot \tau} \cdot \beta(Q) \cdot Q - effect_{,j} \cdot Q \tag{8A.18}$$

$$\frac{dP}{dt} = -\gamma \cdot P + \beta(Q) \cdot Q - e^{-\gamma \cdot \tau} \cdot \beta(Q) \cdot Q - effect_{,j} \cdot P \tag{8A.19}$$

$$\beta(Q) - \beta_o \cdot \frac{\theta^n}{\theta^n + Q^n} \tag{8A.20}$$

where γ is the death rate in the proliferative phase, δ is the death rate in the non-proliferative phase, τ is the duration of proliferation, β_o is the maximum recruitment rate, θ is the cell population of the growth phase when $\beta = \frac{\beta_o}{2}$ and n is a positive parameter depicting the sensitivity of the transition rate to the cell population of the growth phase.

8A.6 Drug Subcutaneous Route

For the model of the SC route, Equation (8A.4) will be replaced by Equations (8A.4a) and (8A.4b), which account for drug bioavailability (k_b) and absorption delay (k_a):

$$\frac{dS}{dt} = Inflow - k_a \cdot k_b \cdot S \tag{8A.4a}$$

$$V_B \cdot \frac{dC_B}{dt} = \sum_{i:H,Li,M,Le,K} Q_i \cdot C_{i,j} - Q_B \cdot C_{B,j} + k_a \cdot k_b \cdot S \tag{8A.4b}$$

Appendix 8B Patient Data

Patient number: 001
Disease status: Secondary
Patient characteristics:
Age: 75 years Sex: F Height: 152 cm
Body weight: 56 Kg BSA: 1.54 m^2

I. Baseline characteristics

Pre-treatment data	
Bone marrow aspirate	
% Blasts in BM aspirate	21%
Prognostic category	Intermediate
Full blood count	**Date: 12/03/2010**
WBC (×10^9 L)	8.7

II. Chemotherapy treatment schedule

Cycle	Date	Drugs	Doses	Dose reduction	Route	Number of days and schedule given
1	Day 1	Ara-C	20 mg		SC	10 days, twice a day every 12 h
2	Day 37	Ara-C	20 mg		SC	10 days, twice a day every 12 h
3	Day 69	Ara-C	20 mg		SC	10 days, twice a day every 12 h
4	Day 105	Ara-C	20 mg		SC	10 days, twice a day every 12 h

III. Response to treatment

Completion of course	Cycle 1	Repeat marrow	Cycle 2	Repeat marrow	Cycle 3	Repeat marrow	Cycle 4	Repeat marrow
Date	Day 36		Day 70		Day 91		Day 146	
Cellularity (1 = hypo, 2 = normo, 3 = hyper)	3		3		2		3	
Blasts (%)	14		4		5		15	
Marrow response	PR		CR		CR		Relapse	

Patient number: 002
Disease status: Secondary
Patient characteristics:
Age: 72 years Sex: F Height: 150 cm
Body weight: 47 Kg BSA: 1.4 m^2

I. Baseline characteristics

Pre-treatment data	
BM aspirate	
% Blasts in BM aspirate	83%
Prognostic category	Intermediate
Full blood count	**Date: 06/02/2008**
WBC (×10^9 L)	46.5

II. Chemotherapy treatment schedule

Cycle	Date	Drugs	Doses	Dose reduction	Route	Number of days and schedule given
1	Day 1	Ara-C	20 mg		SC	10 days, twice a day every 12 h
2						
3						
4						

III. Response to treatment
Marrow examinations

Completion of course	Cycle 1	Repeat marrow	Cycle 2	Repeat marrow	Cycle 3	Repeat marrow	Cycle 4	Repeat marrow
Date	Day 48							
Cellularity (1 = hypo, 2 = normo, 3 = hyper)	3							
Blasts (%)	4							
Marrow response	CR							

Patient number: 006
Disease status: De novo
Patient characteristics:
Age: 71 years Sex: F Height: 160 cm
Body weight: 57 Kg BSA: 1.59 m^2

I. Baseline characteristics

Pre-treatment data	
BM aspirate	
% Blasts in BM aspirate	36 %
Prognostic category	–
Full blood count	**Date: 06/02/2008**
WBC ($\times 10^9$ L)	1.6

II. Chemotherapy treatment schedule

Cycle	Date	Drugs	Doses	Dose reduction	Route	Number of days and schedule given
1	Day 1	Ara-C	20 mg		SC	10 days, twice a day every 12 h
2	Day 42	Ara-C	20 mg		SC	10 days, twice a day every 12 h
3	Day 74	Ara-C	20 mg		SC	10 days, twice a day every 12 h
4	Day 109	Ara-C	20 mg		SC	10 days, twice a day every 12 h

III. Response to treatment
Marrow examinations

Completion of course	Cycle 1	Repeat marrow	Cycle 2	Repeat marrow	Cycle 3	Repeat marrow	Cycle 4	Repeat marrow
Date	Day 42		Day 70		Day 110		Day 145	
Cellularity (1 = hypo, 2 = normo, 3 = hyper)	1		1		3		1	
Blasts (%)	3		2		2		0	
Marrow response	CR		CR		CR		CR	

Patient number: 011
Disease status: Secondary
Patient characteristics:
Age: 24 years Sex: M Height: 170 cm
Body weight: 59.5 Kg BSA: 1.68 m^2

I. Baseline characteristics

Pre-treatment data	
BM aspirate	
% Blasts in BM aspirate	56 %
Prognostic category	–
Full blood count	**Date: 18/10/2011**
WBC ($\times 10^9$ L)	0.9

II. Chemotherapy treatment schedule

Cycle	Date	Drugs	Doses	Dose reduction	Route	Number of days and schedule given
1	Day 1	Ara-C	168 mg		IV	10 days, twice a day every 12 h
		DNR	100 mg		IV	1 h dose on days 1, 3 and 5
2						
3						
4						

III. Response to treatment
Marrow examinations

Completion of course	Cycle 1	Repeat marrow	Cycle 2	Repeat marrow	Cycle 3	Repeat marrow	Cycle 4	Repeat marrow
Date	Day 48							
Cellularity (1 = hypo, 2 = normo, 3 = hyper)	1							
Blasts (%)	3							
Marrow response	CR							

Patient number: 016
Disease status: Secondary
Patient characteristics:
Age: 80 years Sex: M Height: 167.5 cm
Body weight: 79.3 Kg BSA: 1.92 m^2

I. Baseline characteristics

Pre-treatment data	
BM aspirate	
% Blasts in BM aspirate	90 %
Prognostic category	–
Full blood count	**Date: 30/06/2010**
WBC ($\times 10^9$ L)	3.3

II. Chemotherapy treatment schedule

Cycle	Date	Drugs	Doses	Dose reduction	Route	Number of days and schedule given
1	Day 1	DNR	95 mg		IV	1 h dose on days 1, 3 and 5
		Ara-C	190 mg		IV	10 days, twice a day every 12 h
2	Day 66	Ara-C	20		SC	10 days, twice a day every 12 h
3						

III. Response to treatment
Marrow examinations

Completion of course	Cycle 1	Repeat marrow	Cycle 2	Repeat marrow	Cycle 3	Repeat marrow	Cycle 4	Repeat marrow
Date	Day 45		Day 101					
Cellularity (1 = hypo, 2 = normo, 3 = hyper)	3		2					
Blasts (%)	1		1					
Marrow response	CR		CR					

Patient number: 026
Disease status: De novo
Patient characteristics:
Age: 45 years Sex: F Height: 169.3 cm
Body weight: 94.8 Kg BSA: 2.11 m^2

I. Baseline characteristics:

Pre-treatment data	
BM aspirate	
% Blasts in BM aspirate	71 %
Prognostic category	–
Full blood count	**Date: 26/05/2011**
WBC (×10^9 L)	1.2

II. Chemotherapy treatment schedule

Cycle	Date	Drugs	Doses	Dose reduction	Route	Number of days and schedule given
1	Day 1	DNR	150 mg		IV	1 h dose on days 1, 3 and 5
		Ara-C	170 mg		IV	10 days, twice a day every 12 h
2	Day 56	DNR	85 mg		IV	1 h dose on days 1, 3 and 5
		Ara-C	170 mg		IV	8 days, twice a day every 12 h
3						

III. Response to treatment: marrow examinations

Completion of course	Cycle 1	Repeat marrow	Cycle 2	Repeat marrow	Cycle 3	Repeat marrow	Cycle 4	Repeat marrow
Date	Day 48		Day 116					
Cellularity (1 = hypo, 2 = normo, 3 = hyper)	2		2					
Blasts (%)	0		0					
Marrow response	CR		CR					

References

Basse, B., Baguley, B.C., Marshall, E.S., Joseph, W.R., Brunt, B.V., Wake, G. and Wall, D.J.N. (2003) A mathematical model for analysis of the cell cycle in cell lines derived from human tumours. *Mathematical Biology*, 47, 295–312.

BC Cancer Agency (2007) Cancer drug manual: cytarabine. http://www.bccancer.bc.ca/HPI/DrugDatabase/DrugIndexPro/Cytarabine.htm

Bischoff, K.B. and Dedrick, R.L. (1968) Thiopental pharmacokinetics. *Journal of pharmaceutical sciences*, 57(8), 1346–1351.

Brody, S. (1945) *Bioenergetics and Growth*. Reinhold Publications Corp., New York.

Brown, E., Hopper, J., Hodges, J.L., Bradley, B., Wennesland, R. and Yamauchi, H. (1962) Red cell, plasma, and blood volume in healthy women measured by radiochromium cell labeling and hematocrit. *Journal of Clinical Investigation*, 41(12), 2182–2190.

Caldwell, J.C., Evans, M.V. and Krishnan, K. (2012) Cutting edge PBPK models and analyses: providing the basis for future modelling efforts and bridges to emerging toxicology paradigms. *Journal of Toxicology*, 1–10.

Cancer Research UK (2008) Leukaemia. http://www.cancerresearchuk.org/health-professional/cancer-statistics/statistics-by-cancer-type/leukaemia

Chaturvedi, P.R., Decker, C.J. and Odinecs, A. (2001) Prediction of pharmacokinetic properties using experimental approaches during early drug discovery. *Current Opinion in Chemical Biology*, 5, 452–463.

Chen, H.G. and Gross, J.F. (1979) Physiologically based pharmacokinetic models for anticancer drugs. *Cancer Chemotherapy and Pharmacology*, 2, 85–94.

Chouker, A., Martignoni, A., Dugas, M., Eisenmenger, W., Schauer, R., Kaufmann, I., Schelling, G., Lohe, F., Jauch, K., Peter, K. and Thiel, M. (2004) Estimation of liver size for liver transplantation: the impact of age and gender. *Liver Transplantation*, 10, 678–685.

Dedrick, R.L., Forrester, D.D. and Ho, D.H.W. (1972) In vitro–in vivo correlation of drug metabolism-deamination of 1-b-D-arabinofuranosylcytosine. *Biochemical Pharmacology*, 21, 1–16.

Dua, P., Dua, V. and Pistikopoulos, E.N. (2008) Optimal delivery of chemotherapeutic agents in cancer. *Computers & Chemical Engineering*, 32(1–2), 99–107.

Essers, M.A.G. and Trumpp, A. (2010) Targeting leukemic stem cells by breaking their dormancy. *Molecular Oncology*, 4(5), 443–450.

Fister, K.R. and Panetta, J.C. (2000) Optimal control applied to cell-cycle-specific cancer chemotherapy. *Siam Journal on Applied Mathematics*, 60(3), 1059–1072.

Garattini, S. (2007) Pharmacokinetics in cancer chemotherapy. *European Journal of Cancer*, 43, 271–282.

Gillies, H.C., Herriott, D., Liang, R., Ohashi, K., Rogers, H.J. and Harper, P.G. (1987) Pharmacokinetics of idarubicin (4-demethoxydaunorubicin; IMI-30; NSC 256439) following intravenous and oral administration in patients with advanced cancer. *British Journal of Clinical Pharmacology*, 23, 303–310.

Harrold, J.M. and Parker, R.S. (2009) Clinically relevant cancer chemotherapy dose scheduling via mixed-integer optimisation. *Computers and Chemical Engineering*, 33, 2042–2054.

Hijiya, N., Panetta, J.C, Zhou, Y., Kyzer, E.P., Howard, S.C., Jeha, S., Razzouk, B.I., Ribeiro, R.C., Rubnitz, J. E., Hudson, M.M., Sandlund, J.T., Pui, C.H. and Relling, M.V. (2006) Body mass index does not influence pharmacokinetics or outcome of treatment in children with acute lymphoblastic leukemia. *Childhood: A Global Journal of Child Research*, 108, 3997–4002.

Himmelstein, K.J. and Lutz, R.J. (1979) A review of the applications of physiologically based pharmacokinetic modelling. *Journal of Pharmacokinetics and Biopharmaceutics*, 7(2), 127–145.

Holford, N.H.G. and Sheiner, L.B. (1982) Kinetics of pharmacologic response. *Pharmacology & Therapeutics*, 16, 143–166.

Jones, H.M., Gardner, I.B. and Watson, K.J. (2009) Modelling and PBPK simulation in drug discovery. *The AAPS Journal*, 11, 155–166.

Kimmel, M. and Swierniak, A. (2003) Using control theory to make cancer chemotherapy beneficial from phase dependence and resistant to drug resistance. *Mathematical Biosciences*, 1–44.

Komarova, N.L. and Wodarz, D. (2005) Drug resistance in cancer: principles of emergence and prevention. *Proceedings of the National Academy of Sciences of the United States of America*, 102, 9714–9719.

Krieger, A., Panoskaltsis, N., Mantalaris, A., Georgiadis, M. and Pistikopoulos, E.N. (2013) Modelling and analysis of individualized pharmacokinetics and pharmacodynamics for volatile anaesthesia. *IEEE Transactions on Biomedical Engineering*, 61(1), 25–34.

Lacayo, N.J., Lum, B.L., Becton, D.L., Weinstein, H., Ravindranath, Y., Chang, M.N., Bomgaars, L., Lauer, S.J., Sikic, B.I. and Dahl, G.V. (2002) Pharmacokinetic interactions of cyclosporine with etoposide and mitoxantrone in children with acute myeloid leukemia. *Leukemia*, 920–927.

Ledzewicz, L. and Schättler, H. (2002) Optimal bang-bang controls for a two-compartment model. *Optimisation*, 114, 609–637.

Ledzewicz, U. and Schättler, H. (2007) Optimal controls for a model with pharmacokinetics maximizing BM in cancer chemotherapy. *Mathematical Biosciences*, 206, 320–342.

Lewin, B., Cassimeris, L., Lingappa, V.R. and Plopper, G. (2007) *Cells*, 2nd ed. Jones and Bartlett, Burlington, MA.

Looby, M., Linke, R. and Weiss, M. (1997) Pharmacokinetics and tissue distribution of idarubicin and its active metabolite idarubicinol in the rabbit. *Cancer Chemotherapy & Pharmacology*, 39, 554–556.

Lutz, R.L., Galbraith, W.M., Dedrick, R.L., Shrager, R. and Mellett, L.B. (1977) A model for the kinetics of distribution of actinomycin-d in the beagle dog. *The Journal of Pharmacology and Experimental Therapeutics*, 200, 469–478.

Michor, F. (2008) Mathematical models of cancer stem cells. *Journal of Clinical Oncology: Official Journal of the American Society of Clinical Oncology*, 26, 2854–2861.

Morrison, P.F., Lincoln, T.L. and Aroesty, J. (1975) *Rand Report: The Disposition of Ara-C and Its Metabolites: A Pharmacokinetic Simulation*. Rand Corporation, Santa Monica, CA.

Panetta, J.C., Wilkinson, M., Pui, C.H. and Relling, M.V. (2002) Limited and optimal sampling strategies for etoposide and etoposide catechol in children with leukemia. *Journal of Pharmacokinetics and Pharmacodynamics*, 29(2), 171–188.

Panoskaltsis, N., Mantalaris, A. and Wu, J.H.D. (2005) Engineering a mimicry of bone marrow tissue ex vivo. *Journal of Bioscience & Bioengineering*, 100(1), 28–35.

Panoskaltsis, N., Reid, C.D.L. and Knight, S.C. (2003) Quantification and cytokine production of circulating lymphoid and myeloid cells in acute myelogenous leukemia (AML). *Leukemia*, 17, 716–730.

Parker, R.S. and Doyle, F.J. (2001) Control relevant modelling in drug delivery. *Advanced Drug Delivery Reviews*, 48, 211–228.

Pefani, E., Panoskaltsis, N., Mantalaris, A., Georgiadis, M.C. and Pistikopoulos, E.N. (2013) Design of optimal patient-specific chemotherapy protocols for the treatment of acute myeloid leukemia (AML). *Computers & Chemical Engineering*, 57, 187–195.

Pefani, E., Panoskaltsis, N., Mantalaris, A., Georgiadis, M.C. and Pistikopoulos, E.N. (2014) Chemotherapy drug scheduling for the induction treatment of patients diagnosed with acute myeloid leukemia. *Transactions of Biomedical Engineering Journal*, 61(7), 2059–2056.

Peyret, T. and Krishnan, K. (2011) QSARs for PBPK modelling of environmental contaminants. *SAR and QSAR in Environmental Research*, 22, 129–169.

Pichardo, J.C., Trindade, A.A., Brindle, J.M. and Bolch W.E. (2007) Method for estimating skeletal spongiosa volume and active marrow mass in adult male and adult female. *The Journal of Nuclear Medicine*, 48, 1880–1888.

Quartino, A., Karlsson, M.O., Freijs, A., Jonsson, N., Nygren, P., Kristensen, J., Lindhagen, E. and Larsson, R. (2007) Modelling of in vitro drug activity and prediction of clinical outcome in acute myeloid leukemia. *Journal of Clinical Pharmacology*, 47, 1014–1021.

Ratain, M. and Plunkett, W.K. (2003) Pharmacology. In *Holland-Frei Cancer Medicine*, 6th ed. BC Decker, Hamilton, ON, Canada.

Raza, A., Preisler, H.D., Day, R., Yasin, Z., White, M., Lykins, J., Kukla, C., Barcos, M., Bennett, J. and Browman G. (1990) Direct relationship between remission duration in acute myeloid leukemia and cell cycle kinetics: a leukemia intergroup study. *Blood*, 76, 2191–2197.

Reich, S.D., Bachur, N.R., Goebel, R.H. and Berman, M. (1977) A pharmacokinetic model for high-dose methotrexate infusions in man. *Journal of Pharmacokinetics and Biopharmaceutics*, 5(5), 421–433.

Relling, M.V., Yanishevski, Y., Nemec, J., Evans, W.E., Boyett, J.M., Behm, F.G. and Pui, C.H. (1998) Etoposide and antimetabolite pharmacology in patients who develop secondary acute myeloid leukemia. *Leukemia*, 12(3), 346–352.

Richard, B., Launay-Iliadis, M.C., Iliadis, A., Just-Landi, S., Blaise, D., Stoppa, A.M., Viens, P., Gaspard, M.H., Maraninchi, D., Cano, J.P. and Carcassonne, Y. (1992) Pharmacokinetics of mitoxantrone in cancer patients treated by high-dose chemotherapy and autologous bone marrow transplantation. *British Journal of Cancer*, 65, 399–404.

Saltzman, W.M. (2001) *Drug Delivery: Engineering Principles for Drug Therapy*. Oxford University Press, Oxford.

Schmitt, A., Li, L., Giannopoulos, K., Greiner, J., Reinhardt, P., Wiesneth, M. and Schmitt, M. (2008) Quantitative expression of Toll-like receptor-2, -4, and -9 in dendritic cells generated from blasts of patients with acute myeloid leukemia. *Transfusion*, 48(5), 861–870.

Sherer, E., Hannemann, R.E., Rundell, A. and Ramkrishna, D. (2006) Analysis of resonance chemotherapy in leukemia treatment via multi-staged population balance models. *Journal of Theoretical Biology*, 240, 648–661.

Sidoli, F.R., Asprey, S.P. and Mantalaris, A. (2006) A coupled single cell-population-balance model for mammalian cell cultures. *Industrial and Engineering Chemistry Research*, 45, 5801–5811.

Swierniak, A., Kimmel, M. and Smieja J. (2009) Mathematical modelling as a tool for planning anticancer therapy, *European Journal of Pharmacology*, 625, 108–121.

UCLH (2009) Dosage adjustment for cytotoxics in renal impairment, Version 3. http://www.eastmidlandscancernetwork.nhs.uk/Library/RenalDosageAdjustments.pdf

Undevia, S.D., Gomez-Abuin, C.G. and Ratain, M.J. (2005) Pharmacokinetic variability of anticancer agents. *Nature Reviews*, 5, 447–458.

Wennesland, R., Brown, E., Hopper, J., Hodges, J.L., Guttentag, O.E., Scott, K.G., Tucker, I.N. and Bradley, B. (1959) Red cell, plasma and blood volume in healthy men measured by radiochromium (Cr51) cell tagging and hematocrit: influence of age, somatotype and habits of physical activity on the variance after regression of volumes to height and weight combined. *Journal of Clinical Investigation*, 38(7), 1065–1077.

Williams, W.J., Beutler, E., Erslev, A.J. and Lichtman, M.A. (1983) *Hematology*, 3rd ed. McGraw-Hill, New York.

Yun, Y.E. and Edginton, A.N. (2013) Correlation-based prediction of tissue-to-plasma partition coefficients using readily available input parameters. *Xenobiotica*, 1–14.

Index

Modelling Optimization and Control of Biomedical Systems, First Edition.
Edited by Efstratios N. Pistikopoulos, Ioana Naşcu, and Eirini G. Velliou.

k

l

m

www.ingramcontent.com/pod-product-compliance
Lightning Source LLC
LaVergne TN
LVHW011527200926
848484LV00001B/173